Molecular Glycobiology

EDITED BY

Minoru Fukuda
*La Jolla Cancer Research Foundation,
La Jolla, California, USA and Institute of Medical Science,
University of Tokyo, Tokyo, Japan*

and

Ole Hindsgaul
*Department of Chemistry, University of Alberta,
Edmonton, Canada and La Jolla Cancer Research Foundation,
La Jolla, California, USA*

at
OXFORD UNIVERSITY PRESS
Oxford New York Tokyo

Oxford University Press, Walton Street, Oxford OX2 6DP

Oxford   New York   Toronto
Delhi   Bombay   Calcutta   Madras   Karachi
Kuala Lumpur   Singapore   Hong Kong   Tokyo
Nairobi   Dar es Salaam   Cape Town
Melbourne   Auckland   Madrid
and associated companies in
Berlin   Ibadan

Oxford is a trade mark of Oxford University Press

Published in the United States
by Oxford University Press Inc., New York

© Oxford University Press, 1994

All rights reserved. No part of this publication may be
reproduced, stored in a retrieval system, or transmitted, in any
form or by any means, without the prior permission in writing of Oxford
University Press. Within the UK, exceptions are allowed in respect of any
fair dealing for the purpose of research or private study, or criticism or
review, as permitted under the Copyright, Designs and Patents Act, 1988, or
in the case of reprographic reproduction in accordance with the terms of
licences issued by the Copyright Licensing Agency. Enquiries concerning
reproduction outside those terms and in other countries should be sent to
the Rights Department, Oxford University Press, at the address above.

This book is sold subject to the condition that it shall not,
by way of trade or otherwise, be lent, re-sold, hired out, or otherwise
circulated without the publisher's prior consent in any form of binding
or cover other than that in which it is published and without a similar
condition including this condition being imposed
on the subsequent purchaser.

A catalogue record for this book is available from the British Library

Library of Congress Cataloging in Publication Data
Molecular glycobiology / edited by Minoru Fukuda and Ole Hindsgaul.
(Frontiers in molecular biology)
Includes bibliographical references and index.
1. Carbohydrates.   2. Glycoproteins.   3. Glycolipids.   4. Lectins.
I. Fukuda, Minoru, 1945-   .   II. Hindsgaul, Ole.   III. Series.
QP701.M65 1994   574.19'2482–dc20   93–41975
ISBN 0 19 963387 8 (h/b)
ISBN 0 19 963386 X (p/b)

Typeset by
Footnote Graphics, Warminster, Wilts.
Printed in Great Britain by
The Bath Press, Avon

Molecular Glycobiology

Frontiers in Molecular Biology

SERIES EDITORS

B. D. Hames

Department of Biochemistry and Molecular Biology
University of Leeds, Leeds LS2 9JT, UK

AND

D. M. Glover

Department of Biochemistry
University of Dundee, Dundee DD1 4HN, UK

OTHER TITLES IN THE SERIES

Human Retroviruses
Bryan R. Cullen

Steroid Hormone Action
Malcolm G. Parker

Mechanisms of Protein Folding
Roger H. Pain

To our mentors and teachers, particularly:
Fujio Egami,
Raymond Lemieux

Preface

The past few years has seen a tremendous surge in interest in glycobiology. This dramatic increase in the significance of glycobiology is mainly derived from the recent findings that cell surface carbohydrates are critically involved in cell adhesion, and thus cell–cell interaction.

This book is intended to provide the reader with current knowledge on the roles of carbohydrates in cell adhesion, the historical background to recent discoveries, and the prospects in this field. We have collected excellent articles from outstanding scientists, making this book representative of this field.

Historically, three fields in glycobiology were crucial to recent dramatic developments. First, the structural studies in glycoproteins and glycolipids revealed that there are carbohydrates which are unique to certain cell types. This concept is crucial to the understanding of cell surface carbohydrates as cell-type specific recognition molecules (Chapter 1).

The second critical notion came from the vast knowledge of animal lectins. It was discovered twenty years ago that glycoproteins in blood circulation are taken up quickly if their sialic acids are removed. Further studies revealed that this clearance is caused by the binding of asialoglycoproteins to a carbohydrate-binding protein that recognizes terminal galactose. Once animal cells were known to have carbohydrate-binding proteins, lectins, a large number of animal lectins were characterized. Those studies demonstrated that there is a conserved motif in the amino acid sequence; carbohydrate recognition domain (CRD) in the lectins. This notion was critical to the understanding of carbohydrate-binding capability in other adhesive molecules (Chapter 2).

The third critical notion came from the isolation and characterization of the glycosyltransferases that form carbohydrates. These studies showed that carbohydrate moieties are usually built one by one and each reaction is carried out by a glycosyltransferase that forms only a specific linkage. The advent of molecular biology in this field enables us to manipulate the expression of carbohydrates and ask the function of the formed carbohydrates (Chapter 3).

Based on these critical advances in this field, the most recent studies demonstrated that oligosaccharides uniquely present in leukocytes act as a ligand for adhesion molecules in endothelia and platelets. When these adhesive molecules, selectins, were cloned, it was immediately obvious that they contained carbohydrate recognition domains. Thus there was a clear synergy of the studies on cell-type specific carbohydrates and animal lectins. Moreover, these studies were preceded by the findings that lymphocyte-endothelial interaction is dependent on carbohydrates (Chapter 4).

Once this carbohydrate ligand, sialyl-Le^X, was recognized, it was a matter of time before this oligosaccharide was synthesized and used for therapeutic purposes.

This line of research has become much easier because enzymatic synthesis of carbohydrates is now possible thanks to the availability of glycosyltransferases generated by cloned cDNAs (Chapter 5).

Continuing this line of research, it has become critical to develop a drug which is different from natural compounds but identical or similar in steric structure. In order to achieve this goal, it is essential to understand the sterical structure of carbohydrates, particularly in solution (Chapter 6).

These research efforts thus allow us to understand the role of molecular mechanisms in carbohydrates and apply this knowledge to biological as well as clinical usage. I hope that such research directions will put glycobiology at the centre of biology and medicine. In order to facilitate this goal, we have published one book on methodology and one on the perspectives in glycobiology. I sincerely hope that *Glycobiology: a practical approach* and *Molecular glycobiology* serve well to advance our progress in glycobiology.

I thank all of the authors for contributing extremely well-written articles. I also would like to thank the staff of the Oxford University Press for encouragement and various efforts, and Ms Leslie DePry at the La Jolla Cancer Research Foundation for excellent editorial assistance. Without these persons, this book would not have been published.

La Jolla, California M.F.
April 1994

Contents

List of contributors xv

1 Cell surface carbohydrates: cell-type specific expression 1
MINORU FUKUDA

1. Introduction 1
2. Structure of *N*-glycans 3
    2.1 Diversity of *N*-glycans 4
    2.2 Commonality of *N*-glycans 8
3. Cell-type specific expression of *N*-glycans 11
    3.1 Poly-*N*-acetyllactosamines—glycans with *N*-acetyllactosamine repeats 11
    3.2 Modification of poly-*N*-acetyllactosamines 16
    3.3 Cell-type specific modification of poly-*N*-acetyllactosamines 17
    3.4 Poly-*N*-acetyllactosamines in blood group antigens 19
    3.5 Poly-*N*-acetyllactosamines in cell adhesion 21
    3.6 Poly-*N*-acetyllactosamines in tumour cells 26
4. Mucin-type glycans (*O*-glycans) 29
    4.1 Mucin-type glycans—distribution and role 29
    4.2 Mucin-type glycoproteins in the plasma membrane—glycophorin A and leukosialin 30
    4.3 Structures of mucin-type oligosaccharides 32
    4.4 Changes of mucin-type oligosaccharides attached to leukosialin during differentiation 33
    4.5 Changes in mucin-type oligosaccharides in tumour and immunodeficiency 34
5. β-1,6-*N*-acetylglucosaminyltransferases 38
6. Glycobiology—a new frontier in cell biology and molecular biology 40
Acknowledgements 43
References 43

2 Molecular structure of animal lectins — 53
KURT DRICKAMER

1. Introduction — 53
2. Classification of animal lectins — 54
3. Lectins containing P-type CRDs: mannose 6-phosphate receptors — 55
   3.1 Targeting of lysosomal hydrolases — 56
   3.2 Structural features — 56
4. Diversity of C-type animal lectins — 58
   4.1 Biological functions of C-type animal lectins — 58
   4.2 Evolution of C-type CRDs — 70
   4.3 Structure of a prototype C-type CRD — 71
5. Soluble β-galactoside-binding (S-type) lectins — 74
   5.1 Structural features of S-type lectins — 74
   5.2 Biological functions of S-type lectins — 76
6. Other groups of animal lectins — 77
7. Conclusions — 78
Acknowledgements — 79
References — 79

3 Molecular cloning of glycosyltransferase genes — 88
HARRY SCHACHTER

1. Introduction — 88
2. Glycosyltransferases — 89
   2.1 Biosynthesis of *N*-glycan antennae — 90
   2.2 Biosynthesis of *O*-glycans — 96
   2.3 Biosynthesis of human blood group carbohydrate epitopes — 97
3. Cloning of glycosyltransferase genes — 98
   3.1 Galactosyltransferases — 98
   3.2 Sialyltransferases — 118
   3.3 Fucosyltransferases — 128
   3.4 *N*-Acetylgalactosaminyltransferases — 137
   3.5 *N*-Acetylglucosaminyltransferases — 138
4. Conclusions — 146
   4.1 Genomic organization — 146

4.2	Glycosyltransferase gene families	147
4.3	The domain structure of glycosyltransferases	147
4.4	Control of glycosyltransferase activities in time and space	148
	Acknowledgements	149
	References	149

4 Carbohydrate recognition in cell–cell interaction 163
JOHN B. LOWE

1.	**Introduction**	163
2.	**Oligosaccharides in lymphocyte homing**	163
	2.1 General aspects of lymphocyte homing	163
	2.2 Lymphocyte–PNHEV adhesive interactions are oligosaccharide-dependent	165
	2.3 The MEL-14 homing receptor contains a carbohydrate recognition domain	166
	2.4 Functions of the domains of L-selectin	169
	2.5 GlyCAM-1 is part of an endogenous ligand for L-selectin	170
	2.6 Sialic acid and sulphate are essential to the adhesive activity of GlyCAM-1	171
	2.7 L-selectin expression and function in neutrophil adhesion to endothelium	172
3.	**Oligosaccharide-dependent cell adhesion mediated by E-selectin**	174
	3.1 E-selectin-dependent adhesion of myeloid cells to activated endothelium	174
	3.2 E-selectin contains a carbohydrate recognition domain similar to the L-selectin CRD	175
	3.3 Ligands for E-selectin—sialylated, fucosylated oligosaccharides as candidates	175
	3.4 Ligands for E-selectin—biosynthesis of sialylated, fucosylated oligosaccharides	177
	3.5 Ligands for E-selectin—the sialyl-Le^X determinant as a myeloid cell counter-receptor for E-selectin	179
	3.6 Ligands for E-selectin—the sialyl-Le^a determinant as an E-selectin counter-receptor	183
	3.7 Ligands for E-selectin—potential as anti-inflammatory pharmaceuticals	185
4.	**Oligosaccharide-dependent cell adhesion mediated by P-selectin**	186
	4.1 P-selectin in platelets and endothelial cells	186

  4.2 P-selectin contains a carbohydrate recognition domain similar to the L-selectin and P-selectin CRDs   187
  4.3 P-selectin-dependent adhesion of myeloid cells to activated platelets and endothelium   188
  4.4 Ligands for P-selectin—sialylated, fucosylated oligosaccharides as candidates   189
  4.5 Ligands for P-selectin—evidence for a protein component in addition to sialylated, fucosylated oligosaccharides   191
  4.6 Ligands for P-selectin—evidence that P-selectin-dependent cell adhesion may also operate through sulphatides   191
 5. The sequential, multimolecule model for leukocyte rolling, adhesion, and transmigration   192
Acknowledgements   194
References   194

5 Chemical synthesis of oligosaccharides   206
SHAHEER H. KHAN and OLE HINDSGAUL

1. **Introduction**   206
 1.1 Our current capability in oligosaccharide synthesis   207
2. **The problems in oligosaccharide synthesis**   207
 2.1 Protection—deprotection in oligosaccharide synthesis   209
 2.2 Glycosylation methods   209
 2.3 Convergent block synthesis   214
3. **Synthesis of acceptor substrates for glycosyltransferases**   215
 3.1 Synthesis of acceptor analogues   218
4. **Inhibitors for glycosyltransferases**   220
5. **Enzyme-assisted oligosaccharide synthesis**   220
6. **Conclusion**   222

Acknowledgements   222
References   223

6 Conformational studies on oligosaccharides   230
STEVEN W. HOMANS

1. **Introduction**   230

2. Conformational parameters of oligosaccharides	230
3. Sterical structures of oligosaccharides	231
3.1 Theoretical predictions	232
3.2 Nuclear magnetic resonance	234
3.3 X-ray crystallography	236
3.4 Conclusions	236
4. Dynamics of oligosaccharides	236
4.1 Theoretical predictions	236
4.2 Nuclear magnetic resonance	240
4.3 Conclusions	242
5. Sterical structure of oligosaccharide-protein complexes	243
5.1 Glycoproteins	243
5.2 Lectin–oligosaccharide interactions	244
5.3 Antibody–oligosaccharide interactions	248
6. Functional implications	249
6.1 Oligosaccharide biosynthesis	249
6.2 Oligosaccharide-mediated recognition phenomena	251
References	253
Appendix to Chapter 3	254
Index	260

Contributors

KURT DRICKAMER
Department of Biochemistry and Molecular Biophysics, College of Physicians and Surgeons of Columbia University, 630 W. 168th St, New York, NY 10032, USA.

MINORU FUKUDA
Glycobiology Program, La Jolla Cancer Research Foundation, 10901 North Torrey Pines Road, La Jolla, CA 92037, USA, and Department of Biochemistry, Institute of Medical Science, University of Tokyo, 4-6-1 Shirokanedai, Minato-ku, Tokyo 108, Japan.

OLE HINDSGAUL
Department of Chemistry, University of Alberta, Edmonton, Alberta T6G 2G2, Canada, and Glycobiology Program, La Jolla Cancer Research Foundation, 10901 North Torrey Pines Road, La Jolla, California, 92037, USA.

STEVEN W. HOMANS
The University of Dundee, Carbohydrate Research Centre, Medical Sciences Institute, Dundee, Scotland DD1 4HN, UK.

SHAHEER H. KHAN
Carbohydrate Research and Development, Perkin-Elmers Applied Biosystems Division, 850 Lincoln Drive, Foster City, CA 94404, USA.

JOHN B. LOWE
Howard Hughes Medical Institute, University of Michigan Medical School, MSRB I, Room 3510, 1150 W. Medical Center Drive, Ann Arbor, MI 48109-0650, USA.

HARRY SCHACHTER
Biochemistry Research Division, Hospital for Sick Children, University of Toronto, 555 University Avenue, Toronto, Ontario M5G 1X8, Canada.

1 | Cell surface carbohydrates: cell-type specific expression

MINORU FUKUDA

1. Introduction

Cell surface carbohydrates are major components of the outer surface of mammalian cells and these carbohydrates are very often characteristic of cell types. Carbohydrate structures change dramatically during mammalian development. Specific sets of carbohydrates are expressed at different stages of differentiation and in many instances these carbohydrates are recognized by specific antibodies, thus providing differentiation antigens (1–4). In mature organisms, expression of distinct carbohydrates is eventually restricted to specific cell types, providing cell-type-specific carbohydrates. Aberrations in these cell surface carbohydrates are associated with various pathological conditions, including malignant transformation.

It is assumed that cell type-specific carbohydrates are probably involved in cell–cell interaction, in particular as molecules that are recognized by carbohydrate-binding proteins. The latter is likely include lectins and some glycosyltransferases. It is thus possible that the diversity of carbohydrate structures in different cell types provides specific signals that are recognized by counter-receptor molecules. It is then plausible to assume that alterations of cell surface carbohydrates provide cells with different recognition molecules, which are necessary when cells interact with different cells as they differentiate. Alteration of cell surface carbohydrates in malignant transformation similarly facilitates tumour cells in their interaction with other cells in order to achieve tumour cell colonization and metastasis.

Carbohydrates are unique in the complexity of their structures. In contrast to nucleic acids and proteins, carbohydrates are linked together in more than one form. First, one sugar (residue) can be bound to three or four different positions of hydroxyl groups of neighbouring sugar residues. Second, the linkage between two sugar residues can be two isomers, α- and β-linkages. Third, the above two variations allow carbohydrates to have branches. This characteristic separates carbohydrates from the rest of the biological macromolecules, which contain almost exclusively linear structures. Because of this complexity, carbohydrates can provide almost unlimited variations in the structure. Thus, it is safe to say that carbohydrates are well-suited to providing various recognition signals, that serve as ligands for recognition by other molecules. On the other hand, it is also noticeable that the actual variation in carbohydrate structure is rather limited and only a fraction of the theoretically possible structures are found in nature. Actual variation, therefore, appears to be borne out of necessity and it is likely that most cell type-specific carbohydrates are functional, although we have not yet succeeded in determining their functions in all of the cases. This limitation in variation is caused mainly by two factors. Firstly, carbohydrates are not primary gene products. They are synthesized by glycosyltransferases which are protein products of the gene. Thus, it is necessary to provide a gene for every new structure of carbohydrates. It is then reasonable that living cells have a limited number of variations in carbohydrates since necessity of gene diversification is involved. Secondly, carbohydrates are very often synthesized through a common precursor. Commonality in the precursor structure thus also limits the variation of the final products.

In addition to this, carbohydrates are diverse in another manner. Carbohydrates can be present without being attached to other molecules. However, the majority of carbohydrates present in cells are attached to proteins or lipids and the latter are also called glycolipids. The carbohydrates attached to proteins can be further divided into proteoglycans and glycoproteins. In proteoglycans, the carbohydrates are composed of repeats of disaccharide units and they are relatively large (>3000 dalton). In glycoproteins, a repeating structure is usually absent except for poly-*N*-acetyllactosamines (see also section 3.2). In this chapter, I will mainly discuss glycoproteins and the other complex carbohydrates will be mentioned only when necessary.

Before I start to describe the cell-type-specific oligosaccharides, I mention the contribution of carbohydrates to the whole molecule. Figure 1 illustrates the

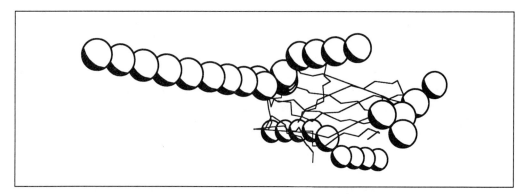

Fig. 1 Schematic representation of Thy-1. Each sugar residue is represented by a sphere of 0.608 nm in diameter. The polypeptide portion is represented by a continuous zigzag bar. Although the native molecule contains a glycosyl-phosphatidylinositol anchor, this is deleted in the figure for simplicity. This figure is taken from reference 5

molecular shape of the Thy-1 glycoprotein (5). In this figure, the sugar residues are expressed by a sphere, while a polypeptide bond is indicated by a continuous folded bar. Since a carbohydrate residue contains more carbon and oxygen atoms than an amino acid residue, the size of sugar residues is much larger than that of amino acid. Figure 1 thus illustrates that the majority of the surface area of this molecule is occupied by sugar residues, although Thy-1 contains only three N-glycans in 1 1 1 amino acid residues. In an extreme case such as lysosomal membrane glycoprotein-1, 18 N-glycans are attached to 354 amino acid residues (6). Such a molecule can be imagined as having a protein floated within a shell composed of oligosaccharides. These results strongly suggest that the contribution of carbohydrate moieties to glycoproteins are extremely significant, pointing toward the possibility that the oligosaccharide moiety can be the first to encounter molecules that recognize the glycoprotein.

In this chapter, I will try to summarize our current knowledge on cell type-specific expression of oligosaccharides. I will first describe structural variation in N-glycans (section 2). In particular, I will dissect the N-glycans into the core and terminal structure in order to better understand the heterogeneity of carbohydrates. In the following section (section 3) I will describe cell type-specific expression of N-glycans, focusing on terminal structures and poly-N-acetyllactosamines and their functionality. Section 4 deals with cell type-specific expression of O-glycans. I will then briefly describe the importance of the β-1,6-N-acetylglucosaminyl structures (section 5). Finally, I will describe the prospects in glycobiology in relation to cell biology and molecular biology (section 6). These descriptions should serve as an introduction to the rest of the book.

2. Structure of N-glycans

Carbohydrates attached to proteins can be classified into two groups, N-glycans and O-glycans. In O-glycans, saccharides are attached to hydroxyl groups of amino acids. In N-glycans, the reducing terminal N-acetylglucosamine is linked to the amide group of asparagine, thus forming an aspartylglycosylamine linkage. Since the hydroxyl group originally attached to C-1 of the N-acetylglucosamine residue is removed during the formation of this linkage, the amino group is directly attached to C-1. By examining N-glycan structures elucidated in the past two decades, it is possible to classify N-glycans into three different groups; high mannose type, complex type, and hybrid type. All of these three types contain a common structure, Manα1 → 6(Manα1 → 3)Manβ1 → 4GlcNAcβ1 → 4GlcNAc → Asn (see dotted line in Figure 2). This common structure is called the 'core', or the trimannosyl core. This core portion is composed of the di-N-acetylditobiose and three mannose residues, in which the latter form branches with α1→6 or α1→3 linkage. In high mannose-type oligosaccharides, two to six more mannoses are attached to this core (Figure 2). The largest high mannose oligosaccharide thus contains nine mannose and two N-acetylglucosamine residues, which were originally discovered in bovine thyroglobulin (7).

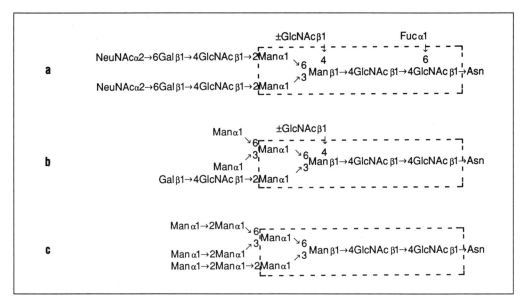

Fig. 2 Three different types of N-glycans. (a) Bi-antennary complex-type; (b) hybrid-type oligosaccharides; (c) high mannose-type oligosaccharides. Many of the high mannose-type oligosaccharides lack some of α-1,2-linked mannose residues, thus containing 5–8 mannose residues per molecule. Complex-type oligosaccharides could have many variations (see also Table 1). The structures encircled by dotted lines represent the 'core' structure that is common in all those different types

Complex-type oligosaccharides always contain three mannose residues. To the outer two α-mannose residues, N-acetyllactosamine, Galβ1 → 4GlcNAc are attached. This is followed by sialic acid residues or additional N-acetyllactosamines. In the latter case, one side chain contains repeats of N-acetyllactosamine and these glycans are called poly-N-acetyllactosamines. Poly-N-acetyllactosamines will be discussed further later in this chapter. Hybrid-type N-glycans contain more than three mannose residues, but yet also contain an N-acetyllactosamine side chain which is attached to α1→3 linked mannose (Figure 2b). This type of oligosaccharide is formed because the high mannose oligosaccharide is partially processed and then additional sugars are attached. These structural variations from high mannose-type to complex-type oligosaccharide can be fully understood once the biosynthetic mechanisms of N-glycans are understood (for details see section 2.2).

2.1 Diversity of N-glycans

The general profile of N-glycans can be summarized as described above. However, N-glycans, in particular those of complex type, have enormous diversity. In order to understand this diversity, it is useful to look separately at the core portion and the outer chain portion (8). The core portion of N-glycans is composed of three mannose and two N-acetylglucosamine residues. Complex-type oligosaccharides contain side chains elongating from α-mannose residues. The first residue of this side chain is *N*-acetylglucosamine and variation arises from the difference in the number of *N*-acetylglucosamine residues attached to α-mannoses and in the positions to which *N*-acetylglucosamines attached. So far, six different structures have been reported on this variation, as shown in Table 1. The first diversity in complex-type *N*-glycans is thus provided by a varying number of side chains that attach to

Table 1 Variation in numbers of side chains of complex-type *N*-glycans

Monoantennary	±GlcNAcβ1 ±Fucα1 ↓ ↓ Manα1↘⁶ 4 6 　　　³Manβ1→4GlcNAcβ1→4GlcNAc→Asn ---GlcNAcβ1 → 2Manα1↗
Biantennary	±GlcNAcβ1 ±Fucα1 ↓ ↓ ---GlcNAcβ1 → 2Manα1↘⁶ 4 6 　　　　　　　　　³Manβ1→4GlcNAcβ1→4GlcNAc→Asn ---GlcNAcβ1 → 2Manα1↗
Triantennary 2,4-branched	±GlcNAcβ1 ±Fucα1 ↓ ↓ ---GlcNAcβ1 → 2Manα1↘⁶ 4 6 　　　　　　　　　³Manβ1→4GlcNAcβ1→GlcNAc→Asn ---GlcNAcβ1↘⁴ 　　　　　²Manα1↗ ---GlcNAcβ1↗
Triantennary 2,6-branched	±GlcNAcβ1 ±Fucα1 ↓ ↓ ---GlcNAcβ1↘⁶ 4 6 　　　²Manα1↘⁶Manβ1→4GlcNAcβ1→4GlcNAc→Asn ---GlcNAcβ1 → 2Manα1↗³ ---GlcNAcβ1
Tetrantennary	±GlcNAcβ1 ±Fucα1 ↓ ↓ ---GlcNAcβ1↘⁶ 4 6 　　　²Manα1↘⁶Manβ1→4GlcNAcβ1→4GlcNAc→Asn ---GlcNAcβ1 → 2Manα1↗³ ---GlcNAcβ1 ---GlcNAcβ1
Pentantennary	---GlcNAcβ1↘ ±GlcNAcβ1 ±Fucα1 ---GlcNAcβ1→⁶₄Manα1 ↓ ↓ ---GlcNAcβ1↗² ↘ 4 6 　　　　　　　　　⁶Manβ1→4GlcNAcβ1→GlcNAc→Asn ---GlcNAcβ1↘⁴ ↗³ 　　　　　²Manα1 ---GlcNAcβ1↗

the α-mannose residues in the core. This side chain is also called antennary and, for instance, biantennary oligosaccharides have two side chains separately linked to C-2 of α-mannose residues. It is also necessary to mention that not all of these variations are frequently found. Monoantennary (9), 2,6-branched triantennary (10), and pentantennary N-glycans (11) rarely exist. In contrast, biantennary, 2,4-branched triantennary and tetrantennary N-glycans are frequently found.

The diversity of N-glycans is also caused by variations in the side chains. The side chains constitute the portion including the nonreducing ends. The majority of the functionality of carbohydrates lies in the portion which includes the nonreducing terminus. It is this side chain that is important to the functions of carbohydrates.

There are tremendous variations in side chains and some of them are described in Table 2. The side chains can be roughly classified into five groups. The first

Table 2 Variation in side chains of complex-type N-glycans

Group 1	Galβ1→4GlcNAcβ1→
Group 2	Galβ1→3GlcNAcβ1→
	NeuNAcα2→6Galβ→4GlcNAcβ1→
	NeuNAcα2→3Galβ→4GlcNAcβ1→
	Galα1→3Galβ1→4GlcNAcβ1→
	Fucα1→2Galβ1→4GlcNAcβ1→
Group 3	±NeuNAcα2
	↓
	6
	NeuNAcα2→3Galβ1→3GlcNAcβ1→
	Fucα1
	↓
	3
	Galβ1→4GlcNAcβ1→
	Fucα1
	↓
	4
	Galβ1→3GlcNAcβ1→
	Fucα1
	↓
	3
	NeuNAcα2→3Galβ1→4GlcNAcβ1→
	Fucα1
	↓
	4
	NeuNAcα2→3Galβ1→3GlcNAcβ1→
Group 4	R→(Galβ1→4GlcNAcβ1→3)$_n$Galβ1→4GlcNAcβ1→
	R→(Galβ1→4GlcNAcβ1
	↘6
	R→(Galβ1→4GlcNAcβ1→3)$_n$Galβ1→4GlcNAcβ1→
	(NeuNAcα2→8)$_n$NeuNAcα2→3Galβ1→4GlcNAcβ1→
Group 5	SO_4^- (3 or 4) GalNAcβ1→4GlcNAcβ1→
	NeuNAcα2→3GalNAcβ1→4GlcNAcβ1→

group is N-acetyllactosamine. This side chain is frequently found and still has the potential for either further elongation to form poly-N-acetyllactosamine, termination by sialylation, or other modification. Thus, premature termination might yield this side chain. The second group is the one terminated by sialic acid or fucose attached to the subterminal galactose residues. The addition of α1→2 linked galactose to N-acetylglucosamine also halts further elongation. There is thus no further elongation in side chains once group 2 modification takes place. Group 3 is the combination of fucose and sialic acid or addition of two sialic acids. The addition of fucose apparently halts the chain elongation in nonsialylated, fucosylated form. The addition of sialic acid to the subterminal galactose is followed by sialylation or fucosylation, resulting in disialyl structure, NeuNAcα2→3Galβ1→3(NeuNAcα2→6)GlcNAcβ1→, or sialyl-Le^x, NeuNAcα2α→3Galβ1→4(Fucα1→3)GlcNAcβ1→ or sialyl-Le^a, NeuNAcα2→3Galβ1→3(Fucα1→4)GlcNAcβ1→. Group 4 contains long side chains. Either linear or branched poly-N-acetyllactosamine contain two or more N-acetyllactosamine repeats. On the other hand, α2→8 linked sialic acid can be repeatedly added to α2→3 linked sialic acid, forming polysialyl side chains. Group 5 is unique in that GalNAc is added instead of galactose. Further sulphation or sialylation of this structure has also been found (12). More variations in this type of side chain may be found in future studies.

The combination of these diversities in the side chains and the antennary provides a vast array of heterogeneity in N-glycans. Figure 3 illustrates one such example found in N-glycans present in granulocytes (13). Notice that all of the side

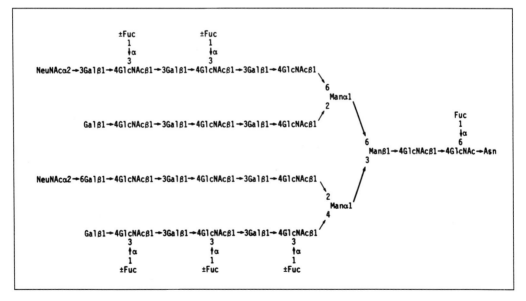

Fig. 3 Sialylated fucosyl lactosaminoglycans isolated from human granulocytes. A representative structure is shown here. There is a tremendous heterogeneity among different side chains. Although each side chain is composed of three N-acetyllactosamine units, this can be variable as well (from reference 13)

chains could have different terminal structures and the length of each antennary is also variable from 0 to 3.

2.2 Commonality of N-glycans

Although the structures of N-glycans vary enormously, it is possible to define the commonality in the structures by examining their core portions. As shown in Figure 2, complex-type N-glycans, high mannose-type, and hybrid-type N-glycans share a common core structure, which is denoted by dotted lines.

This commonality can be understood easily once the biosynthetic process of N-glycans is understood. The biosynthesis of N-glycans is initiated by the addition of a precursor sugar which is attached to dolichol. This lipid-linked oligosaccharide is composed of nine mannoses, three glucoses, and two N-acetylglucosamine residues. The structure of this lipid intermediate has been elucidated partly because the structure of high mannose type from bovine thyroglobulin, which also contains nine mannoses and two N-acetylglucosamines, had been established (7). This dolichol-linked intermediate is then transferred as a block to the Asn residue in a polypeptide. This step is common also for complex-type N-glycans, thus all N-glycans are synthesized from the common precursor. The conversion of this high mannose oligosaccharide to complex-type oligosaccharide thus requires multiple steps, involving trimming and then addition. This elaborate process is called 'processing'.

Processing of N-glycans is first initiated by the removal of three glucose residues at the endoplasmic residue (ER) (Figure 4, step ii). After one $\alpha 1 \rightarrow 2$ linked mannose is removed, the glycoprotein is then transported to cis-Golgi. If the first steps are inhibited by, for example, α-deoxynojirimycin or castanospermine, the exit of some glycoproteins from the ER is considerably slowed down (for review see reference 14). The oligosaccharide is then processed to remove three more $\alpha 1 \rightarrow 2$ linked mannoses by α-mannosidase I (step iii). The resultant oligosaccharide can now be an acceptor for N-acetylglucosaminyltransferase I, forming product e. This step is one of the most critical steps in N-glycan biosynthesis. Unless product e is formed, there is no further processing including the addition of fucose to the innermost N-acetylglucosamine or the addition of N-acetylglucosamine for the formation of antennary structures. In fact, it was shown that a CHO mutant cell line, Lec I mutant, lacks N-acetylglucosaminyltransferase I and as a result accumulates $(Man)_5(GlcNAc)_2$ high mannose oligosaccharides (8). When product e is further processed by α-mannosidase II, intermediate f is formed. This structure is usually absent but can be found in bovine rhodopsin (15, 16). This protein was found to also contain Manα1 $\rightarrow$ 6(Manα $\rightarrow$ 3)Manα1 $\rightarrow$ 6[GlcNAcβ1 $\rightarrow$ 2Manα1 $\rightarrow$ 3]Manβ1 $\rightarrow$ 4R, structure e, which is one step upstream from the product f.

Once product f is formed, it is converted to biantennary saccharide by the addition of galactose and sialic acid, or to triantennary and tetrantennary saccharide by more branches being formed by N-acetylglucosaminyltransferase IV and V followed by the addition of galactose and sialic acid. The hybrid-type oligosaccharide

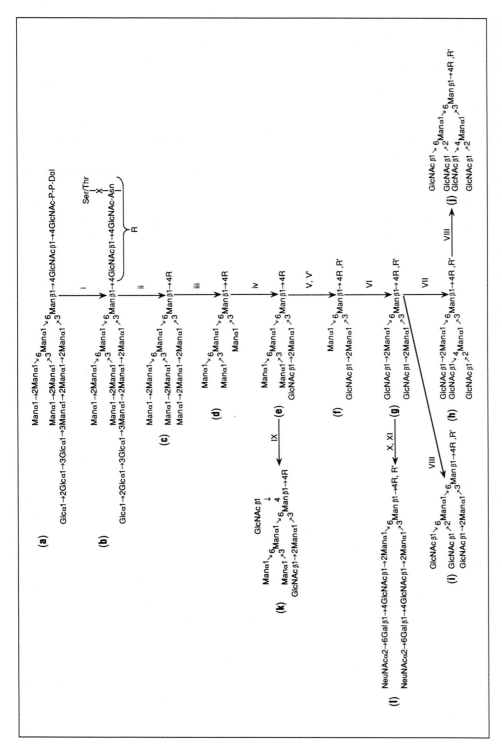

Fig. 4 The processing pathway of *N*-glycans. High mannose oligosaccharides containing glucose residues are first assembled to form a lipid-linked intermediate (a). This lipid-linked oligosaccharide is then transferred to the Asn residue in a polypeptide (b). This oligosaccharide attached to the polypeptide now undergoes many steps forming complex-type oligosaccharides (h, i, j, k, l): i, oligosaccharyltransferase; ii, ER glucosidases I and II; iii, ER α-1,2-mannosidase I; iv, *N*-acetylglucosaminyltransferase I; v, Golgi α-mannosidase II; v', (Fuc to GlcNAc) α-1,6-fucosyltransferase; vi, *N*-acetylglucosaminyltransferase II; vii, *N*-acetylglucosaminyltransferase IV; viii, *N*-acetylglucosaminyltransferase V; ix, *N*-acetylglucosaminyltransferase III; x, β-1,4-galactosyltransferase; xi, α-2,6-sialyltransferase (based on reference 162)

is formed when removal of α-mannose residues from product e does not take place and at the same time N-acetylglucosamine, and possibly galactose, is added (product k). This can be associated with the addition of β1→4 linked GlcNAc (bisecting N-acetylglucosamine) as shown in reaction IX. The product k can be, however, further processed. In this case, complex-type N-glycans with bisecting N-acetylglucosamine can be formed.

If there is no addition of N-acetylglucosamine (shown as step IV), the oligosaccharide remains high mannose type. Since high mannose-type oligosaccharides are the result of partial processing, they usually contain a varying number (5–9) of mannose residues. There is no evidence yet if any particular form of high mannose is functional.

It is common for one glycoprotein to contain more than one N-glycan attached to different sites of a polypeptide. It is also common in such cases that N-glycan attached to one position remains as high mannose oligosaccharides, while the other is processed to complex-type oligosaccharides. What determines such a difference in the processing dependent on the N-glycosylation sites, called site-specific glycosylation? The studies showed that the processing is heavily dependent on the accessibility of oligosaccharides in a given site to carbohydrate processing enzymes (17, 18). If mutant cell lines that synthesize only high mannose-type are used, glycoproteins with only high mannose type can be synthesized. If you treat this glycoprotein by endo-β-N-acetylglucosaminidase H, N-glycans can be susceptible to the enzyme only where complex-type oligosaccharides be formed in wild-type cells. This can be understood reasonably well since later processing takes place after a polypeptide is folded. High-mannose oligosaccharides remain as high mannose type because it is difficult for processing enzymes to act on them.

I have so far described the processing of N-glycans attached to secretory and plasma membrane glycoproteins. Lysosomal enzymes are processed in the same way before N-acetylglucosaminyltransferase I adds N-acetylglucosamine. Instead of this reaction, N-acetylglucosaminyl-1-phosphotransferase adds a GlcNAc-1-phosphase to C-6 of mannose residues in high mannose oligosaccharides present in lysosomal enzymes. This step is specific for lysosomal enzymes because the enzymes recognize the common domain structure exerted by folded lysosomal enzymes (19). Since such structure is absent in nonlysosomal proteins, other proteins do not accept N-acetylglucosamine-1-phosphate. After N-acetylglucosaminyl-1- phosphate is attached to lysosomal enzymes, an a-N-acetylglucosaminylphosphodiesterase, an 'uncovering enzyme,' removes N-acetylglucosamine leaving phosphate attached to high mannose oligosaccharides present in the lysosomal enzymes. This newly formed mannose-6-phosphate is recognized by mannose-6-phosphate receptors, and the receptor and lysosomal enzyme form a complex. This complex is then transported to prelysosomes, where the complex is dissociated and the lysosomal enzymes are now enclosed in the lysosomal membrane. The mannose-6-phosphate receptor, on the other hand, returns to *trans*-Golgi and they will again form a complex with a lysosomal enzyme and the cycle will be repeated (20).

The significance of N-glycan processing can be summarized as follows. Firstly, the processing unifies all the necessary biosynthesis of N-glycan into one set of consecutive reactions. Thus, it provides unified reactions on a vast array of different glycoproteins.

Secondly, the processing of N-glycan provides a unified specific modification (for instance, mannose-6-phosphate) of specific kinds of proteins (lysosomal enzymes), which are structurally quite different. This illustrates the functional significance of glycosylation of glycoprotein in general, that is to provide a unified modification in apparently unrelated proteins. The unified processing of N-glycans is also vital in providing a common structure in N-glycans which is shared among a vast variety of peptide structures.

3. Cell-type specific expression of N-glycans

As described above, the variation of N-glycans is controlled by (1) the degree of branches (antennary) and (2) the variation in terminal structures in the side chains. One of the recent critical findings is that these two events often correlate with each other. In particular, it has been increasingly evident that the formation of certain terminal structures is preferentially formed on particular side chains. In this section, I mainly describe poly-N-acetyllactosamines, since the formation of poly-N-acetyllactosamine is heavily dependent on the formation of a particular antennary and then poly-N-acetyllactosamines, in turn, provide a preferable backbone for many of the cell-type specific glycosylation.

3.1 Poly-N-acetyllactosamines—glycans with N-acetyllactosamine repeats

As described above, complex type N-glycans usually contain only one N-acetyllactosamine in each antennary. It has been shown that some N-glycans are distinct from the usual complex-type N-glycans because they have more than one N-acetyllactosamine unit in one side chain. Poly-N-acetyllactosamines were initially discovered in N-glycans present on human erythrocytes (for a review see references 1 and 2). It was found that the majority of N-glycans in human erythrocytes is much larger than typical N-glycans present in other cells, and they were found to be enriched in N-acetylglucosamine and galactose. Further studies demonstrated that these glycans contain side chains consisting of N-acetyllactosaminyl repeats. Because such side chains are much longer than a typical side chain consisting of only one N-acetyllactosamine, poly-N-acetyllactosamines containing N-glycans are much larger than the typical complex-type N-glycans. In fact, the majority of poly-N-acetyllactosamines in human erythrocytes contain 5–15 N-acetyllactosaminyl repeats in one side chain (21). Immediately after the discovery of poly-N-acetyllactosamines in human erythrocytes, it was recognized that they contain the majority of ABO blood group antigens in human erythrocytes. In parallel, it was discovered

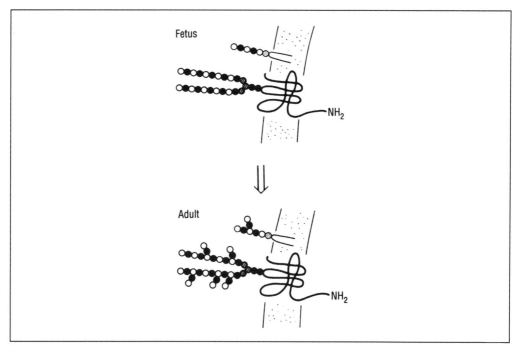

Fig. 5 Changes in poly-N-acetyllactosamines during the development from fetus to adult erythrocytes. In fetal erythrocytes, a linear unbranched poly-N-acetyllactosamine is linked to Gal→Glc→Cer or to (Man)$_3$(GlcNAc)$_2$ core of band 3. When fetal erythrocytes develop into adult erythrocytes, the linear chains are converted to those having branches. (○) galactose; (●) N-acetylglucosamine; (◐) mannose; (◉) glucose (from reference 28)

that the chemical determinants of I/i antigens are branches (I) and linear (i) poly-N-acetyllactosamines, respectively (22–24). The Ii antigens represent the adult (I) and fetal (i) human erythrocyte antigens and the conversion from i to I antigen takes place during development within a year after birth (Figure 5). These two notable initial findings that poly-N-acetyllactosamines can be modified to carry blood group and differentiation antigens made this kind of glycans the favourite subject of various researchers as seen in the following sections. The studies on poly-N-acetyllactosamines were greatly aided by the discovery that endo-β-galactosidase can preferentially cleave the poly-N-acetyllactosamines (25), as depicted below.

$$R_1 \to 4 \text{ or } 3 \text{ GlcNAc(or GalNAc)}\beta 1 \to 3\text{Gal}\beta 1 \to 4\text{GlcNAc(or Glc)} \to R_2$$
$$\downarrow$$
$$R_1 \to 4 \text{ or } 3 \text{ GlcNAc(or GalNAc)}\beta 1 \to 3\text{Gal}\beta + {}^*\text{GlcNAc(or Glc)} \to R_2$$

In many studies, the susceptibility to endo-β-galactosidase was useful to detect poly-N-acetyllactosaminyl structure. In fact, a method was developed to detect poly-N-acetyllactosamines by endo-β-galactosidase treatment followed by the addition of radiolabelled galactose to newly formed nonreducing N-acetylglucosamine residues, catalysed by β-1,4-galactosyltransferase (26). However, this method has severe limitations in the structural characterization of poly-N-acetyllactosamines in F9 embryocarcinoma cells, where structural studies provided clear evidence that there are abundant poly-N-acetyllactosamines. Since F9 cells contain exclusively branched (I) type poly-N-acetyllactosamines and the degree of branching is much higher than that in human erythrocytes, they are scarcely hydrolysed by endo-β-galactosidase and thus barely detectable by the above method. In addition to branchings at galactose residues, branchings at fucose residues also hinder the hydrolysis (For details of the specifying of endo-β-galactosidase, please refer to reference 27.) These facts thus indicate that one needs to be careful about the results obtained by one method and those studies need to be corroborated by other studies, in particular by structural studies.

On the other hand, it has been shown in human erythrocytes that glycoproteins carrying poly-N-acetyllactosamines can be detected by digestion of erythrocytes with endo-β-galactosidase after cell surface-labelling. SDS-gel electrophoresis of labelled glycoproteins before and after endo-β-galactosidase treatment clearly demonstrated that band 3 and band 4.5 are the major carriers for poly-N-acetyllactosamines (28), for which the evidence by structural studies was also obtained. Again it may not always be true that this method yields clear-cut data because the results are clearly dependent on the nature of poly-N-acetyllactosamines present in a given cell.

Poly-N-acetyllactosaminyl repeats are not uniformly distributed at different antennae arising from α-mannose. As initially discovered in human erythrocyte poly-N-acetyllactosamines, poly-N-acetyllactosamine extensions can be found more often in the side chains arising from the α-mannose residue which is attached to C-6 of β-mannose (1→6 side) (21). In particular, tetrantennary N-glycans are the most preferable acceptor for poly-N-acetyllactosamines extension, where the side chain arising from C-6 and that from C-2 in 1→6 side contain more poly-N-acetyllactosaminyl extension than the other side chains (see Figure 5). Secondly, α2→3 linked sialic acid residues are present in almost the same preferential manner as poly-N-acetyllactosaminyl extension. Thus a majority of sialylated, long poly-N-acetyllactosaminyl side chains are terminated by α2 → 3 linked sialic acid but not by α2 → 6 linked sialic acid. Third, α2 → 6 linked sialic acid, in contrast, most frequently found in the side chain arising from C-2 of α-mannose, which is attached to C-3 of β-mannose (1→3 side) Figure 6 summarizes these structural rules and provides a representative structure containing poly-N-acetyllactosaminyl repeats, α2→3 linked and α2→6 linked sialic acid. In this figure, in general, the number of m, n, o is in the order of $m>n>o$.

These characteristic structural features exist because glycosyltransferases have unique specificity toward acceptor structures. It was shown that β-1,3-N-acetylglucosaminyltransferase preferentially adds a N-acetylglucosamine residue to non-reducing terminal galactose in the side chains, with the same order, $m>n>o$ in Figure 6 (29). The β-1,3-N-acetylglucosaminyltransferase is the key enzyme forming poly-N-acetyllactosamine extension. N-Acetyllactosamine is formed as long as β-1,3 linked N-acetylglucosamine is added for extension since β-1,4-galactosyl-

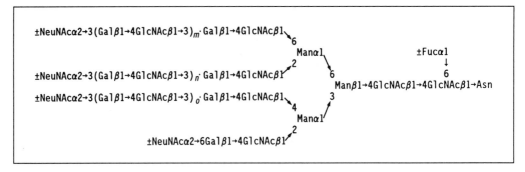

Fig. 6 Representative structures of tetrantennary N-glycan. The structure is based on the N-glycan structures reported in various cells. Poly-N-acetyllactosamine extension takes place in the order of $m>n>o$, and α2→3 sialylation takes place in a similar preference as poly-N-acetyllactosamine extension, whereas α2→6 sialylation preferentially takes place at the side chain attached to C-3 of the α-mannose

transferase is almost ubiquitously present. Thus, the amount of the β-1,3-N-acetylglucosaminyltransferase is apparently the limiting factor for poly-N-acetyllactosamine extension in a given cell.

α-2,6-Sialyltransferase, which prefers mostly nonreducing terminal galactose residues on the side chains arising from C-2 of 1→3 side, apparently has the opposite preferential side chains (30). In fact, the majority of N-glycans present on erythropoietin produced in CHO cells lack sialic acid residues in this side chain, because CHO cells do not express α2 → 6 sialyltransferase (31). This intriguing property of the preference in different side chains as acceptors is crucial to obtain distinctly different side chains depending on where the side chains are arising. Such an unique property can also be found in α-1,3-galactosyltransferase and most likely in α-1,3-fucosyltransferase (13). Such specificity of the gycosyltransferases on different side chain branches is called 'branch specificity' (30).

When glycoproteins present in a given cell are analysed, it is noticed that only a limited number of glycoproteins contain poly-N-acetyllactosamines. For instance, in human erythrocytes band 3 and band 4.5 contain a significant amount of poly-N-acetyllactosamines (21, 25), whereas glycophorins A, B, and C appear to lack poly-N-acetyllactosamines and instead contain an almost exclusively typical, short N-glycans (32, 33). This raises the question in regard to how poly-N-acetyllactosamine is preferentially formed on certain proteins but not on the others. There are several points to be considered. First, band 3 and band 4.5 (in the latter the major component is glucose transporter) have a configuration of type III membrane protein (Figure 7). The polypeptides of these glycoproteins thus traverse the lipid bilayer more than once. Two of the major lysosomal membrane glycoproteins, CD63 and Limp II also have a type III configuration. All of these glycoproteins contain a significant amount of poly-N-acetyllactosamines, as evidenced by their heterogeneity and much higher molecular weights of mature proteins compared to those of the precursor polypeptides (34). It is thus possible that this kind of topology in relation to the membrane may facilitate poly-N-acetyllactosamine

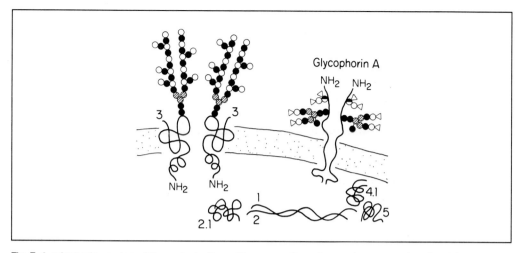

Fig. 7 A schematic version of the organization of human erythrocyte membrane proteins. Band 3 and glycophorin A represent type III and type I membrane protein, respectively. Band 3 has a long carbohydrate chain, poly-N-acetyllactosamine, in its N-glycans. Glycophorin A, on the other hand, has short, typical N-glycan and 15 short O-glycans (only one O-glycan is shown in the figure). (●) N-acetylgalactosamine; (△) sialic acid; other symbols are the same as in Figure 5 (from reference 1)

modification. The second critical observation so far is that poly-N-acetyllactosamines are preferentially formed on membrane proteins and secretory proteins contain only a small amount of poly-N-acetyllactosamines. Even when they contain N-acetyllactosaminyl repeats, they have at most one or two additional N-acetyllactosamines in a side chain (35). The lysosomal membrane glycoproteins, lamp-1 and lamp-2, are the major sialoglycoproteins in nucleated cells which contain poly-N-acetyllactosamines. When the amino acid sequences that were attached by poly-N-acetyllactosamines were elucidated, it was discovered that there was no clear consensus amino acid sequence that appeared to be a signal for poly-N-acetyllactosamine formation (36). These results suggest the third possibility that the tertiary structure is likely to be the key factor to determine if a given glycosylation site is modified by poly-N-acetyllactosamine. This is almost reminiscent of the previous discovery that the degree of N-glycan processing is the direct result of how far processing enzyme can approach the N-glycan in a given site (see above). In order to obtain poly-N-acetyllactosamine modification, it is likely that the N-glycan in that site should be fully accessible to β-1,4-galactosyltransferase and β-1,3-N-acetylglucosaminyltransferase. It is possible that glycophorins may not contain poly-N-acetyllactosamines simply because the surrounding O-glycans hamper efficient glycosylation at the N-glycan site.

Fifthly, we still need to understand why only membrane proteins, in particular those with type III configuration, are preferentially modified by poly-N-acetyllactosamines. When hCG-α chain is expressed in COS-1 cells, this secretory protein contains typical complex-type N-glycans. hCG-α chain was then fused with the transmembrane and cytoplasmic domains of vesicular stomatitis virus-G protein (VSV-G), making the chimeric protein a membrane glycoprotein. N-Glycans attached to this chimeric protein were extensively modified by poly-N-acetyl-lactosamines (37). Thus one possibility is that membrane anchoring is a prerequisite for poly-N-acetyllactosamine formation of glycoproteins. During these studies, we also found that the hCG-α/VSV-G chimeric protein is processed much more slowly than the hCG-α chain. In order to form poly-N-acetyllactosamine, the protein needs to be in contact with β-1,4-galactosyltransferase and β-1,3-N-acetylglucosaminyltransferase for a sufficient amount of time at the Golgi complex. Otherwise, the next step of glycosylation, such as sialylation, terminates the N-acetyllactosamines extension. In order to test this hypothesis, HL-60 cells were incubated at 37 or 21 °C and N-glycans attached to lamp molecules were examined (38). It was shown previously that the incubation at 21 °C prevents the existence of glycoproteins from the Golgi complex, allowing the glycoproteins to stay longer in the Golgi complex (39). The results clearly show that delaying the exit from the Golgi complex increases the poly-N-acetyllactosamine content of lamp molecules (38). These results thus suggest that there are two critical factors for the formation of poly-N-acetyllactosamines.

First, the tertiary structure of proteins to be glycosylated must allow for access to the glycosylation site by N-glycan processing enzymes, including β-1,4-galactosyltransferase and β-1,3-N-acetylglucosaminyltransferase. Secondly, it is essential for proteins to go through the Golgi complex with a relatively slow speed. It is likely that most of the glycoproteins containing poly-N-acetyllactosamines satisfy the above criterion. Further studies in this line are expected in order to provide a more clear-cut picture of this aspect.

3.2 Modification of poly-N-acetyllactosamines

By having N-acetyllactosamine repeats, poly-N-acetyllactosamines are modified more extensively than those with only one N-acetyllactosamine unit. First, it was found in granulocyte glycolipids that glycolipids with N-acetyllactosamine repeats are substituted with α1→3 fucose more significantly than those with short glycolipids. The longer they are, they are present as Le^x glycolipids, Galβ1 → 4(Fucα1 → 3)GlcNAcβ1 → 3(Galβ1 → 4GlcNAcβ1 → 3)$_n$Galβ1 → 4Glc → Cer but not those without a fucose residue, Galβ1 → 4GlcNAcβ1 → 3(Galβ1 → 4GlcNAc β1 → 3)$_n$Galβ1 → 4Glc → Cer(40). Secondly, O-glycans containing poly-N-acetyllactosamine side chains are more fucosylated than those with one N-acetyllactosamine unit (41). Thirdly, fucosylation of sialylated poly-N-acetyllactosamine is much more favourable than fucosylation of sialylated N-acetyllactosamine (42). These combined results suggest that α1→3/4 and α1→3 fucosyltransferases preferentially add fucose to poly-N-acetyllactosamines than N-acetyllactosamine. The I branching enzyme seems to have a similar property. When branched glycolipids are examined, all of the branches are formed only after N-acetyllactosamine plus lactose units from the ceramide; Galβ1→ 4GlcNAcβ1→ 3(Galβ1→ 4GlcNAc β1→ 6)Galβ1→ 4GlcNAcβ1→ 3Galβ1 → 4Glc → Cer. Galβ1 → 4GlcNAcβ1 → 3(Galβ1 → 4GlcNAc β1 → 6)Galβ1 →

4Glc→Cer has not been detected (43). In human band 3 poly-N-acetyllactosamines, branches are formed from the second N-acetyllactosamine unit from α-mannose, but not from the N-acetyllactosamine unit directly attached to α-mannose (21). These combined results thus suggest that branchings and fucosylation and possibly other modification along the side chains appear to take place more preferentially on poly-N-acetyllactosamines than one N-acetyllactosamine. It can be argued that poly-N-acetyllactosamines are preferentially modified because they have more acceptor sites. However, many of the above reactions take place at a nonreducing terminus so that essentially there is no increase in the number of acceptor sites in poly-N-acetyllactosamines. These results strongly suggest that poly-N-acetyllactosamines are better acceptors for glycosyltransferases, either by increasing the affinity to the acceptor or by increasing the velocity of the reaction. The former is more likely to cause this increase.

3.3 Cell-type specific modification of poly-N-acetyllactosamines

Not only are poly-N-acetyllactosamines extensively modified, but these modifications take place very often in a cell-type specific manner. The well established examples for this specific expression are probably those in human erythrocytes and granulocytes. Erythrocytes and granulocytes directly differentiate from the same precursor stem cells. In adult human erythrocytes, poly-N-acetyllactosamines are branched at galactose residues, forming Galβ1 → 4 GlcNAcβ1 → 3(Galβ1 → 4GlcNAcβ1 → 6)Gal, the I-branch. The termini of these branches (and linear chains) are then modified by α1→2 fucose, forming an H-structure. This H-structure now serves as an acceptor for α-1,3-N-acetylgalactosaminyltransferase or α-1,3-galactosyltransferase, forming A or B blood group antigen. In granulocytes, poly-N-acetyllactosamines have a continuously linear structure and no branching at galactose residues takes place. This linear poly-N-acetyllactosamine is then modified by α-1,3-N-fucosyltransferase, forming Le^x, or X antigen (Figure 8) (see reference 2 also). If sialylation takes place before fucosylation, such sequential reaction results in the formation of sialyl-Le^x, NeuNAcα2 → 3Galβ1 → 4(Fucα1 → 3)GlcNAc → R (see the next section). The formation of these Le^x and sialyl-Le^x structures are restricted to myeloid cell lineage and only granulocytes and monocytes are enriched with these structures in blood cells. In contrast, the presence of ABO blood group antigens is restricted to erythroid cell, and no other blood cells contain ABO blood group antigens (for review see reference 1).

This restricted expression of Le^x or ABO antigenic structures are caused by cell-type specific expression of glycosyltransferases. Apparently, α-1,2-fucosyltransferase (H enzyme), α-1,3-N-acetylgalactosaminyltransferase (A enzyme), and α-1,3-N-galactosyltransferase (B enzyme) are present only in the erythroid cell lineage while α-1,3-fucosyltransferase is present only in the myeloid cell lineage

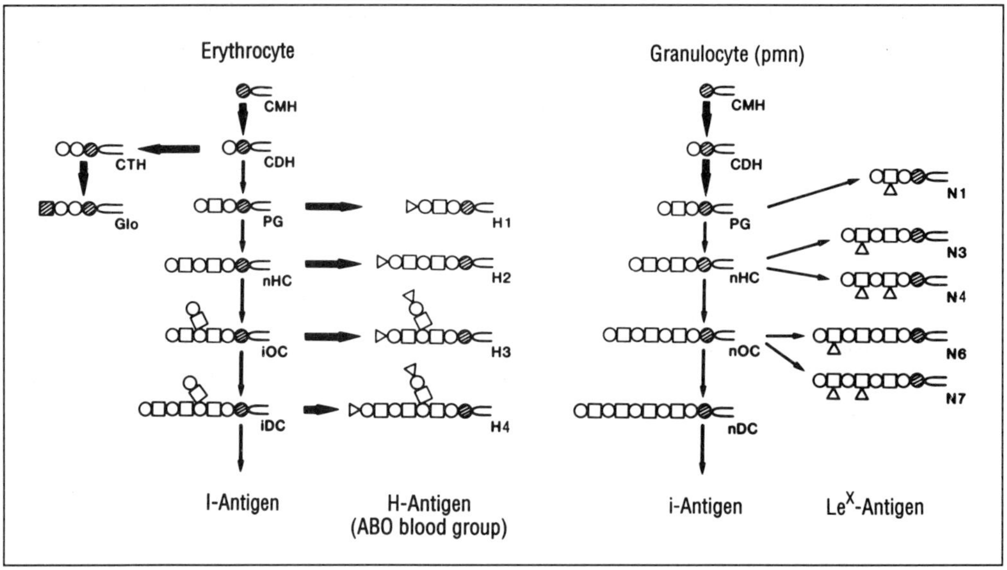

Fig. 8 Glycolipids present in normal adult erythrocytes and granulocytes. The carbohydrate structures of glycolipids are schematically shown. In erythrocytes, glycolipids of the lactoneo series are branched by 'branching enzyme' (N-acetyllactosamine: β-1,6-N-acetylglucosaminyltransferase). This branched polylactosaminyl backbone expresses I-antigen. The branched polylactosaminyl structure is then fucosylated at terminal galactose by α-1,2-fucosyltransferase, and fucosylated polylactosaminoglycan serves as a precursor for A and B blood antigens. In granulocytes, glycolipids of lactoneo-series are continuously extended by 'extension enzyme' (N-acetyllactosamine; β-1,3-N-acetylglucosaminyltransferase). This linear polylactosaminyl backbone expresses i-antigen. The linear polylactosaminyl structure is then fucosylated at N-acetylglucosamine by α-1,3-fucosyltransferase, and fucosylated polylactosamine expresses Le^x-antigen (from reference 2)

(44). These results point toward general rules in glycosylation. First, there is cell-type specific glycosylation. Second, such expression is determined by the presence of key glycosyltransferase, of which expression is restricted to certain cell types. Another good example in cell-type specific glycosylation is linear (i) versus branched (I) poly-N-acetyllactosamines. This structural change was first discovered in human erythrocytes. It was first shown that a patient's haemolytic anaemia was caused by antibodies which agglutinated the erythrocytes. This agglutination occurred only when the patient's body was exposed to cold temperature. It turned out that the antibodies agglutinate the erythrocytes from the majority of normal individuals expressing this antigen, designated I. Of the 22 000 random donors, only five were found to be I-negative, i (45). Further studies indicate that some of the antiserum in haemolytic anaemia patients contains antibodies agglutinated with these i erythrocytes, but not the I erythrocytes. Moreover, it was later found that human fetal and neonatal erythrocytes express this i antigen (46). The i antigen is then replaced with I antigen within a few months after birth. Two later independent studies provided the evidence that i antigens and I antigens are expressed by linear and branched poly-N-acetyllactosamines, respectively (22–24). In particular, the structural analysis of N-glycans attached to human band 3 clearly indicates that poly-N-acetyllactosamines from fetal and neonatal erythrocytes have linear structure, whereas adult erythrocytes have a branched structure (22) (see also Figure 5). This change is due to the appearance of the I-branching enzyme, β-1,6-N-acetylglucosaminyltransferase, which is absent in fetal and neonatal erythroid precursor cells. It is not known why this change takes place. However, the presence of linear poly-N-acetyllactosamine may be critical in suppressing the agglutination of erythrocytes by antibodies when the mother and fetus have incompatible blood groups. It has been shown that erythrocytes of the neonate reacts weakly with anti-A or anti-B IgG antibodies when the erythrocytes of the neonate suffers from haemolytic disorders due to ABO-incompatible pregnancies. Antibodies directed to carbohydrate antigens are, in general, of low or moderate affinity. The low affinity of carbohydrate–anticarbohydrate antibody will be significantly increased if two antigenic determinants are linked to two neighbouring branches. This is because such bivalent determinants could bind to two binding sites of the same IgG antibody (termed 'monogamous bivalency') (47). Although it has not been proven, it is assumed that lack of branched, bivalent ABH determinants in the fetus may have a protective effect on an ABO-incompatible pregnancy.

Poly-N-acetyllactosamine is also expressed in mouse cells (48–50). During mouse embryogenesis, it was shown that the I antigen is expressed throughout the preimplantation period, whereas the i antigen is first detected in the 5-day embryo. This expression of the i antigen is more pronounced in the primary endoderm, and the increase in the i antigen is associated with a decrease in the I antigen during development.

3.4 Poly-N-acetyllactosamines in blood group antigens

Poly-N-acetyllactosamines in human erythrocytes carry ABO-blood group antigens. First, the terminal galactose is substituted with α1→2 fucose, forming the H antigen. This formation is catalysed by α-1,2-fucosyltransferase, H-antigen and this enzyme adds fucose exclusively on a type 2 chain, Galβ1 → 4GlcNAc → R(51). However, a type 1 chain, Galβ1 → 3GlcNAc → R can serve as an acceptor for α-1,2-fucosyltransferase in other tissues, in particular epithelial tissues of gastrointestinal ducts and lung (44). The enzyme responsible for this reaction is different from the H-enzyme. This enzyme was originally described as a regulatory, Se gene, but later found to code an α1 → 2 fucosyltransferase itself (52).

The resultant H-structure, Fucα1 → 2Galβ1 → 4(or 3)GlcNAc → R, is present in O-type blood group individuals. In A and B individuals, the H-antigen is further converted to the A, or B antigen. Blood group A and B antigens are formed by the addition of α1 → 3 N-acetylgalactosamine and α1 → 3 galactose on H-structure, respectively (Figure 9). These reactions are carried out by α-1,3-N-acetylgalactosaminyltransferase and α-1,3-galactosyltransferase, and requires the H-structure for the acceptor. The A and B blood group phenotype is dominant over the O blood group phenotype. Recently, cDNA encoding the H enzyme, α-1,2-fucosyltransferase, has been cloned (51). The gene for this cDNA was found to be located in

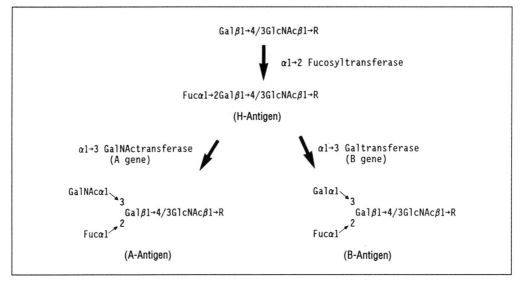

Fig. 9 Biosynthetic pathway of A, B, and H antigen formation. Almost every individual contains α1→2 fucosyltransferase, whereas the presence of A or B enzyme depends on the genomes of each individual

chromosome 19, which was shown to also be the chromosome for H-locus. At this point, cDNA for Sec-type α-1,2-fucosyltransferase has not been cloned yet, thus no comparison in sequences can be made for the two different α-1,2-fucosyltransferases. cDNAs encoding α-1,3-N-acetylgalactosaminyltransferase and α-1,3-galactosyltransferase were also isolated (53, 54). The studies indicate that the nucleotide sequences of the two cDNAs are identical, except for seven positions. O-type blood group express cDNA that is very similar to that encoding the A or B enzyme. However, the mRNA transcript present in O-type individuals has one nucleotide deletion, so that the amino acid sequence of the resulting translation product is entirely different from that of A or B enzymes. Moreover, the translation product is truncated because of the appearance of a stop codon long before the corresponding stop codon in A or B enzymes. These results strongly suggest that the translation product in O-individuals is nonfunctional. These results explain well why the phenotype of the A or B gene is dominant over that of the O gene.

The difference in A and B enzymes is due to whether UDP-GalNAc or UDP-Gal is utilized for the enzymatic action. Thus the difference in affinity toward these nucleotide sequences determines if a particular protein becomes the A enzyme or the B enzyme. The difference in the amino acid sequences between A and B enzymes exists only in four amino acids (residues 176, 235, 266, and 268). The wild-type A enzyme can be expressed as AAAA (arginine, glycine, leucine, glycine) and the B enzyme as BBBB (glycine, serine, methionine, alanine). By individual mutation of these amino acids, it was discovered that the difference in two amino acids is particularly important to determine specificity. The amino acid sequence with the combination of AAAA, AAAB, ABAA, BAAA, BAAB, BBAA encodes a protein with A transferase activity, while AABB, ABBB, BABB, BBAA encode a protein with B transferase activity. On the other hand, the combination of AABA, ABAB, ABBA, BABA, BBAB, and BBBA encodes an enzyme with both A and B transferase activities. Thus, the last group of hybrid protein can utilize either UDP-GalNAc or UDP-Gal, presumably in a compromised position. The results explain the rare phenotype, so-called cis-AB, where the same gene product is apparently responsible for both A and B enzyme activities (54). These results clearly indicate that the specificity of nucleotide sugar binding is very precisely determined. On computer analysis, the last two substitutions in B transferase were found to decrease the flexibility of the protein around the amino acids. Because there is a difference in size between Gal and GalNAc, this decrease of flexibility likely determines which sugar can be utilized by the enzyme. Further studies by X-ray crystallography are expected to reveal the difference in the structures of these two enzymes. It is important to keep in mind that such a subtle difference actually determines what kind of sugars are transferred.

In addition to the B enzyme, there is additional α-1,3-galactosyltransferase which has homology with the B enzyme. This α-1,3-galactosyltransferase does not require a fucose attached to N-acetyllactosamine. The enzyme is present in New World monkeys and many primate mammals but is absent in Old World primates, anthropoid apes, or man (55). In man, a genomic sequence very homologous to α-1,3-galactosyltransferase in nonprimate mammals can be found (56). However, the sequence does not encode a functional α-1,3-galactosyltransferase, because it has two separate deletions of a nucleotide, which results in frame-shift and at the same time, truncation of the protein product by a stop codon (56). The genes for these two enzymes, B-enzyme and N-acetyllactosamine α-1,3-galactosyltransferase, were probably produced by gene duplication of a primordial gene. In fact, the genes coding for both enzymes were found to be at the same loci, human chromosome 9 q33–34 (57). Frame-shift may be the most common mechanism to produce nonfunctional glycosyltransferase. As a rare phenotype, Bombay type, these individuals lack H-antigen, α-1,2-fucosyl residue attached to Galβ1 → 4GlcNAc → R. By comparison of genomic DNA from normal and Bombay individuals, Lowe et al. discovered that the H-gene in the Bombay individuals have several mutations (58). By replacing the normal H-gene cDNA with the nucleotides found in the Bombay, it was discovered that all of the proteins translated by cDNAs lacked the activity of α-1,2-fucosyltransferase.

These results as a whole reveal that the genetics of carbohydrate antigens such as the ABO blood group can be well-understood once information on the genes are obtained for the enzymes responsible for key reactions. In the future, I expect that this kind of development will continuously emerge.

3.5 Poly-N-acetyllactosamines in cell adhesion

A main functional significance of carbohydrates is the role they play in cell–cell interactions. It has been suggested, for instance, that cell surface carbohydrates play critical roles in neural adhesion (59), sperm–egg interaction (60), and bacterial adhesion (61). However, progress made in the last three years has demonstrated that some mammalian cell adhesion is mediated by carbohydrate–protein interaction. In order to introduce this exciting new development, I would like to provide some historical background in this particular field. One of the early observations, which indicates the role of carbohydrates, showed that plasma glycoproteins are cleared from the blood circulation once sialic acid residues are removed from the glycoproteins. Ashwell and his colleagues then elucidated that this clearance is due to binding of asialoglycoproteins to galactose-binding protein in liver hepatocytes (62). Removal of sialic acid thus exposes terminal galactose in glycoproteins, and then the latter is bound to hepatic lectin. Since then, a number of lectins present in the animal kingdom were isolated and characterized, and their amino acid sequences were elucidated by isolating and sequencing cDNAs coding for the lectins. These results reveal that animal lectins can be classified into two groups: C-type lectin, which requires Ca^{2+} for the function, and the S-type lectin, which requires the sulfide group. By comparing the amino acid sequences obtained, it is possible to define the amino acid residues which appear to be determining the *c*arbohydrate *r*ecognition *d*omain, CRD. Such domains in C-type and S-type lectins are different and unique for each type of animal lectin. These studies, pioneered by Drickamer, thus provided a basis for evaluating whether or not a given protein binds to carbohydrates through the carbohydrate recognition domain (63, see also Chapter 2). Recently, a group of adhesive molecules has been identified in leukocyte–endothelial interaction. Those molecules participate in adhesion between circulating leukocytes and vascular endothelia or platelets. The first group was initially called ELAM-1 (endothelial leukocyte adhesion molecule-1). This molecule appears on vascular endothelial cells after stimulation by cytokines during inflammation. The molecule thus traps neutrophils and monocytes at inflammatory sites, leading into extravasation of leukocytes. The second group was called PADGEM (platelet activation-dependent granule-external membrane protein), or GMP-140, also known as CD62. PADGMP is also a cell adhesion molecule that mediates the interaction of stimulated platelets with monocytes and platelets. The third group of the molecules is called MEL-14, or lymphocyte honing receptor or LEC-CAM. This molecule is the principal adhesion molecule present in leukocytes when they need to adhere to high endothelial venules (HEV) of peripheral lymph nodes. When lymphocytes are taken out from the peripheral blood and reinjected to the blood circulation, it was discovered that those injected lymphocytes accumulate in lymph nodes or Peyers' patches, depending on the cell type. This phenomenon, called lymphocyte homing, was discovered by Gesner and Ginsburg (64), to be dependent on carbohydrates of lymphocytes. It was then found that there are homing receptors that facilitate the binding of lymphocytes and granulocytes to endothelial cells, and these are called lymphocyte homing receptors. Rosen and his colleagues systematically studied this interaction and discovered that adhesion of leukocytes is dependent on carbohydrate–protein interaction (65). In particular, they found that fucosylated, sulphated polysacchar-

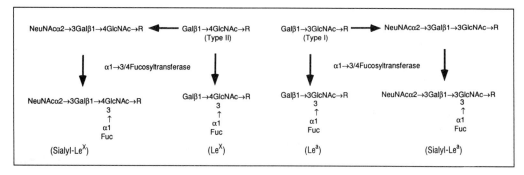

Fig. 10 Structure and biosynthesis of Lewis and Lewis-related oligosaccharide determinants. Le^a constitutes a part of Lewis blood group antigens, while Le^X constitutes a differentiation antigen. Blood cells express only type II chains while other tissues such as gastrointestinal duct express type I chains as well. Type II oligosaccharides in blood cells are fucosylated by $\alpha1\rightarrow3$ fucosyltransferase, while type I and type II oligosaccharides in gastrointestinal tissues are mostly fucosylated by Lewis enzyme, $\alpha1\rightarrow3/4$ fucosyltransferase

ides are the best inhibitors for this interaction. Just three years ago, molecular anatomy of the above three different adhesive molecules were elucidated by the isolation of cDNAs. When these amino acid sequences were examined, it was immediately recognized that the NH$_2$-terminal domains of all of these three molecules are highly homologous to the carbohydrate recognition domain, CRD (66–68). Right next to this domain, epidermal growth factor-like domain is present. This domain is followed by two to nine complement regulatory domains. This domain organization is completely conserved in the above adhesive molecules. These molecules are thus now called, E-(endothelia), P-(platelets), and L-(leukocytes) selectin, respectively (see also Chapter 4).

When these molecules were discovered to contain a lectin-like domain, it was thus natural to test immediately if they bound to carbohydrates. Fortunately, we and others have shown that granulocytes and HL-60 cells (13, 69, 70), both of which bind strongly to E- and P-selectin, are enriched with Le^X and sialyl-Le^X structures, in particular at the termini of poly-N-acetyllactosamines. Three independent groups thus obtained the evidence that E- or P-selectin binds strongly to sialyl-Le^X structure and leukocyte–endothelium or leukocyte–platelet interaction can be inhibited by oligosaccharides or glycolipids having sialyl-Le^X structures (71–74). In particular, Lowe's group demonstrated that Chinese hamster ovary (CHO) cells expressing sialyl-Le^X bind to E-selectin, while wild-type CHO lacking the same structure do not (71). This approach was possible because his group cloned cDNA encoding α-1,3/4-fucosyltransferase which forms sialyl-Le^X (75). By transfecting CHO cells with α-1,3/4-fucosyltransferase, CHO cells were transformed to express sialyl-Le^X (as well as Le^X structures). However, another $\alpha1\rightarrow3$ fucosyltransferase could express only Le^X and those transfected cells did not adhere to activated endothelial cells. These results clearly established that cell surface carbohydrates in leukocytes play critical roles in cell–cell interaction.

In static assays, neutrophils and monocytes bind to those cells expressing E- or P-

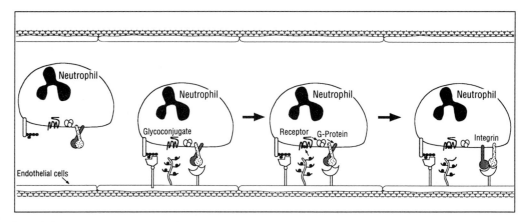

Fig. 11 Schematic representation of neutrophil and monocyte adhesion to endothelial cells. In unstimulated state, leukocytes are rapidly flowing through the blood vessel. Once endothelial cells are activated by inflammatory agents, E-selectin (and P-selectin) are transported to the cell surface and bind to leukocytes resulting in the slow down leukocytes, or 'rolling effect'. Once leukocytes are close to endothelial cells because of rolling chemoattractants such as MIP, originally bound to cell surface heparan sulphate, are transferred to the receptor (which is most likely a member of the seven transmembrane receptors). The chemoattractant then induces signals, probably through G protein, which then induces upregulation or conformational change of integrins. Those active integrins now bind to the ICAM-1 in endothelial cells, establishing tight binding to endothelials. The last step then leads to the penetration of neutrophils between the boundary of endothelial cells, and vascular extravasation

selectin. In the blood stream, leukocytes are extremely rapidly flowing through the vessel before E- or P-selectin mediated-adhesion takes place. Once E- or P-selectin appears on the cell surface of endothelium, neutrophils and monocytes are slowed down, although they are not necessarily bound tightly to the endothelial cells (76, 77). This slowing down, called the 'rolling effect,' leads to contact with chemoattractants, such as MIP (78), originally bound to the endothelial cell surface heparan sulphate, which is then transferred to receptors (Figure 11). Once neutrophils and monocytes receive a sufficient amount of chemoattractants, cells undergo a transformation that apparently leads into the activation of integrins consisting of β_2-subunit. Such activation of integrins now allows neutrophils and monocytes to become firmly attached to endothelial cells, which leads to penetration of leukocytes between interendothelial cell boundaries, and then to the extravasation of leukocytes (79, see Chapter 4 also). It was demonstrated that the first step is critical to successful extravascularization. If 'rolling' is inhibited either by antibodies against E-or P-selectin, or by sialyl Le^X-containing oligosaccharides or glycopeptides (80), the whole process can be inhibited (see Figure 11). These results clearly indicate how synergistic and ordered interactions are critical for successful inflammatory response. Further detailed molecular mechanisms during this process will undoubtedly yield exciting discoveries, which will most likely have a profound impact upon the understanding of cell–cell interaction in general.

Although the above descriptions are suitable for general description in cell–cell interaction, there are several issues to be solved. One aspect is the carrier molecules that present ligands to L-selectin. There are several reports that appear to identify glycoprotein molecules that carry ligands. First, Butcher and colleagues isolated monoclonal antibodies that inhibit the lymphocyte adhesion to high endothelial venules. These monoclonal antibodies recognized a series of molecules on endothelial cells, which they call adressin (81). These addressins can be classified into two groups: one present in peripheral lymph nodes and the other present in Pyer's patch, mucosal tissue. Very recently, cDNA for one of these mucosal addressins, MAdCAM-1, was cloned (82). The predicted amino acid sequence indicates that this receptor contains three globular domains, ICAM-1, VCAM-1, and IgA1. In addition, VCAM-1 and IgA1 domains are separated by Ser/Thr rich segment, resembling mucin-type sequence. It is noteworthy that this protein, in an environment enriched with protease, has a domain homologous to IgA1, which is in a similar harsh environment. This reminds us that lamp-1 and lamp-2, which are also in a harsh environment, also have a hinge region that is homologous to the IgA1 hinge-region (34). Lasky et al., on the other hand, discovered molecules that bind to L-selectin, which can be secreted from cultured lymph nodes. Their major molecular species is similar to one of adressins. However, this molecule is not anchored to the plasma membrane and thus differs from adressins. The cDNA sequence reveals that the molecule, termed GlyCAM-1, is a mucin-like protein, and the presentation of multiple O-glycans, some of which are the ligands for L-selectin, may be important (83). As described above, carbohydrate–protein interaction is generally not so strong that multiple interactions between ligands and binding molecules are usually necessary for binding to occur. Because O-glycans are present as clusters, (see also section 4.2 of this chapter) mucin type O-glycans may be a more favourable ligand presentor than N-glycans. In another study, a carrier molecule of P-selectin ligand was identified. This molecule, P-120, also contains O-glycans (84). More recently, cDNA encoding CD68 has been cloned. The deduced amino acid sequences indicate that this molecule is a variant of lamp-1 molecules (85). Lamp-1 contains two intralumenal domains that are separated by a hinge-like region (34). CD68 appears to have a homology with lamp-1 in the hinge-like domain, the second intralumenal domain followed by transmembrane and cytoplasmic domains. The NH_2-terminal domain of CD68 is a mucin-like domain, containing multiple O-glycans, different from that of lamp-1. It is possible that P-120 is CD68 or related to that.

In contrast to the accumulated knowledge of carbohydrate ligands for E- and P-selectin, the carbohydrate ligand for L-selectin is still not known. By using various inhibitors, however, it is inferred that L-selectin binds to carbohydrates containing sulphate and fucose (86). Further studies will be of significance to reveal the actual ligand structure for L-selectin. More of the molecules containing the carbohydrate recognition domain may be found in other proteins. In fact, a proteoglycan was found to contain the carbohydrate recognition domain (87), although this domain is present in the internal part of the molecule, and it is not certain if it is actually functional in binding to carbohydrates. When only the carbohydrate binding domain is expressed by in vitro translation, it was found that the protein segment binds to specific monosaccharides (88). Future studies will be of significance to determine the nature of carbohydrate ligands if intact molecules bind the carbohydrates.

3.6 Poly-N-acetyllactosamines in tumour cells

Cell surface carbohydrates of tumour cells were reported to have an altered structure when parent cells and tumour virus-transformed cells are compared. These initial reports were then extended by Waren and Glick, showing that transformed cell lines express more glycopeptides with high molecular weight than the parent cells. This increased size was initially concluded to have a higher degree of sialylation, or an increased amount of sialic acid in tumour cells (for review see reference 89). In contrast, Ogata *et al.* discovered that tumour cells express more glycopeptides which are not bound to concavalin A conjugated to Sepharose. It was shown that ConA-Sepharose binds to high mannose oligosaccharides and biantennary oligosaccharides but not tri- and tetrantennary saccharides (90). This phenomenon was further analysed and Takasaki *et al.* concluded that there is an increase of tri- and tetrantennary saccharides in transformed cells (91). In the same analysis, it was found that some of the *N*-glycans in tumour cells contain more than four *N*-acetyllactosamine units, suggesting that there may be *N*-glycans with more than four antennas. Yamashita *et al.* then elucidated the structures of those glycans with high numbers of *N*-acetyllactosamine units and found that they contain *N*-acetyllactosamine repeats, poly-*N*-acetyllactosamine side chains (10). Hereby, it was established that tumour cells synthesize *N*-glycans with an increased number of antennas and at the same time also with poly-*N*-acetyllactosamine side chains. These results were confirmed by other studies in the following years. In particular, Dennis *et al.* found that the increased amount of tetrantennary *N*-glycans and poly-*N*-acetyllactosamine are also associated with tumour cells showing high metastatic capacity, compared to those with low metastatic capacity (92). It was then shown that there is a preferential site for poly-*N*-acetyllactosamine extension. The antennary increase in tumour cells, that is from the C-6 of 2,6-linked mannose, is the most preferential site for poly-*N*-acetyllactosamine extension (see 'branch specificity' section 3.1). This is the reason why the increase of tetrantennary saccharides which contain the side chain elongating for the C-6 of α-mannose, results in the increase of poly-*N*-acetyllactosamines. Thus it may be concluded that tumour cells, in particular metastatic tumour cells are enriched with tetrantennary *N*-glycans containing poly-*N*-acetyllactosamine repeats.

In relation to these findings, it was reported that some glycosylation inhibitors inhibit tumour metastasis when administered to animals. For example, such an effect was shown by tunicamycin (93), which completely blocks *N*-glycan synthesis, and by castanospermine and swainsonine (94, 95). The latter two drugs presumably inhibit the processing leading to complex-type carbohydrates, and thereby inhibit the formation of poly-*N*-acetyllactosamines as well as *N*-acetyllactosamine. Thus, these results apparently point toward the possibility that the amount of tetrantennary saccharides and poly-N-acetyllactosamines are directly related to the tumourigence of tumour cells.

Why are such oligosaccharides important for tumourigenicity? In the previous section, I described that poly-N-acetyllactosamines provide a preferable backbone for sialyl-Le^X formation. It is thus possible to speculate that the increase of poly-N-acetyllactosamine actually leads to an increased amount of sialyl-Le^X structures, in the long backbone. It is likely that such presentation of carbohydrate ligands is more accessible to binding proteins, such as selectins, than to sialyl-Le^X in short side chains. Furthermore, it is expected that the amount of terminal Le^X itself is increased in tetrantennary saccharide with poly-N-acetyllactosamine backbones, since they are a preferable acceptor for α-1,3-fucosyltransferase and possibly for α-1,3/4-fucosyltransferase (13, 69). The process of tumour metastasis is reminiscent of the process for the leukocyte adhesion to inflammatory sites. In fact, many carcinoma cells are found to have an increased amount of sialyl-Le^X (96–98) and sialyl-Le^a (99) structures which were shown to be ligands for E-selectin (100). It is also known that tumour cells are attached to platelets, forming an agglutination of tumour cells surrounded by platelets (101). It is thus possible that blood-borne metastatic tumour cells are first bound to platelets, through tumour cell carbohydrate–platelet P-selectin interaction. Such aggregates can be easily trapped in capillary veins, whereby tumour cells release some of the cytokines that activate endothelial cells. Alternatively, activated platelets, upon multiple binding to tumour cells, release cytokines, which activate endothelial cells. Once endothelial cells are activated, tumour cells are bound to E-selectin (and P-selectin) expressed on endothelial cells. Such binding to selectins on endothelial cells leads to firm attachment to endothelial cells, and then eventual extravasation. This process thus establishes tumour cell growth in metastatic sites (see also reference 101).

This hypothesis is extremely attractive since it unifies many findings so far accumulated on the property of tumour cells, in particular metastatic tumour cells. Further studies will be exciting if they can show that this is the actual mechanism, or at least one of the mechanisms, for successful establishment of metastasis. In relation to this hypothesis, it is noteworthy that myelogenous leukaemic cells are enriched with sialyl-Le^X and sialyl-di-Le^X structures (102, 103). In this case, it is possible that leukaemic cells utilize sialyl-Le^X–E-selectin interaction to attach and cross the endothelial barrier, eventually establishing the perfusion of tumour cells in the body fluid. If the above hypothesis is correct, it is assumed that the amount of sialyl-Le^X in a preferable carrier molecule should be increased. Results from various laboratories indicate that this is the case. First, it was shown that tumour cells were found to adhere to endothelial cells by E-selectin mediated adhesion (104, 105). The results thus seem to be direct evidence that E-selectin mediated adhesion is the major force in tumour cell adhesion to endothelial cells. Second, the cell surface expression of lamp-1 and lamp-2 was compared between highly metastatic and low metastatic colonic carcinoma cell lines (106). The results indicate that the highly metastatic cells express more lamp-1 and lamp-2 on their cell surface than the low metastatic cells. At the same time, it was shown that the sialyl-

Le^x structure with high affinity to the antisialyl-Le^x antibody, presumably attached to poly-N-acetyllactosamines, is increased on highly metastatic cells. The colonic carcinoma cells were then tested for adhesion to human and mouse endothelial cells. The results demonstrated that highly metastatic tumour cells exhibit a better efficiency of E-selectin mediated adhesion. This was also true for the adhesion to CHO cells stably expressing E-selectin (107). A third series of experiments increased the amount of cell surface lamp-1 of the low metastatic colon cardinoma cells by genetic manipulation. It was then asked if those cells were more adhesive to E-selectin expressing cells. The results showed that the amount of cell-surface lamp-1 is directly proportional to the extent of E-selectin mediated adhesion of those tumour cells. The level of the cell-surface lamp-1 is also closely correlated to the amount of cell-surface sialyl-Le^x. Moreover, such adhesion can be effectively inhibited by soluble lamp-1 containing sialyl-Le^x structures (108). No inhibition was obtained when a soluble lamp-1, used as an inhibitor, did not contain sialyl-Le^x structures. These results strongly suggest that soluble lamp-1, or its related molecules, is useful as an inhibitor for therapeutic purposes (Figure 12). In relation to these findings, it is noteworthy to mention the studies on the prognosis of different groups of patients who differed in the amount of cell surface sialyl-Le^x structures in tumour cells (109). The results indicate that patients with a higher expression of sialyl-Le^x show poorer prognosis than those with a lower expression of sialyl-Le^x. These studies thus strongly suggest that one of the key factors in metastatic spread is the amount of sialyl-Le^x structure, the ligand for E- and P-selectin in tumour cells. Further studies will be important if those cells expressing more cell-surface lamp-1 become highly metastatic cells. Although it is a challenging task, it also may be possible to interfere with tumour metastasis by injecting soluble lamp-1, or by injecting cells continuously expressing soluble lamp-1 that contain sialyl-Le^x structures.

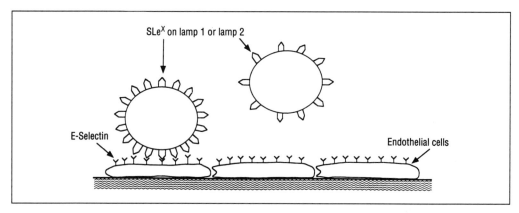

Fig. 12 Increase of lamp-1 molecules on the cell surface of SP colonic carcinoma cells results in increased adhesion to E-selectin-expressing cells. Low metastatic colonic carcinoma SP cells express a small amount of sialyl-Le^x on cell surface lamp molecules (see the *right* cell). When the same cells were transfected with lamp-1 expression vectors, the number of the cell surface lamp-1 molecules was increased. This increase was accompanied by an increased amount of cell surface sialyl-Le^x determinants and the increased efficiency of adhesion to E-selectin-expressing cells (see the *left* cell) (from reference 108)

In relation to poly-*N*-acetyllactosamine, it is noteworthy that the defect in poly-*N*-acetyllactosamine results in haemolytic anaemia, HEMPAS. The disease is apparently caused by incomplete synthesis of poly-*N*-acetyllactosamine in erythroid cells. At this point, it is likely that this disease is a collection of genetic defects in glycosylation, which by one way or another results in incomplete synthesis of *N*-glycans, particularly poly-*N*-acetyllactosamine. Further studies on genetic defects in each case will likely yield critical information on the biosynthetic mechanisms of poly-*N*-acetyllactosamines in erythroid cell development (for review see reference 110).

4. Mucin-type glycans (*O*-glycans)

So far I have described *N*-glycans which are attached to asparagine residues of polypeptides. As a second abundant form of carbohydrates attached to proteins, there is a group of glycans in which the reducing terminal *N*-acetylgalactosamines are attached to serine and threonine residues of polypeptides. Since this group of glycoproteins and carbohydrates was initially found in a mucus substance, called mucin, this type of carbohydrate is called mucin-type glycans. Mucin-type glycans belong to a member of the group of glycoproteins in which the reducing terminal sugars are attached to hydroxyl groups of polypeptides. This superfamily is called *O*-glycans. *O*-Glycans can be classified into different subgroups, depending on the nature of sugar and amino acid residues which are involved in the linkage between carbohydrate and proteins. They are classified as the following: (a) *N*-acetylgalactosamine linked to serine or threonine (mucin-type carbohydrates) (111), (b) *N*-acetylglucosamine linked to serine or threonine (*O*-GlcNAc type) (112), and (c) xylose linked to serine or threonine in proteoglycans (113). These three groups are relatively abundant in animal cells, although *O*-GlcNAc type has so far been found only in cytoplasmic or nuclear proteins. Less often encountered are (d) galactose linked to hydroxylsine in collagen (114) and proteins containing collagen-like domain such as C1q (115), (e) xylosyl glucose or glucose linked to serine or threonine in clotting factors (116), and (f) fucose linked to plasma glycoproteins (117). In addition, in plants, there are (g) arabinose linked to hydroxyproline (118) and (h) galactose linked to serine, and in fungi, mannose linked to threonine or serine (119). As you can see, *O*-glycans include a wide variety of glycoproteins, but in this chapter I would like to restrict myself to mucin-type glycoproteins, because they are abundantly present in animal cells.

4.1 Mucin-type glycans—distribution and role

Mucin-type saccharides are characterized by their wide variety of distribution and by the vast heterogeneity in their structures. Mucin-type glycoproteins are initially found in mucus covering the surface of epithelial cells in the gastrointestinal system and respiratory ducts. It is likely that mucin proteins are produced from various types of cells present in mucus. However, very few studies showed how many different cell types are producing mucus glycoproteins. If one analyses the mucin proteins from a particular cell, one can still find three different mucin proteins in certain cases (120). Since no studies have been made so far, on a single mucin-producing cell, heterogeneity of oligosaccharides can be due to the fact that they are synthesized in more than one kind of cell. It is also likely that heterogenous saccharides are attached to even a single mucin protein. Thus, not only the heterogeneity of core protein, but also that of carbohydrate moieties provide heterogeneity of mucin glycoproteins.

For what reason are mucin-type glycoproteins so heterogenous? The predominant and critical role of mucin proteins is to provide a smooth surface layer on epithelial cells. Because O-glycans can be attached in a high density as clusters, mucin-type glycans provide a highly hydrated surface. Secondly, these saccharides can have highly heterogenous structures so that they provide a barrier between the body fluid, which may contain proteases, and the epithelial surfaces. Moreover, complex and heterogenous carbohydrates are more resistant to various digestive enzymes because the digestion of those carbohydrates requires many different hydrolases (121). Thirdly, it is possible that mucin-type carbohydrates serve as a trap for bacterial adhesion. Many bacterium infecting the duct contain lectins, which bind to the carbohydrates of mucin. By having a variety of carbohydrates, there are more chances that any bacterium may bind to some portion of the mucin-type saccharides (122). Such adhesion then could lead to the excretion of bacteria from the duct either by coughing or in the stool.

4.2 Mucin-type glycoproteins in the plasma membrane — glycophorin A and leukosialin

In mucin glycoproteins, a great number of saccharides are attached to polypeptides as a cluster and these polypeptides are covered by a continuous layer of oligosaccharides. This type of oligosaccharide attachment is not unique to so-called mucin-type glycoproteins but also can be seen in glycoproteins in the plasma membrane. Here I would like to describe two such glycoproteins. One is glycophorin A and the other is leukosialin (CD43). Glycoprotein A and leukosialin are the major sialoglycoproteins of human erythrocytes and leukocytes, respectively. Glycophorin A has an extracellular domain consisting of 92 amino acids, among which 16–19 are occupied by mucin-type saccharides (for review see reference 123). Leukosialin contains 234 amino acid residues in the extracellular domain, among which 93 residues are serine or threonine. Since about 90% of these hydroxyl amino acids are occupied by oligosaccharides in leukosialin, one out of the three amino acids in the extracellular domain is attached by oligosaccharides (for review see reference 124) (Figure 13).

One of leukosialin's functions is to cover the surface of leukocytes with a large amount of negative charges, preventing the adhesion of leukocytes to each other

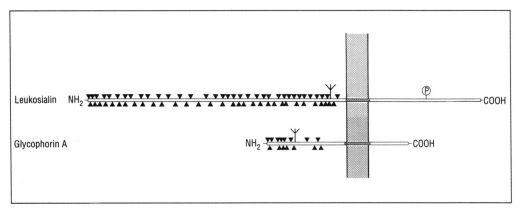

Fig. 13 Schematic structure of leukosialin and glycophorin A. Leukosialin and glycophorin A contain ~ 80 and 15 O-linked oligosaccharides, respectively, and these oligosaccharides are denoted by ▼, while one N-glycan is denoted by Y. The phosphorylation site in leukosialin is indicated by Ⓟ (163). The size of each domain is shown in proportion to sizes

and to endothelial cells in the blood stream. Moreover, it was shown that leukosialin plays some role in cell–cell interaction. For instance, it was demonstrated that T and B lymphocyte interaction can be facilitated by the presence of leukosialin in T lymphocytes (125). It was also demonstrated that the addition of antibody specific to leukosialin activates monocytes or T lymphocytes (124). Since the mucin-type oligosaccharides attached to leukosialin change dramatically in T-cell development or T-cell activation (see next section), it is likely that those changes in O-glycans modulate the interaction mediated by leukosialin.

Isolated cDNA sequence revealed that the amino acid sequence of human leukosialin has tandem repeats of short amino acid sequences that provide O-glycan attachment sites (126). Similarly, the amino acid sequences of bovine submaxillary apomucin (127) and large intestine apomucin (128) also contain tandem repeats of amino acid sequences that contain multiple O-glycan attachment sites. One of the mechanisms to produce such tandem repeats could be duplication of exons coding such amino acid sequences. When the genomic sequences of human leukosialin were revealed, however, the whole coding sequence of leukosialin was unexpectedly found to lie on one exon (129). Similarly, the whole coding region of apomucin from submaxillary gland and large intestine was found to be coded by only a single exon (130, 131). This includes 25 tandem repeats of the sequence with 81 amino acids in submaxillary gland apomucin.

These results suggest that there may be some common genetic mechanisms for generation of the polypeptide backbone that are heavily decorated by mucin-type oligosaccharides. It is possible that these sequences were first made by duplication of one exon, and then the duplicated exons were connected together by eliminating introns. On the other hand, a similar mucin-like glycoprotein, glycophorin A, is coded by several exons (132) and the amino acid sequences containing O-glycan attachment sites are coded by multiple exons (Figure 14). A most recent report

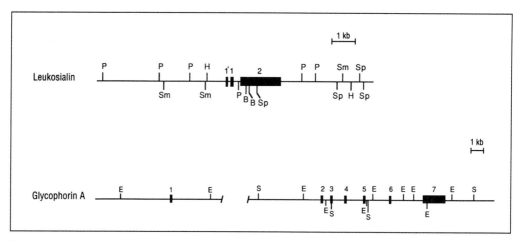

Fig. 14 The organization of the leukosialin and glycophorin A genes. The exons are depicted by solid boxes in proportion to their sizes. Exon 1' is utilized in only one case, while exon 1 is utilized in all the other cases for expression of leukosialin. The restriction enzyme sites shown are: E, *EcoR*I; P, *Pst*I; Sm, *Sma*I; H, *Hind*III; B, *BamH*I; Sp, *Sph*I. The distance between exons 1 and 2 in glycophorin is more than 15 kb (from reference 124)

showed that another mucin-like glycoprotein, GlyCAM-1 is also coded by several exons (133). These results indicate that mucin and mucin-like glycoproteins can be classified into two groups, depending on their genomic organization.

4.3 Structures of mucin-type oligosaccharides

The structures of mucin-type oligosaccharides appear to be extremely heterogenous. However, it is also possible to classify these oligosaccharides, according to the structures of the core portions of their oligosaccharides. This classification, proposed by Schachter and his colleagues is based on the sugar residues attached to *N*-acetylgalactosamine, which is linked to threonine or serine residues (134). First, the addition of β-1,3-linked galactose to *N*-acetylgalactosamine forms 'core 1'. Core 1 can be converted to 'core 2' by β-1,6-*N*-acetylglucosaminyltransferase (Figure 15). On the other hand, the addition of β-1,3-linked *N*-acetylglucosamine to *N*-acetylgalactosamine forms 'core 3'. Core 3 can be converted to 'core 4' by another β-1,6-*N*-acetylglucosaminyltransferase. If α-2,6-sialyltransferase acts on *N*-acetylgalactosamine, NeuNAcα2→6 *N*-acetylgalactosamine is formed. This product is no longer modified, becoming a final product, although all of the other core structures (core 1–4) can be further extended (for review see reference 134). In addition to these frequently encountered structures, NeuNAcα2→6(GalNAcα1→3) GalNAc was found in human rectum tumour cells (135). So far, there are no additional findings on this particular core structure but it is tentatively called 'core 5'. This type of classification for mucin-type glycans, based on their core structures, is extremely useful. This is particularly true when mucin-type oligosaccharides are analysed from different sources of tissues or cells. The glycosyltransferases respon-

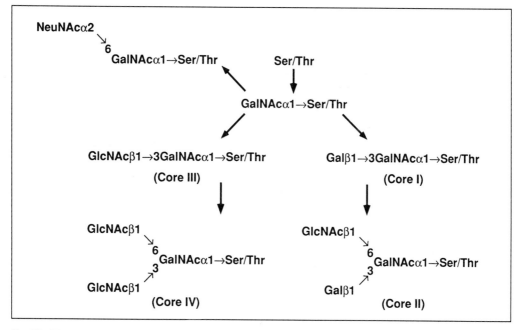

Fig. 15 Biosynthesis steps of the core portions of oligosaccharides (O-glycans). As the first step, N-acetylgalactosamine is added to serine or threonine residues in apomucin proteins. This is then converted to core I by β-1,3-galactosyltransferase. Core I can then be converted to core II by β-1,6-N-acetylglucosaminyltransferase. GalNAcα1→Ser(Thr) is also converted to Core III by β-1,3-N-acetylglucosaminyltransferase and then to core IV by another β-1,6-N-acetylglucosaminyltransferase. GalNAcα1→Ser(Thr) can also be a substrate for α-2,6-sialyl-transferase, forming NeuNAcα2→6GalNAcα1→Ser(Thr). This oligosaccharide can no longer be an acceptor for other glycosyltransferase and thus becomes a final product

sible for core 2, core 3, or core 4 formation are present in relatively restricted tissues or cells so that the formation of these structures are characteristic of different cell types (see also reference 136). Thus, cell-type specific expression of mucin-type oligosaccharides also exists. In order to describe this point, the glycosylation of leukosialin is described in the next few sections.

4.4 Changes of mucin-type oligosaccharides attached to leukosialin during differentiation

Because leukosialin is expressed in various leukocytes, oligosaccharides attached to leukosialin can be analysed in various cell lineages. First, it was discovered that the molecular weight of leukosialin varies significantly depending on the cell. Leukosialins from K562 (erythroleukaemia), HL-60 (promyelocyte) and HSB-2 (T lymphocyte) display apparent molecular weights of 105, 130, and 120 kDa, respectively. On the other hand, leukosialin in the peripheral blood display apparent molecular weights of 105 kDa (in resting T lymphocytes) or 130 kDa (in granulocytes, monocytes, and platelets). The difference in molecular weights was judged to be caused by the difference in mucin-type oligosaccharides, because the precursor polypeptide, detected by ^{35}S-methionine pulse labelling, exhibited an identical molecular weight irrespective of mucin-type oligosaccharides (O-glycans) attached to leukosialin (137). The analyses of the oligosaccharides showed that K562 leukosialin almost exclusively contain O-glycans with core 1, NeuNAcα2 → 3Galβ1 → 3(NeuNAcα2 → 6)GalNAc. In contrast, HL-60 leukosialin contains those with core 2, NeuNAcα2 → 3Galβ1 → 3 (NeuNAcα2 → 3Galβ1 → 4GalNAcβ1 → 6)GalNAc. It is thus obvious that leukosialin with more complex O-glycans exhibits a larger molecular weight. HSB-2 leukosialin was found to contain a mixture of O-glycans with core 1 and core 2, and it showed an intermediate molecular weight between HL-60 and K562 leukosialins. These results clearly indicate that the molecular weight of mucin-type glycoproteins such as leukosialin is determined by the complexity of attached O-glycans (136). It was shown that the size of a major sialoglycoprotein of human and mouse T lymphocytes increased from 110 to 130 kDa, upon mitogenic, T-cell activation (138, 139). This major sialoglycoprotein was found to be leukosialin by immunoprecipitation with leukosialin-specific antibodies, and the oligosaccharides attached to leukosialin were found to change (140). Thus, O-glycans present in resting T lymphocytes were found to be mainly those with core 1, NeuNAcα2 → 3Galβ1 → 3(NeuNAcα2 → 6)GalNAc. After T-cells are activated, those cells now express O-glycans with core 2, NeuNAcα2 → 3Galβ1 → 3(NeuNAcα2 → 3Galβ1 → 4GlcNAcβ1 → 6)GalNAc. This conversion is caused by the appearance of β-1,6-N-acetylglucosaminyltransferase, core 2 β-1,6-N-acetylglucosaminyltransferase, which is absent in resting T lymphocytes (140) (Figure 16a).

As described above, leukosialin is apparently involved in the interaction between T and B lymphocytes. The changes of oligosaccharides attached to leukosialin thus likely modulate the leukosialin function, considering a large number of O-glycans cover the leukosialin polypeptides. Moreover, it is possible that some of the O-glycans may serve as ligands (see below).

4.5 Changes in mucin-type oligosaccharides in tumour and immunodeficiency

The amount of the oligosaccharides with core 2 is also increased in various pathologic conditions. In particular, knowledge has been accumulated in patients with leukaemia and immunodeficiency syndrome. O-glycans with core 1, NeuNAcα2 → 3Galβ1 → 3(NeuNAcα2 → 6)GalNAc, are expressed on T lymphocytes in the peripheral blood of normal individuals. Leukocytes from patients with T-cell leukaemia express a large amount of the oligosaccharides with core 2, NeuNAcα2 → 3Galβ1 → 3(NeuNAcα2 → 3Galβ1 → 4GlcNAcβ1 → 6)GalNAc (141). As described above, this hexasaccharide is increased in activated T lymphocytes under normal conditions. The same hexasaccharide is, however, substantially increased in cases of acute leukaemia and moderately in cases of chronic lymphocytic

MUCIN-TYPE GLYCANS (O-GLYCANS) | 35

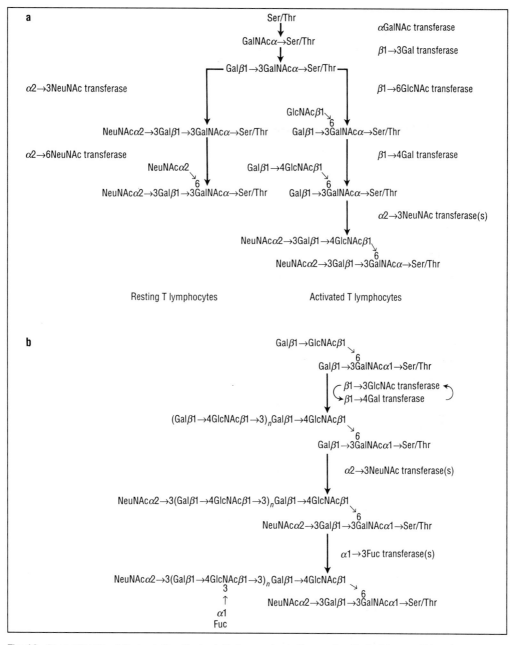

Fig. 16 Biosynthesis of leukosialin-attached O-glycans in resting and activated human T lymphocytes and poly-N-acetyllactosaminyl formation in O-glycans. (a) Resting T lymphocytes express no detectable Galβ1→3GalNAc→R (GlcNAc to GalNAc)β-1,6-N-acetylglucosaminyltransferase, whereas activated T lymphocytes express a significant amount of the enzyme (from reference 140). (b) Poly-N-acetyllactosaminyl extension can be extended from the GlcNAcβ1→6 linkage formed in core 2. Poly-N-acetyllactosaminyl extension can be further modified by α-1,3-fucosyltransferase; forming sialyl-Le^x termini (from reference 153)

leukaemia. It appears that the amount of the hexasaccharides is more pronounced when more immature T cells are produced in leukaemia. This increase in the amount of hexasaccharides is caused by the increase of β-1,6-N-acetylglucosaminyltransferase that forms core 2 branchings (141). Similarly, it was reported that the core 2 β-1,6-N-acetylglucosaminyltransferase is increased in myelogenous leukaemia cells (142), although it should be noted that normal granulocytes are also enriched with the hexasaccharide.

The increase in the amount of hexasaccharide has also been found in patients with immunodeficiency diseases, such as Wiskott–Aldrich syndrome (143, 144) and acquired immunodeficiency syndrome, AIDS (141). Wiskott–Aldrich syndrome is characterized by thrombocytopenia with reduced platelet volume and number, immunological defects affecting both humoral and cellular responses, and eczema (145). This disease is a congenital one and the defective gene is localized at chromosome X, band p11-q12 (146). Boys are thus inflicted with this disease and the patients die at a young age due to recurrent infection or leukaemia. Remold-O'Donnell *et al.* first reported that a major sialoglycoprotein in T lymphocytes is defective in this disease using L10 monoclonal antibody (147). This major sialoglycoprotein was identified to be leukosialin and it was discovered that its molecular weight was increased in the leukocytes of patients with the syndrome (143, 144). This increase was found to be due to the increased amount of hexasaccharide in core 2, the same oligosaccharides that are increased in activated T lymphocytes under normal conditions. It is possible that L10 monoclonal antibody, which detected an abnormality in leukosialin, binds to leukosialin in a carbohydrate-dependent manner. This may be the reason why leukosialin was not detected in leukocytes from patients with Wiskott–Aldrich syndrome when this monoclonal antibody was used. The increase of the hexasaccharide is clearly associated with the increase of core 2 β-1,6-N-acetylglucosaminyltransferase (143, 144). It is noteworthy that these leukocytes, despite the fact that they express the hexasaccharides, are not functional. They do not respond to various mitogens, which induce T-cell activation under normal conditions. Similar changes in oligosaccharides, being dissociated from functional maturation, are observed in the leukocytes of AIDS patients. The apparent molecular weight of leukosialin was found to increase as the symptoms of AIDS progressed (141). Moreover, it was reported that the antibodies reactive with leukosialin of AIDS patients are newly formed (148). It is possible that the changes in O-glycans attached to leukosialin provoked the antibody production in AIDS patients. These phenomenon can be summarized as follows. T lymphocytes in pathological conditions acquire the hexasaccharides which are present in activated T lymphocytes, despite the fact that they do not have functions associated with activated T-cells under normal conditions. This phenomenon, is thus called 'pseudoactivation'. It was shown that leukosialin is coded by the gene present in chromosome 16, band p11.2 (126). In contrast, the gene defective in Wiskott–Aldrich syndrome is found on the X chromosome, band p11-q12 (146). It will be of interest to determine if one of the transcription factors which activates the expression of core 2 enzyme gene, is related to the protein coded by the X chromosome, band p11-q12.

The above results may be related to the fact that thymocytes, immature T lymphocytes, contain leukosialin that is heterogeneous in molecular weight. Some of them are almost as large as that in activated T lymphocytes. Consistent with this result, the oligosaccharides attached to leukosialin in thymocytes are a heterogenous mixture of the tetrasaccharide with core 1 structure and the hexasaccharide with core 2 (141). Fox et al. isolated a monoclonal antibody which specifically reacts with the hexasaccharide attached to leukosialin. When this monoclonal antibody, T305, was used to detect the hexasaccharides attached to leukosialin, the immature cortical region of thymus is strongly stained while medullary thymocytes are not (149). These results indicate that leukosialin with the hexasaccharide is present in immature thymocytes, but replaced with those with the tetrasaccharide during T-cell maturation in thymus (see also reference 150). It is thus possible that the hexasaccharide is increased in leukaemia and immunodeficiency because immature T lymphocytes are increased in these diseases. The hexasaccharides thus can be regarded as an oncodifferentiation antigen.

The increase of the hexasaccharide was reported in other tumour cells. Amano et al. (151) analysed the carbohydrate structures attached to chorionic gonadotropin and found much more branched hexasaccharides in those isolated from patients with choriocarcinoma than in those isolated from normal individuals. Although the normal individuals also synthesized a core 2 structure, its amount was much less than that of core 1. Yousefi et al. provided the evidence that malignant transformation of rodent cells results in a significant increase of core 2 β-1,6-N-acetylglucosaminyltransferase. This increase resulted in an increase of poly-N-acetyllactosamines in O-glycans (152), as Galβ1 → 4GlcNAcβ1 → 6 side chain is a preferential site for poly-N-acetyllactosamine extension (41) (Figure 16). Previously, the same group found in the same system that the increase of N-acetylglucosaminyltransferase V is associated with the increase in poly-N-acetyllactosamine extension (92). Apparently, it can be generalized that the increases in β-1,6-N-acetylglucosaminyl linkages and poly-N-acetyllactosamines are associated with malignant transformation. As described above, poly-N-acetyllactosamine provides preferable backbones for sialyl-Le^X formation (Figures 16 and 17). In fact, sialyl-Le^X is present in leukosialin from HL-60 cells but absent in that from K562 cells, and the former contains core 2 enzyme while the latter does not (153). These results suggest that O-glycans present on leukosialin or other mucin-type glycoproteins can present ligands to E- and P-selectin. In this context, it is noteworthy that a soluble form of leukosialin is present in the human plasma membrane at a relatively high concentration (10 μg/ml). This soluble form of leukosialin, apparently produced by proteolytic digestion, contains the branched hexasaccharides, indicating that it is produced from neutrophils or activated T lymphocytes (154). It is possible that this molecule may be acting as a modulator in inflammation by binding to E- or P-selectin. It will thus be of significance in determining whether poly-N-acetyllactosamine in O-glycans or those in N-glycans are determining factors for providing sialyl-Le^X and sialyl-Le^a structures as ligands for selectins. In relation to this, it has been demonstrated that leukosialin inhibits LFA-1 mediated binding to

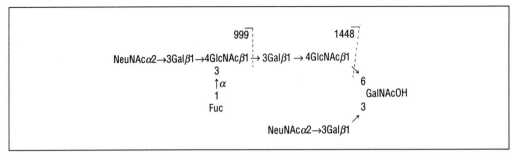

Fig. 17 The structure of a branched and poly-*N*-acetyllactosaminyl *O*-glycan isolated from human granulocytes. The *O*-glycan contains sialyl-Le^x structure at the terminus and has one repeat of *N*-acetyllactosamine. The *O*-glycan with three *N*-acetyllactosaminyl repeats was also isolated from human granulocytes. Numbers indicate the mass fragments formed during the mass spectrometry analysis of the permethylated *O*-glycan (from reference 41)

ICAM-1 (155). This inhibitory activity by leukosialin is, however, abolished by neuramidase treatment of leukosialin. These results strongly suggest that the carbohydrate moiety of mucin-type glycoproteins may play a critical role in modulating cell adhesion.

In leukaemia cells, the increased amount of core 2 enzyme and sialyltransferase was reported (141, 142, 156). Because of this increase, the amount of sialyl-Le^x is higher in leukaemia cells than in their normal counterparts (41, 102), which likely facilitates the adhesion of leukaemia to endothelial cells. This may be the reason why leukaemia cells cross the boundary of endothelial cells which results in perfusion into the body fluid.

5. β-1,6-*N*-acetylglucosaminyltransferases

Poly-*N*-acetyllactosamines were initially discovered in *N*-glycans and glycosphingolipids (see references 1–3). Sulphated forms of poly-*N*-acetyllactosamines were found as keratansulphates, even before the discovery of poly-*N*-acetyllactosamines in *N*-glycans and glycolipids. Since poly-*N*-acetyllactosamines can also be found in *O*-glycans, poly-*N*-acetyllactosamines unify all different kinds of cell surface glycoconjugates in one category. At the same time, the distinct forms of poly-*N*-acetyllactosamines exist and such distinctive features differentiate various forms of poly-*N*-acetyllactosamines. I believe that structural classification and biosynthetic mechanisms in these different forms of poly-*N*-acetyllactosamines are likely to yield useful information which will help us to understand how different glycolylations play distinct roles.

As discussed already, poly-*N*-acetyllactosamine extension is determined by β-1,6-*N*-acetylglucosaminyl branches in both *N*- and *O*-glycans. We have recently cloned a cDNA encoding core 2 β-1,6-*N*-acetylglucosaminyltransferase that forms a branch in *O*-glycans (157). When the protein was expressed by transfecting this cDNA, it was discovered that this enzyme, isolated from HL-60 cDNA library, catalyses only the reaction of core 2 formation. In O-glycans, core 3 can be converted to core 4 by another β-1,6-N-acetylglucosaminyltransferase (Figure 15). Since this enzymatic reaction is not carried out by core 2 β-1,6-N-acetylglucosaminyltransferase present in blood cells, there must be another enzyme that forms core 4 in tissues such as gastrointestinal ducts.

Linear poly-N-acetyllactosamine (i) can be converted to branched ones (I) by another β-1,6-N-acetylglucosaminyltransferase. By cloning cDNA for this enzyme from PA-1 teratocarcinoma cDNA library, we discovered that this enzyme is distinctly different from all the other β-1,6-N-acetylglucosaminyltransferase (158). Despite the difference in acceptor specificity between core 2 and I β-1,6-N-acetylglucosaminyltransferase, however both enzymes have a striking homology in both their nucleotides and their translated amino acid sequences. In particular, three different regions in the catalytic domains of two enzymes have 60% identity in the amino acid sequences (Figure 18). These results suggest that there is a family

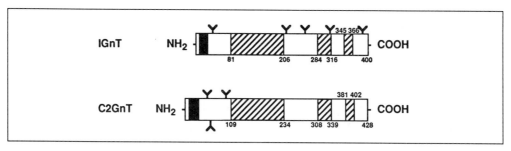

Fig. 18 Comparison of the domain structures and homologous regions of two β-1,6-N-acetylglucosaminyltransferases, IGnT and C2GnT. Solid boxes denote signal/anchor domain; Y denotes potential N-glycosylation sites. From the NH$_2$-terminus, these structures are characterized by a short cytoplasmic domain followed by a single membrane anchoring domain and a long intralumenal domain. The amino acid sequence in each hatched box has 60% identity between the two enzymes. This homologous region lies in the catalytic domain of the enzymes (from reference 158)

of β-1,6-N-acetylglucosaminyltransferase which has a distinctly different acceptor specificity (Table 3). In fact, the Southern blot analysis of restricted human genomic DNA revealed many bands that were cross-hybridized with a cDNA fragment of either core 2 or 1 enzyme. Since preliminary studies showed that the coding region of the core 2 enzyme lies on one exon, the results can be interpreted as showing that there is more than one gene related to core 2 (as well as 1 enzyme) gene. Moreover, the genes encoding two enzymes were found to be at the same locus, chromosome 9, band q21 (158). These results strongly suggest that all of β-1,6-N-acetylglucosaminyltransferases are related to each other, and possibly produced by gene duplication and divergence of a primordial gene. Further studies will be exciting to reveal the genesis of various β-1,6-N-acetylglucosaminyltransferases during evolution.

Table 3 Structure and distribution of different β-1,6-N-acetylglucosaminyl linkages

Acceptors and linkages formed	HL-60	K562	Erythrocytes	Colon
GlcNAcβ1↘6 Galβ1→3GalNAc (core 2)	+++[a]	−	−	+
GlcNAcβ1↘6 GlcNAcβ→3GalNAc (core 4)	−	−	−	++
GlcNAcβ1↘6 GlcNAcβ1→2Man (GnTV)	+++	++	++	++
GlcNAcβ1↘6 GlcNAcβ→3Gal (i)	−	−	+++	+

[a] The presence and absence of structures are denoted by + and −, respectively.

6. Glycobiology—a new frontier in cell biology and molecular biology

So far I have described the structures of carbohydrates and their synthesis in biology and diseases. The field of carbohydrate biochemistry evolved from carbohydrate chemistry to biochemistry and eventually to biology of carbohydrates, or glycobiology. At this point, I would like to summarize the significance of this development and then provide some perspectives on how this field should be developed.

When we measure the significance of scientific development, it is important that the obtained knowledge is essential for one to explore a new direction in that particular field. However, that particular development becomes much more significant if that knowledge critically influences research in other fields and thus has a wider influence on science in general. Such a development therefore becomes an integral part of scientific knowledge at large. In order to explain this well, I would like to take, for example, the processing of N-glycans from high mannose-type to complex-type oligosaccharides.

First, it was important to recognize that the conversion of high mannose-type to complex-type oligosaccharides is associated with the conversion from susceptibility to resistance to the treatment by endo-β-N-acetylglucosaminidase-H. The next critical findings were that the processing of N-glycans takes place step-wisely as newly synthesized glycoproteins are transported from endoplasmic reticulum to the Golgi complex and finally to the *trans*-Golgi network. It was shown that this processing is carried out by a series of reactions, in which the product made in an earlier compartment becomes an acceptor for the glycosyltransferase responsible for the next step of synthesis. Thirdly, it was shown that the glycosyltransferases responsible for the earlier steps of biosynthesis reside in earlier compartments of intracellular organelles. Thus, the orderly addition of sugar residues is controlled by the requirement of an acceptor for a given glycosyltransferase and by the restricted presence of a given glycosyltransferase in the place where such enzymatic reaction is required. These discoveries by carbohydrate biochemists, as well as by some cell biologists such as Rothman (159), allow an investigator to utilize this fact in many instances of research. In particular, a simple experiment to test the susceptibility to endo-β-*N*-acetylglucosaminidase-H has been carried out in numerous studies. Moreover, knowledge on *N*-glycan processing has been extremely useful in order to elucidate the signals for intracellular trafficking, in particular trafficking to lysosomes. This enriched knowledge was accumulated by various carbohydrate biochemists, including those that determined the specificity of endo-β-*N*-acetylglucosaminidase-H. Thus, it is safe to say that the knowledge on *N*-glycans processing critically stimulates the advancement of scientific knowledge in general. Another example is in the field of selectins. Carbohydrate biochemists believed that the vastly different carbohydrates could be good candidates for molecules recognized by binding proteins. Since certain carbohydrates are expressed in a cell-type specific manner, such an expectation has been widely held by researchers. It is rewarding that such an expectation was finally met, at least in the field of selectin-carbohydrate interaction. This rapid development in the studies on selectins also tells us that carbohydrates are likely involved in cell recognition in many other biological processes. The discovery in selectins is considered to be just the beginning in this promising field. Glycobiology is thus expected to have a significant impact on scientific development in general, through the understanding of the roles of carbohydrates in cell recognition.

Having mentioned the exciting potential of glycobiology, how then can we facilitate its scientific development? I feel that there are three research efforts which should be emphasized.

First, it is essential for us to realize that protein–carbohydrate interaction may be playing a role in many of the cases under study. It is then essential to elucidate the carbohydrate structures that are functional as ligands for carbohydrate-binding proteins. For this, it may be useful to utilize affinity chromatography using a binding protein conjugated to insoluble matrix. Once cDNA for the carbohydrate-binding protein is isolated, it is possible to determine which amino acids are critically involved by examining the expression of cDNA after various mutations. Such research efforts can be extended to obtain a three-dimensional structure by X-ray crystallography, as exemplified in serum mannose binding protein (160, see also Chapter 2).

The second research effort is the understanding of the sterical structure of carbohydrates. Such a research direction will be strengthened once the oligosaccharides are synthesized. Our progress in carbohydrate chemistry revealed that there is tremendous heterogeneity in carbohydrate structures. It is now essential to understand how such subtle differences in carbohydrate structure can work as ligands. For this it is critical to synthesize the oligosaccharides and then test the oligosaccharide and their derivatives as ligands. Once the structure most suited for the ligand is known, it is possible to design a synthetic compound(s), which is(are) very similar in sterical structure to that of a ligand, but are less metabolized because they differ from natural compounds (161). This example can be seen in cyclic peptides for peptide analogies. These research efforts will be facilitated by determining the sterical structure of a protein complexed with or without a ligand carbohydrate. X-ray crystallography and 2- or 3-dimensional NMR for solution structure will be imperative for this line of studies (see Chapter 6). In addition, the ability to synthesize a given carbohydrate would be essential for these research efforts.

Thirdly, it is essential to isolate genes or cDNA for all of the enzymes involved in glycosylation (see Chapter 3). This research direction is important in three different aspects. First, the isolation of cDNAs encoding glycosyltransferases revolutionized the chemical synthesis of oligosaccharides. Synthetic carbohydrate chemistry solely based on chemical reaction has been an enormously difficult task, close to art. This is mainly because the functional groups of carbohydrates are very alike. Except for the carboxyl group of sialic acid, and *N*-acyl groups of sialic acid and hexosamines, the difference is only between the primary alcohol and secondary alcohol. Chemical reactions become extremely laborious in order to differentiate such a subtle difference. Thus, it is essential to replace such rather nonspecific chemical reactions with specific reactions catalysed by glycosyltransferases (see Chapter 5). This is possible once the cDNA is isolated and glycosyltransferase is expressed in suitable cells.

The second critical significance in the isolation of cDNA is to test directly the functionality of a given carbohydrate in cell biology. If the cDNA is introduced and the glycosyltransferase is expressed in cells where the enzyme was not otherwise expressed, it is possible to ask how newly expressed carbohydrates contribute to biological function. This has been elegantly shown by Lowe *et al.* in the identification of E-selectin ligand (see Chapter 4). On the other hand, it is possible to decrease the expression of glycosyltranferases by using antisense technology or by abolishing its expression through homologous recombination. Such efforts are extremely useful to determine the roles of carbohydrates in many biological systems, including embryonic development.

Thirdly, the isolation of genes coding glycosyltransferases further broadens this field into molecular biology. Since the formation of carbohydrates are dictated by glycosyltransferases, the amount of a given carbohydrate is determined by how the expression of the enzyme is regulated. It has been shown that glycosylation is cell-type specific and may change during differentiation or pathological conditions. By having cDNAs encoding these glycosyltransferases, it is possible to isolate genomic DNA that includes the regulatory element for a glycosyltransferase. By having genes encoding glycosyltransferase, it is now possible to understand how the expression of these molecules in cells is regulated (see Chapter 3). By reaching this stage, glycobiology becomes a part of molecular biology. Glycosylation may be significant as much as phosphorylation since glycosylation changes the functionality of proteins by modulation and by providing ligands for binding proteins. By understanding the mechanisms underlying the regulation of glycosylation, we are much closer to comprehensive understanding of the biology of cells.

Being one of the scientists working on carbohydrates, I cannot help but to think that our time has finally come. It is our responsibility to grasp this potential and put glycobiology at the forefront of cell biology and molecular biology.

Acknowledgements

The work done in our laboratory was supported by grants RO1 CA33000, CA33895, CA48737, and AI33189 from the National Institutes of Health. I would like to thank my colleagues and collaborators, particularly, Dr Michiko N. Fukuda for sharing her enthusiasm for glycobiology with me throughout the years. I also thank my sons, Ko and Shun Fukuda for their support, and Ms Melissa Moore and Ms Leslie DePry for their excellent editorial efforts and secretarial assistance.

References

1. Fukuda, M. and Fukuda, M. N. (1984) Cell surface glycoproteins and carbohydrate antigens in development and differentiation of human erythroid cells. In *The biology of glycoproteins* (ed. R. J. Ivatt), p. 183. Plenum Press, New York.
2. Fukuda, M. (1985) Cell surface glycoconjugates as onco-differentiation markers in hematopoietic cells. *Biochem. Biophys. Acta*, **780**, 119.
3. Feizi, T. (1985) Demonstration by monoclonal antibodies that carbohydrate structures of glycoproteins and glycolipids are onco-developmental antigens. *Nature*, **314**, 53.
4. Fukuda, M. (ed.) (1992) Cell surface carbohydrates in hematopoietic cell differentiation and malignancy. In *Cell surface carbohydrates and cell development*, p. 127. CRC Press, Boca Raton, FL.
5. Rademacher, T. W., Parekh, R. B., and Dwek, R. A. (1988) Glycobiology. *Ann. Rev. Biochem.*, **57**, 785.
6. Carlsson, S. R., Roth, J., Piller, F., and Fukuda, M. (1988) Isolation and characterization of human lysosomal membrane glycoproteins, h-lamp-1 and h-lamp-2: major sialoglycoproteins carrying polylactosaminoglycan. *J. Biol. Chem.*, **263**, 18911.
7. Ito, S., Yamashita, K., Spiro, R. G., and Kobata, A. (1977) Structure of a carbohydrate moiety of a unit A glycopeptide of calf thyroglobulin. *J. Biochem. Tokyo*, **81**, 1621.
8. Kornfeld, R. and Kornfeld, S. (1985) Assembly of asparagine-linked oligosaccharides. *Annu. Rev. Biochem.*, **54**, 631.
9. Endo, Y., Yamashita, K., Tachibana, Y., Tojo, S., and Kobata, A. (1979) Structures of the asparagine-linked sugar chains of human chorionic gonadotropin. *J. Biochem. (Tokyo)*, **85**, 669.
10. Yamashita, K., Ohkura, T., Tachibana, Y., Takasaki, S., and Kobata, A. (1984) Com-

parative study of the oligosaccharides released from baby hamster kidney cells and their polyoma transformant by hydrazinolysis. *J. Biol. Chem.*, **259**, 10 834–40.
11. Yamashita, K., Kamerling, J. P., and Kobata, A. (1982) Structural study of the carbohydrate moiety of hen ovomucoid. *J. Biol. Chem.*, **257**, 12 809.
12. Baenziger, J. U. and Green, E. D. (1988) Pituitary glycoprotein hormone oligosaccharides: structure, synthesis and function of the asparagine-linked oligosaccharides on lutropin, follitropin, and thyrotropin. *Biochim. Biophys. Acta*, **947**, 287.
13. Fukuda, M., Spooncer, E., Oates, J. E., Dell, A., and Klock, J. C. (1984) Structure of sialylated fucosyllactosaminoglycan isolated from human granulocytes. *J. Biol. Chem.*, **259**, 10 925.
14. Elbein, A. D. (1987) Inhibitors of the biosynthesis and processing of N-linked oligosaccharide chains. *Annu. Rev. Biochem.*, **56**, 497.
15. Liang, C.-J., Yamashita, K., Muellenberg, C. G., Shichi, H., and Kobata, A. (1979) Structure of the carbohydrate moieties of bovine rhodopsin. *J. Biol. Chem.*, **254**, 6414.
16. Fukuda, M. N., Papermaster, D., and Hargrave, P. A. (1979) Rhodopsin carbohydrate: structure of small oligosaccharides attached at two sites near the amino-terminus. *J. Biol. Chem.*, **254**, 8201.
17. Hsieh, P., Resner, M. R., and Robbins, P. W. (1983) Selective cleavage by endo-β-N-acetylglucosaminidase H at individual glycosylation sites of sindbis virion envelope glycoproteins. *J. Biol. Chem.*, **258**, 2555.
18. Timble, R. B., Maley, F., and Chu, F. K. (1983) Glycoprotein biosynthesis in yeast: protein conformation affects processing of high mannose oligosaccharides on carboxypeptidase Y and invertase. *J. Biol. Chem.*, **258**, 2562.
19. Baranski, T. J., Faust, P. L., and Kornfeld, S. (1990) Generation of a lysosomal enzyme targeting signal in the secretory protein pepsinogen. *Cell*, **63**, 281.
20. Kornfeld, S. (1989) The biogenesis of lysosomes. *Annu. Rev. Cell Biol.*, **5**, 483.
21. Fukuda, M., Dell, A., Oates, J. E., and Fukuda, M. N. (1984) Structure of branched lactosaminoglycan, the carbohydrate moiety of band 3 isolated from adult human erythrocytes. *J. Biol. Chem.*, **259**, 8260.
22. Fukuda, M., Fukuda, M. N., and Hakomori, S. (1979) The developmental change and genetic defect in carbohydrate structure of Band 3 glycoprotein of human erythrocyte membrane. *J. Biol. Chem.*, **254**, 3700.
23. Watanabe, K., Hakomori, S., Childs, R. A., and Feizi, T. (1979) Characterization of a blood group I-active ganglioside. Structural requirements for I and i specificities. *J. Biol. Chem.*, **254**, 3221.
24. Feizi, T., Childs, R. A., Watanabe, K., and Hakomori, S. (1979) Three types of blood group I specificity among monoclonal and anti-I autoantibodies revealed by analogues of a branched erythrocyte glycolipid. *J. Exp. Med.*, **149**, 975.
25. Fukuda, M. N. and Matsumura, G. (1976) Endo-β-galactosidase of *Escherichia freundii*. Purification and endoglycosidic action on keratin sulfates, oligosaccharides, and blood active glycoprotein. *J. Biol. Chem.*, **251**, 6218.
26. Viitala, J. and Finne, J. (1984) Specific cell-surface labeling of polyglycosyl chains in human erthrocytes and HL-60 cells using endo-β-galactosidase and galactosyltransferase. *Eur. J. Biochem.*, **138**, 393.
27. Fukuda, M. N. (1991) Endo-β-galactosidases. In *Handbook of endoglycosidases and glycopeptides* (ed. T. Muramatsu and N. Takahoshi), pp. 55–103. CRC Press, Boca Raton, FL.
28. Fukuda, M. N., Fukuda, M., and Hakomori, S. (1979) Cell surface modification by

endo-β-galactosidase. Change of blood group activities and release of oligosaccharides from glycoproteins and glycosphingolipids of human erythrocytes. *J. Biol. Chem.*, **254**, 5458.
29. Van den Eijnden, D. N., Koenderman, A. H. L., and Schiphorst, W. E. C. M. (1988) Biosynthesis of blood group i-active polylactosaminoglycans. Partial purification and properties of an UDP-GlcNAc:N-acetyllactosaminide β1→3-N-acetylglucosaminyltransferase from Novikoff tumor cell ascites fluid. *J. Biol. Chem.*, **263**, 12 461.
30. Joziasse, D. H., Schiphorst, W. E. C. M., Van den Eijnden, D. H., Van Kuik, J. A., Van Halbeek, H., and Vliegenthart, J. F. G. (1987) Branch specificity of bovine colostrum CMP-sialic acid: Galβ1→4GlcNAcRα2→6-sialyltransferase. Sialylation of Bi-, tri-, and tetraantennary oligosaccharides and glycopeptides of the N-acetyllactosamine type. *J. Biol. Chem.*, **262**, 2025.
31. Sasaki, H., Bothner, B., Dell, A., and Fukuda, M. (1987) Carbohydrate structure of erythropoietin expressed in Chinese hamster ovary cells by a human erythropoietin cDNA. *J. Biol. Chem.*, **262**, 12 059.
32. Yoshima, H., Furthmayr, H., and Kobata, A. (1980) Structures of the asparagine-linked sugar chains of glycophorin. *J. Biol. Chem.*, **255**, 9713.
33. Irimura, T., Tsuji, T., Tagami, S., Yamamoto, K., and Osawa, T. (1981) Structure of a complex-type sugar chain of human glycophorin A. *Biochemistry*, **20**, 560.
34. Fukuda, M. (1991) Lysosomal membrane glycoproteins. *J. Biol. Chem.*, **266**, 21 327.
35. Yoshima, H., Matsumoto, A., Mizuochi, T., Kawasaki, T., and Kobata, A. (1981) Comparative study of the carbohydrate moieties of rat and human plasma α_1-acid glycoproteins. *J. Biol. Chem.*, **256**, 8476.
36. Carlsson, S. R. and Fukuda, M. (1990) The polyactosaminoglycans of human lysosomal membrane glycoproteins lamp-1 and lamp-2. *J. Biol. Chem.*, **265**, 20 488.
37. Fukuda, M., Guan, J.-L., and Rose, J. K. (1988) A membrane-anchored form but not the secretary form of human chorionic gonadotropin-α chain acquires polyactosaminoglycan. *J. Biol. Chem.*, **263**, 5314.
38. Wang, W.-C., Lee, N., Aoki, D., Fukuda, M. N., and Fukuda, M. (1991) The poly-N-acetyllactosamines attached to lysosomal membrane glycoproteins are increased by the prolonged association with the Golgi complex. *J. Biol. Chem.*, **266**, 23 185.
39. Matlin, K. and Simons, K. (1983) Reduced temperature prevents transfer of a membrane glycoprotein to the cell surface but not terminal glycosylation. *Cell*, **34**, 233.
40. Fukuda, M. N., Dell, A., Oates, J. E., Wu, P., Klock, J. C., and Fukuda, M. (1985) Structures of glycosphingolipids isolated from human granulocytes: the presence of a series of linear poly-N-acetyl lactosaminyl ceramide and its significance in glycolipids of whole blood cells. *J. Biol. Chem.*, **260**, 1067.
41. Fukuda, M., Carlsson, S. R., Klock, J. C., and Dell, A. (1986) Structures of O-linked oligosaccharides isolated from normal granulocytes, chorinic myelogenous leukemia cells, and acute myelogenous leukemia cells. *J. Biol. Chem.*, **261**, 12 796.
42. Holms, E. H., Ostrander, G. K., and Hakomori, S. (1986) Biosynthesis of the sialyl-Le^X determinant carried by type 2 chain glycosphingolipids (IV^3 NeuAcIII^{3}FucnLc$_4$, VI^{3}NeuAcV^{3}FucnLc$_6$, and VI^{3}NeuAcIII^{3}V^{3}Fuc$_2$nLc$_6$) in human lung carcinoma PC9 cells. *J. Biol. Chem.*, **261**, 3737.
43. Hakomori, S. (1981) Glycosphingolipids in cellular interaction, differentiation, and oncogenesis. *Ann. Rev. Biochem.*, **50**, 733.

44. Watkins, W. M. (1980) Biochemistry and genetics of the ABO, Lewis, and P blood group systems. *Adv. Hum. Genet.*, **10**, 1.
45. Wiener, A. S., Unger, L. H., Cohen, L., and Feldman, J. (1956) Type-specific cold autoantibodies as a cause of acquired hemolytic anemia and hemolytic transfusion reactions: biologic test with bovine red cells. *Ann. Intern. Med.*, **44**, 221.
46. Marsh, W. L. (1961) Anti-i: cold antibody defining the Ii relationship in human red cells. *Br. J. Haematol.*, **7**, 200.
47. Romans, D. G., Tilley, C. A., and Dorrington, K. J. (1980) Monogamous bivalency of IgG antibodies. I., Deficiency of branched ABHI-active oligosaccharide chains on red cells of infants causes the weak antiglobulin reactions in hemolytic disease of the newborn due to ABO incompatibility. *J. Immunol.*, **124**, 2807.
48. Muramatsu, T., Gachelin, G., Damonnevelle, M., Delarbre, C., and Jacob, F. (1979) Cell surface carbohydrates of embryonal carcinoma cells: polysaccharidic side chains of F9 antigens and of receptors to two lectins, FBP and PNA. *Cell*, **18**, 183.
49. Kapadia, A., Feizi, T., and Evans, M. J. (1981) Changes in the expression and polarization of blood group i and I antigens in postimplantation embryos and teratocarcinomas of mouse associated with cell differentiation. *Exp. Cell Res.*, **131**, 185.
50. Knowles, B. B., Rappaport, J., and Solter, D. (1982) Murine embryonic antigen (SSEA-1) is expressed on human cells and structurally related human blood group antigen I is expressed on mouse embryos. *Dev. Biol.*, **93**, 54.
51. Larsen, R. D., Ernst, L. K., Nair, R. P., and Lowe, J. B. (1990) Molecular cloning, sequence, and expression of a human GDP-L-fucose: β-D-galactoside 2-α-L-fucosyltransferase cDNA that can form the H blood group antigen. *Proc. Natl Acad. Sci. USA*, **87**, 6674.
52. Oriol, R., Danilovs, J., and Hawkins, B. R. (1981) A new genetic model proposing that the Se gene is a structural gene closely linked to the H gene. *Am. J. Hum. Genet.*, **33**, 421.
53. Yamamoto, F.-I., Clausen, H., White, T., Marken, J., and Hakomori, S.-I. (1990) Molecular genetic basis of the histo-blood group ABO system. *Nature (London)*, **345**, 229.
54. Yamamoto, F.-I. and Hakomori, S.-I. (1990) Sugar-nucleotide donor specificity of Histo-blood group A and B transferases is based on amino acid substitutions. *J. Biol. Chem.*, **265**, 19257.
55. Galili, U., Shohet, S. B., Kobrin, E., Stutts, C. L. M., and Macher, B. A. (1988) Man, apes, and old world monkeys differ from other mammals in the expression of α-galactosyl epitopes on nucleated cells. *J. Biol. Chem.*, **263**, 17755.
56. Larsen, R. D., Rivera-Marrero, C. A., Ernst, L. K., Cummings, R. D., and Lowe, J. B. (1990) Frameshift and nonsense mutations in a human genomic sequence homologous to a murine UDP-gal:β-D-Gal(1,4)-D-GlcNAcα(1,3)-galactosyltransferase cDNA. *J. Biol. Chem.*, **265**, 7055.
57. Shaper, N. L., Lin, S., Joziasse, D. H., Do, Y. K., and Yangfeng, T.L. (1992) Assignment of two human α1,3-galactosyltransferase gene sequences (GGTA1 and GGTA1P) to chromosome 9q33-q43 and chromosome 12q14-q15. *Genomics*, **12**, 613.
58. Kelly, R. J., Ernst, L. K., Larsen, R. D., Bryant, J. G., Robinson, J. S., and Lowe, J. B. (1994) Molecular basis for A, B and H blood group deficiency in Bombay (O_h) and para-Bombay individuals. *Proc. Natl. Acad. Sci.* (in press).
59. Schachner, M. (1989) Families of neural adhesion molecules. *Ciba Foundation Symposium 145*, p. 156. Wiley, New York.

60. Wassarman, P. M. (1992) Cell surface carbohydrates and mammalian fertilization (Chapter 8). In *Cell surface carbohydrates and cell development* (ed. M. Fukuda). CRC Press, Boca Raton, FL.
61. Karlsson, K. (1989) Animal glycosphingolipids as membrane attachment sites for bacteria. *Annu. Rev. Biochem.*, **58**, 309.
62. Ashwell, G. and Hanford, J. (1982) Carbohydrate-specific receptors of the liver. *Ann. Rev. Biochem.*, **51**, 531.
63. Drickamer, L. K. (1988) Two distinct classes of carbohydrate-recognition domains in animal lectins. *J. Biol. Chem.*, **263**, 9557.
64. Gesner, B. M. and Ginsburg, V. (1964) Effect of glycosidases on the fate of transfused lymphocytes. *Proc. Natl Acad. Sci. USA*, **52**, 750.
65. Rosen, S. D., Singer, M. S., Yednock, T. A., and Stoolman, L. M. (1985) Involvement of sialic acid on endothelial cells in organ-specific lymphocyte recirculation. *Science*, **228**, 1005.
66. Lasky, L. A., Singer, M. S., Yednock, T. A., Dowbonko, D., Fennie, C., Rodriguez, H. et al. (1989) Cloning of a lymphocyte receptor reveals a lectin domain. *Cell*, **56**, 1045.
67. Bevilaqua, M. P., Stengelin, S., Gimbrone, M. A., Jr, and Seed, B. (1989) Endothelial leukocyte adhesion molecule is an inducible receptor for neutrophils related to complement regulatory proteins and lectins. *Science*, **243**, 1160.
68. Johnston, G. I., Cook, R. G., and McEver, R. P. (1989) Cloning of GMP-140, a granule protein of platelets and endothelium: sequence similarity to proteins involved in cell adhesion and inflammation. *Cell*, **56**, 1033.
69. Spooncer, E., Fukuda, M., Klock, J. C., Oates, J. E., and Dell, A. (1984) Isolation and characterization of polyfucosylated lactosaminoglycan from human ganulocytes. *J. Biol. Chem.*, **259**, 4792.
70. Mizoguchi, A., Takasaki, S., Maeda, S., and Kobata, A. (1986) Changes in asparagine-linked sugar chains of human promyelocytic leukemic cells (HL-60) during monocytoid differentiation and myeloid differentiation. Decrease of high-molecular-weight oligosaccharides in acidic fraction. *J. Biol. Chem.*, **259**, 11 949.
71. Lowe, J. B., Stoolman, L. M., Nair R. P., Larsen, R. D., Berhend, T. L., and Marks, R. M. (1990) ELAM-1-dependent cell adhesion to vascular endothelium determined by a transfected human fucosyltransferase cDNA. *Cell*, **63**, 475.
72. Phillips, M. L., Nudelman, E., Gaeta, F. C. A., Perez, M., Singhal, A. K., Hakomori, S.-I. et al. (1990) ELAM-1 mediates cell adhesion by recognition of a carbohydrate ligand, sialyl-Le^x. *Science*, **250**, 1130.
73. Waltz, G., Aruffo, A., Kolanus, W., Bevilacqua, M., and Seed, B. (1990) Recognition of ELAM-1 of the sialyl-Le^x determinant on myeloid and tumor cells. *Science*, **250**, 1132.
74. Polley, M. J., Phillips, M. L., Wayner, E., Nudelman, E., Singhal, A. K., Hakomori, S.-I. et al. (1991) CD62 and endothelial cell–leukocyte adhesion molecule 1 (ELAM-1) recognize the same carbohydrate ligand, sialyl-Lewis x. *Proc. Natl Acad. Sci. USA*, **88**, 6224.
75. Kukowska-Latallo, J. F., Larsen, R. D., Nair, R. P., and Lowe, J. B. (1990) A cloned human cDNA determines expression of a mouse stage-specific embryonic antigen and the Lewis blood group α(1,3/1,4) fucosyltransferase. *Genes & Dev.*, **4**, 1288.
76. Lawrence, M. B. and Springer, T. A. (1991) Leukocytes roll on a selectin at physiologic flow rates: distinction from and prerequisite for adhesion through integrins. *Cell*, **65**, 859.
77. Ley, K., Gaehtgens, P., Fennie, C., Singer, M. S., Lasky, L. A., and Rosen, S. D. (1991) Lectin-like cell adhesion molecule 1 mediates leukocyte rolling in mesenteric venules *in vivo*. *Blood*, **77**, 2553.

78. Tanaka, Y., Adams, D. H., Hubscher, S., Hirano, H., Siebenlist, U., and Shaw, S. (1993) T-cell adhesion induced by proteoglycan-immobilized cytokine MIP-1β. *Nature*, **361**, 79.
79. Springer, T. A. (1990) Adhesion receptors of the immune system. *Nature*, **346**, 425.
80. Mulligan, M. S., Lowe, J. B., Larsen, R. D., Walker, L., Maemura, K., Fukuda, M. *et al.* (1993) Protective effects of sialylated oligosaccharides in immune complex-induced acute lung injury. *J. Exp. Med.*, **178**, 623.
81. Streeter, P. R., Rouse, B. T. N., and Butcher, E. C. (1988) Immunohistologic and functional characterization of a vascular addressin involved in lymphocyte homing into peripheral lymph nodes. *J. Cell Biol.*, **107**, 1853.
82. Briskin, M. J., McEvoy, L. M., and Butcher, E. C. (1993) MAdCAM-1 has homology to immunoglobin and mucin-like adhesion receptors and to IgA1. *Nature*, **363**, 461.
83. Lasky, L. A., Singer, M. S., Dowbenko, D., Imai, Y., Henzel, W. J., Grimley, C. *et al.* (1992) An endothelial ligand for L-selectin is a novel mucin-like molecule. *Cell*, **69**, 927.
84. Moore, K. L., Stults, N. L., Diaz, S., Smith, D. F., Cummings, R. D., Varki, A. *et al.* (1992) Identification of a specific glycoprotein ligand for P-selectin (CD62) on myeloid cells. *J. Cell. Biol.*, **118**, 445.
85. Holness, C. L. and Simmons, D. L. (1993) Molecular cloning of CD68, a human macrophage marker related to lysosomal glycoproteins. *Blood*, **81**, 1607.
86. Imai, Y., Lasky, L. A., and Rosen, S. D. (1993) Sulphation requirement for GlyCAM-1, an endothelial ligand for L-selectin. *Nature*, **361**, 555.
87. Zimmermann, D. R. and Ruoslahti, E. (1989) Multiple domains of the large fibroblast proteoglycan, versican. *EMBO J.*, **8**, 2975.
88. Saleque, S., Ruiz, N., and Drickamer, K. (1993) Expression and characterization of a carbohydrate-binding fragment of rat aggrecan. *Glycobiology*, **3**, 185.
89. Warren, L., Buck, C. A., and Tuszynski, G. P. (1978) Glycopeptide changes and malignant transformation a possible role for carbohydrate in malignant behavior. *Biochem. Biophys. Acta*, **516**, 97.
90. Ogata, S., Muramatsu, T., and Kobata, A. (1976) New structural characteristics of the large glycopeptides from transformed cells. *Nature*, **259**, 580.
91. Takasaki, S., Ikehira, H., and Kobata, A. (1980). Increase of asparagine-linked oligosaccharides with branched outer chains caused by cell transformation. *Biochem. Biophys. Res. Commun.*, **92**, 735.
92. Dennis, J. W., Laferté, S., Waghorne, C., Breitman, M. L., and Kerbel, R. S. (1987) β1→6 Branching of Asn-linked oligosaccharides is directly associated with metastasis. *Science*, **236**, 582.
93. Irimura, T., Gonzalez, R., and Nicolson, G. L. (1981) Effects of tunicamycin on B16 metastatic melanoma cell surface glycoproteins and blood-borne arrest and survival properties. *Cancer Res.*, **41**, 3411.
94. Dennis, J. W. (1986) Effects of swainsonine and polyinosinic: polycytidylic acid on murine tumor cell growth and metastasis. *Cancer Res.*, **46**, 5131.
95. Humphries, M. J., Matsumoto, K., White, S. L., and Olden, K. (1986) Inhibition of experimental metastasis by castanospermine in mice: blockage of two distinct stages of tumor colonization by oligosaccharide processing inhibitors. *Cancer Res.*, **46**, 5215.
96. Fukushima, K., Hirota, M., Terasaki, P. I., Wakisaka, A., Togashi, H., Chia, D. *et al.* (1984) Characterization of sialylated Lewis^X as a new tumor-associated antigen. *Cancer Res.*, **44**, 5279.
97. Fukushi, Y., Nudelman, E., Levery, S. B., Hakomori, S., and Rauvala, H. (1984) Novel

fucolipids accumulating in human adenocarcinoma. III. A hybridoma antibody (FH$_6$) defining a human cancer-associated difucoganglioside (VI^{2}NeuAcV^{3}III^{3}Fuc$_2$nLc$_6$). *J. Biol. Chem.*, **259**, 10 511.
98. Kim, Y. S., Itzkowitz, S. H., Yuan, M., Chung, Y.-S., Satake, K., Umeyama, K., and Hakomori, S.-I. (1988) Le^X and Le^Y antigen expression in human pancreatic cancer. *Cancer Res.*, **48**, 475.
99. Magnani, J. L., Nilsson, B., Brockhaus, M., Zopf, D., Steplewski, Z., Koprowski, H., and Ginsburg, V. (1982) A monoclonal antibody-defined antigen associated with gastrointestinal cancer is a ganglioside containing sialylated lacto-N-fucopentaose II. *J. Biol. Chem.*, **257**, 14 365.
100. Berg, E. L., Robinson, M. K., Mansson, O., Butcher, E. C., and Magnani, J. L. (1991) A carbohydrate domain common to both sialyl Le^a and sialyl Le^X is recognized by the endothelial cell leukocyte adhesion molecule ELAM-1. *J. Biol. Chem.*, **266**, 14 869.
101. Nicolson, G. L. (1989) Metastatic tumor cell interactions with endothelium, basement membrane and tissue. *Curr. Opin. Cell Biol.*, **1**, 1009.
102. Fukuda, M., Bothner, B., Ramsamooj, P., Dell, A., Tiller, P. R., Varki, A. *et al.* (1985) Structures of sialylated fucosyl polylactosaminoglycans isolated from chronic myelogenous leukemia cells. *J. Biol. Chem.*, **260**, 12 957.
103. Fukuda, M. N., Dell, A., Tiller, P. R., Varki, A., Klock, J. C., and Fukuda, M. (1986) Structure of a novel sialylated fucosyl lacto-N-nor-hexasosylceramide isolated from chronic myelogenous leukemia cells. *J. Biol. Chem.*, **261**, 2376.
104. Rice, G. E. and Bevilacqua, M. P. (1989) An inducible endothelial cell surface glycoprotein mediates melanoma adhesion. *Science*, **246**, 1303.
105. Hession, C., Osborn, L., Goff, D., Chi-Rosso, G., Vassallo, C., Pasek, M. *et al.* (1990) Endothelial leukocyte adhesion molecule 1: direct expression cloning and functional interactions. *Proc. Natl Acad. Sci. USA*, **87**, 1673.
106. Saitoh, O., Wang, W., Lotan, R., and Fukuda, M. (1992) Differential glycosylation and cell surface expression of lysosomal membrane glycoproteins in sublines of a human colon cancer exhibiting distinct metastatic potentials. *J. Biol. Chem.*, **267**, 5700.
107. Sawada, R., Tsuboi, S., and Fukuda, M. (1994) Differential E-selectin-dependent adhesion efficiency in sublines of a human colon cancer with distinct metastatic potentials. *J. Biol. Chem.*, **269**, 1425.
108. Sawada, R., Lowe, J. B., and Fukuda, M. (1993) E-selectin-dependent adhesion efficiency of colonic carcinoma cells is increased by genetic manipulation of their cell surface lysosomal membrane glycoprotein-1 expression levels. *J. Biol. Chem.*, **268**, 12 675.
109. Nakamon, S., Kameyama, M., Kumaoka, S., Furukawa, H., Ishikawa, O., Kabuto, T. *et al.* (1993) Increased expression of sialyl Le^X antigens correlates with poor survival in patients with colorectal carcinoma: clinicopathologic and immunohistochemical studies. *Cancer Res.*, **53**, 3632.
110. Fukuda, M. N. (1992) Hempas and other congenital dyserythropoietic anemias. In *Protein blood group antigens of the human red cell* (ed. P. C. Agre and J.-P. Cartron), p. 246. Johns Hopkins University Press, Baltimore.
111. Sadler, J. E. (1984) Biosynthesis of glycoproteins: formation of O-linked oligosaccharides. In *Biology of carbohydrates* (ed. V. Ginsburg and P. W. Robbins), Vol. 2, p. 199. Wiley, New York.
112. Hart, G. W., Haltiwanger, R. S., Holt, G. D., and Kelly, W. G. (1989) Glycosylation in the nucleus and cytoplasm. *Ann. Rev. Biochem.*, **58**, 841.
113. Rodén, L. (1980) *The biochemistry of glycoproteins and proteoglycans* (ed. W. J. Lennarz), p. 267. Plenum Press, New York.

114. Spiro, R. G. (1973) Glycoproteins. *Adv. Protein Chem.*, **27**, 349.
115. Reid, K. B. M. (1979) Complete amino acid sequences of the three collagen-like regions present in subcomponent C1q of the first component of human complement. *Biochem. J.*, **179**, 367.
116. Nishimura, H., Kawabata, S., Kisiel, W., Hase, S., Ilkenaka, T., Takao, T. *et al.* (1989) Identification of a disaccharide (Xyl-Glc) and a trisaccharide (Xyl$_2$-Glc)) O-glycosidically linked to a serine residue in the first epidermal growth factor-like domain of human factors VII and IX and protein Z and bovine protein Z. *J. Biol. Chem.*, **264**, 20 320.
117. Harris, R. J., Leonard, C. K., Guzzetta, A. W., and Spellman, M. W. (1991) Tissue plasminogen activator has an O-linked fucose attached to Threonine-61 in the epidermal growth factor domain. *Biochemistry*, **30**, 2311.
118. Allen, A. K., Desai, N. N., Neuberger, A., and Creeth, J. M. (1978) Properties of potato lectin and the nature of its glycoprotein linkages. *Biochem. J.*, **173**, 665.
119. Ballou, C. E. (1976) Structure and biosynthesis of the mannan component of the yeast cell envelope. *Adv. Microbiol. Phys.*, **14**, 93.
120. Irimura, T., Carlson, D. A., Price, J., Yamori, T., Giavazzi, R., Ota, D. M. *et al.* (1988) Differential expression of a sialoglycoprotein with an approximate molecular weight of 900,000 on metastatic human colon carcinoma cells growing in culture and in tumor tissues. *Cancer Res.*, **48**, 2353.
121. Gottschalk, A., Bhargara, A. S., and Murty, V. L. N. (1972) *Glycoproteins: their composition, structure and function* (ed. A. Gottshalk), p. 810. Elsevier, Amsterdam.
122. Ramphal, R., Carnoy, C., Fievre, S., Michalski, J. C., Houdret, N., Lamblin, G. *et al.* (1991) *Pseudomonas aeruginosa* recognizes carbohydrate chains containing type 1 (Galβ1-3GlcNAc) or type 2 (Galβ1-4GlcNAc) disaccharide units. *Infect. Immunol.*, **59**, 700.
123. Fukuda, M. (1993) Molecular genetics of the glycophorin A gene cluster. *Semin. Hematol.*, **30**, 138.
124. Fukuda, M. (1991) Leukosialin, a major O-glycan-containing sialoglycoprotein defining leukocyte differentiation and malignancy. *Glycobiology*, **1**, 347.
125. Park, J. K., Rosenstein, Y. J., Remold-O'Donnell, E., Bierer, B. E., Rosen, F. S., and Burakoff, S. J. (1991) Enhancement of T-cell activation by the CD43 molecule whose expression is defective in Wiskott–Aldrich syndrome. *Nature*, **350**, 706.
126. Pallant, A., Eskenazi, A., Mattei, M. G., Fournier, R. E. K., Carlsson, S. R., Fukuda, M. *et al.* (1989) Identification of cDNAs encoding human leukosialin and localization of the leukosialin gene to chromosome 16. *Proc. Natl Acad. Sci. USA*, **86**, 1328.
127. Timpte, C. S., Eckhardt, A. E., Abernethy, J. L., and Hill, R. L. (1988) Porcine submaxillary gland apomucin contains tandemly repeated, identical sequences of 81 residues. *J. Biol. Chem.*, **263**, 1081.
128. Gum, J. R., Byrd, J. C., Hicks, J. W., Toribara, N. W., Lamport, D. T. A., and Kim, Y. S. (1989) Molecular cloning of human intestinal mucin cDNAs. Sequence analysis and evidence for genetic polymorphism. *J. Biol. Chem.*, **264**, 6480.
129. Shelley, C. S., Remold-O'Donnell, E., Rosen, F. S., and Whitehead, A. S. (1990) Structure of the human sialophorin (CD43) gene. Identification of features atypical of genes encoding integral membrane proteins. *Biochem. J.*, **270**, 569.
130. Eckhardt, A. E., Timpte, C. S., Abernethy, J. L., Zhao, Y., and Hill, R. L. (1991) Porcine submaxillary mucin contains a cystine-rich, carboxyl-terminal domain in addition to a highly repetitive, glycosylated domain. *J. Biol. Chem.*, **266**, 9678.

131. Kim, Y. S., Gum, J., Byrd, J. C., Toribara, N., Ho, S., and Dahiya, R. (1990) Human intestinal mucin core proteins. *Glycoconjugate J.*, **7**, 388.
132. Kudo, S. and Fukuda, M. (1989) Structural organization of glycophorin A and B genes: Glycophorin B gene evolved by homologous recombination at Alu repeat sequences. *Proc. Natl Acad. Sci. USA*, **86**, 4619.
133. Dowbenko, D., Andalibi, A., Young, P. E., Lusis, A. J., and Lasky, L. A. (1993) Structure and chromosomal localization of the murine gene encoding GLYCAM 1. *J. Biol. Chem.*, **268**, 4525.
134. Schachter, H. and Brockhausen, I. (1992) The biosynthesis of serine (threonine)-*N*-acetylgalactosamine-linked carbohydrate moieties. In *Glycoconjugates: composition, structure and function* (ed. H. J. Allen and E. C. Kisailus). Marcel Dekker, New York.
135. Kurosaka, A., Nakajima, H., Funakoshi, I., Matsuyama, M., Nagayo, T., and Yamashina, I. (1983) Structures of the major oligosaccharides from a human rectal adenocarcinoma glycoprotein. *J. Biol. Chem.*, **258**, 11 594.
136. Carlsson, S. H. and Fukuda, M. (1986) Isolation and characterization of leukosialin, a major sialoglycoprotein on human leukocytes. *J. Biol. Chem.*, **261**, 12 779.
137. Carlsson, S. R., Sasaki, H., and Fukuda, M. (1986) Structural variations of *O*-linked oligosaccharides present in leukosialin isolated from erythroid, myeloid and T-lymphoid cell lines. *J. Biol. Chem.*, **261**, 12 787.
138. Kimura, A. K. and Wigzell, H. (1978) Cell surface glycoproteins of murial cytotoxic T lymphocytes. I. T145, a new cell surface glycoprotein selectively expressed on Ly1^{-2+} cytotoxic T lymphocytes. *J. Exp. Med.*, **147**, 1418.
139. Andersson, L. C., Gahmberg, C. G., Kimura, A. K., and Wigzell, H. (1978) Activated human T-lymphocytes display new surface glycoproteins. *Proc. Natl Acad. Sci. USA*, **75**, 3455.
140. Piller, F., Piller, V., Fox, R. I., and Fukuda, M. (1988) Human T-lymphocytic activation is associated with changes in *O*-glycan biosynthesis. *J. Biol. Chem.*, **263**, 15146.
141. Saitoh, O., Piller, F., Fox, R. I., and Fukuda, M. (1991) T-lymphocytic leukemia express complex, branched *O*-linked oligosaccharides on a major sialoglycoprotein, leukosialin. *Blood*, **77**, 1491.
142. Brockhausen, I., Kuhns, W., Schachter, H., Matta, K. L., Sutherland, D. R., and Baker, M. A. (1991) Biosynthesis of *O*-glycans in leukocytes from normal donors and from patients with leukemia: increase in *O*-glycan core 2 UDP-GlcNAc:Galβ-3GalNAcα-R (GlcNAc to GalNAc) β(1–6)-*N*-acetylglucosaminyltransferase in leukemic cells. *Cancer Res.*, **51**, 1257.
143. Piller, F., Le Deist, F., Weinberg, K. I., Parkman, R., and Fukuda, M. (1991) Altered *O*-glycan synthesis in lymphocytes from patients with Wiskott–Aldrich syndrome. *J. Exp. Med.*, **173**, 1501.
144. Higgins, E. A., Siminovitch, K. A., Zhuang, D., Brockhausen, I., and Dennis, J. W. (1991) Aberrant *O*-linked oligosaccharide biosynthesis in lymphocytes and platelets from patients with the Wiskott–Aldrich syndrome. *J. Biol. Chem.*, **266**, 6280.
145. Rosen, F. S., Cooper, M. D., and Wedgewood, R. J. P. (1983) The primary immunodeficiencies. *N. Engl. J. Med.*, **311**, 300.
146. Peacocke, M. and Siminovitch, K. A. (1987) Linkage of the Wiskott–Aldrich syndrome with polymorphic DNA sequences from human X chromosome. *Proc. Natl Acad. Sci. USA*, **84**, 3430.
147. Remold-O'Donnell, E., Kenney, D. M., Parkman, R., Caims, L., Savag, B., and Rosen, F. S. (1984) Characterization of a human lymphocyte surface sialoglycoprotein that is defective in Wiskott–Aldrich syndrome. *J. Exp. Med.*, **159**, 1705.

148. Ardman, B., Sikorski, M. A., Settles, M., and Staunton, D. E. (1990) Human immunodeficiency virus type 1-infected individuals make autoantibodies that bind to CD43 on normal thymic lymphocytes. *J. Exp. Med.*, **172**, 1151.
149. Fox, R. I., Hueniken, M., Fong, S., Behar, S., Royston, I., Singhal, S. et al. (1983) A novel surface antigen (T305) found in increased frequency on acute leukemia cells and in autoimmune disease states. *J. Immunol.*, **131**, 762.
150. Gillespie, W., Paulson, J. C., Kelm, S., Pang, M., and Baum, L. (1993) Regulation of α2,3-sialyltransferase expression correlates with conversion of peanut agglutin (PNA)$^+$ to PNA$^-$ phenotype in developing thymocytes. *J. Biol. Chem.*, **268**, 3801.
151. Amano, J., Nishimura, R., Mochizuki, M., and Kobata, A. (1988) Comparative study of the mucin-type sugar chains of human chorionic gonadotropin present in the urine of patients with trophoblastic diseases and healthy pregnant women. *J. Biol. Chem.*, **263**, 1157.
152. Yousefi, S., Higgins, E., Daoling, Z., Pollex-Krüger, A., Hindsgaul, O., and Dennis, J. W. (1991) Increased UDP-GlcNAc:Galβ1→3GalNAc-R (GlcNAc to GalNAc)β-1→6-N-acetylglucosaminyltransferase activity in metastatic murine tumor cell lines. Control of polylactosamine synthesis. *J. Biol. Chem.*, **266**, 1772.
153. Maemura, K. and Fukuda, M. (1992) Poly-N-acetyllactosaminyl O-glycans attached to leukosialin. *J. Biol. Chem.*, **267**, 24379.
154. Schmid, K., Hediger, M. A., Brossmer, R., Collins, J. H., Haupt, H., Marti, T. et al. (1992) Amino acid sequence of human plasma galactoglycoprotein: identity with the extracellular region of CD43 (sialophorin). *Proc. Natl Acad. Sci. USA*, **89**, 663.
155. Ardman, B., Sikorski, M. A., and Staunton, D. E. (1992) CD43 interferes with T-lymphocyte adhesion. *Proc. Natl Acad. Sci. USA*, **89**, 5001.
156. Kanani, A., Sutherland, D. R., Fibach, E., Matta, K. L., Hindenburg, A., Brockhausen, I. et al. (1990) Human leukemic myeloblasts and myeloblastoid cells contain the enzyme cytidine 5'-monophosphate-N-acetylneuraminic acid: Galβ1→3GalNAcα (2→3)-sialyltransferase. *Cancer Res.*, **50**, 5003.
157. Bierhuizen, M. F. A. and Fukuda, M. (1992) Expression cloning of cDNA encoding UDP-GlcNAc: Galβ1→3GalNAc→R(GlcNAc to GalNAc) β-1, 6-N-acetylglucosaminyltransferase by gene transfer into CHO cells expressing polyoma large T antigen. *Proc. Natl Acad. Sci.*, **89**, 9326.
158. Bierhuizen, M. F. A., Mattei, M.-G., and Fukuda, M. (1993). Expression of the developmental I antigen by a cloned human cDNA encoding a member of a β-1,6-N-acetylglucosaminyltransferase gene family. *Genes Dev.*, **7**, 468.
159. Dunphy, W. G., Brands, R., and Rothman, J. E. (1985) Attachment of terminal N-acetylglucosamine to asparagine-linked oligosaccharides occurs in central cisternae of the golgi stack. *Cell*, **40**, 463.
160. Weis, W. I., Drickamer, K., and Hendrickson W. A. (1992) Structure of a C-type mannose-binding protein complexed with an oligosaccharide. *Nature*, **360**, 127.
161. Vasella, A. (1990) New reactions and intermediates involving the anomenic center. *XVth international carbohydrate symposium*, p. 6, Yokohama.
162. Kornfeld, R. and Kornfeld, S. (1985). Assembly of asparagine-linked oligosaccharides. *Ann. Rev. Biochem.*, **54**, 631.
163. Piller, V., Piller, F., and Fukuda, M. (1989) Phosphorylation of the major leukocyte surface sialoglycoprotein, leukosialin, is increased by phorbol-12-myristate-13-acetate. *J. Biol. Chem.*, **264**, 18824.

2 | Molecular structure of animal lectins

KURT DRICKAMER

1. Introduction

Many effects of the carbohydrate portions of glycoconjugates result from the physical properties of the sugars themselves and from interaction of the saccharides with each other and with the proteins and lipids to which they are covalently attached. Thus, for example, the properties of proteoglycans in various biological matrices are dominated by the interactions of the constituent glycosaminoglycans with water (1). Other effects of carbohydrate conjugated to proteins can be more subtle and varied (2, 3). A striking example is modulation of activation of tissue-type plasminogen activator by the oligosaccharide attached to it (4).

It is widely believed that an additional major function of the carbohydrates in glycoconjugates is to represent information in the form of three-dimensional structures. One important way in which information in oligosaccharide structures can be recognized or 'decoded' is through the binding of lectins, each of which interacts with a selected subset of these structures (5). As novel, sugar-specific interactions are discovered, it is often found that lectins mediate the interactions. Although we are still in the early stages of identifying situations in which sugar–lectin interactions play a critical role, the list of lectin-mediated processes already includes diverse biological phenomena, from intracellular routing of glycoproteins to cell–cell adhesion and phagocytosis.

The state of our understanding of the lectin components of carbohydrate–lectin interactions varies widely. In some cases, the existence of the lectins is known largely by inference: specific binding of natural glycoproteins or neoglycoproteins to various cellular fractions is usually taken as evidence for the presence of endogenous lectins (6). Similarly, the abundance and variety of animal lectins is demonstrated by the large number of polypeptides which can be isolated by affinity chromatography on immobilized sugars (7). The emphasis of this chapter will be on animal lectins which have been characterized at the level of primary structure.

Although the list of sequenced animal lectins is long and growing, it is striking that they can be grouped into a relatively small number of categories based on their primary structures. The first portion of this chapter will be devoted to the molecular systematics of the animal lectins. In the following sections, the biological properties of lectins in each major structural group will be discussed. The tissue- and developmental stage-specific expression of lectins will also be considered.

2. Classification of animal lectins

Animal lectins can be grouped into classes based on a number of properties: the nature of their saccharide ligands, the type of biological processes in which they participate, their subcellular localization, their physical properties such as solubility in water versus detergents, or their dependence on metal co-factors. Since the primary structures of some fifty animal lectins are now known, classification based on shared sequence characteristics proves to be particularly useful. While the overall structures of the lectins vary widely, the carbohydrate-binding activity can often be ascribed to a limited portion of a given lectin. This active segment is often designated the carbohydrate-recognition domain (CRD) (8). Comparison of the sequences of these CRDs reveals that many fall into three groups, that can be designated P, C, and S. The sequence motifs which are the basis for this classification are shown in Figure 1. CRDs in each group share a pattern of invariant and highly conserved amino acid residues at a characteristic spacing.

CRDs in the P, C, and S groups have common properties beyond similarity of primary structure. A number of properties which relate the lectins in each group are summarized in Table 1. The C-type CRDs derive their name from the fact that they require calcium ions for activity. In addition, they are all extracellular, although they bind a diversity of sugars. In contrast, although the S-type lectins are found both inside and outside cells, they often are dependent on reducing agents (thiols) for full activity. They display no requirement for divalent cations and they all bind β-galactosides. P-type CRDs bind mannose 6-phosphate as their primary ligand. The relationship between these shared properties and the sequence

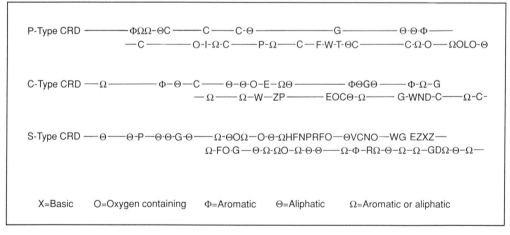

Fig. 1 Sequence motifs characteristic of three classes of carbohydrate-recognition domains

Table 1 Classification of animal lectins

Types of CRDs	Ca^{2+} dependence	State of cysteines	Ligands	Location
P-type	±	SS	Man 6-phosphate	Cell surface Lumenal
C-type	+	SS	Various	Cell surface Lumenal Extracellular
S-type	−	SH	β-Galactosides	Cytoplasm Extracellular

similarities in each group are discussed in more detail below. However, it is important to note that, for the purposes of this review, the criterion used to establish the classification is sequence similarity.

3. Lectins containing P-type CRDs: mannose 6-phosphate receptors

It is useful to begin a discussion of individual lectins by considering the lectins which contain P-type CRDs because the biological properties of these mannose 6-phosphate receptors have been very extensively studied. Two mannose 6-phosphate receptors have been identified in humans, cows, and rodents (9, 10). The properties of the cation-independent (CI) and cation-dependent (CD) receptors are compared in Table 2, and their structures are shown in Figure 2.

Table 2 Characteristics of mannose 6-phosphate receptors

Property	CI receptor	CD receptor
Calcium stimulation	−	+
Subunit M_r	~290 000	~45 000
Oligomeric	Monomer	Dimer
Intracellular pathways	TGN → lysosomes Surface → lysosomes	TGN → lysosome

TGN—*trans*-Golgi network.

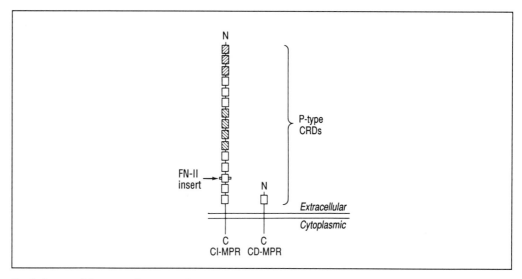

Fig. 2 Structural features of mannose 6-phosphate receptors. P-type CRDs are denoted by boxes, with crosshatching indicating regions where two active domains must be located. Adapted from reference 9

3.1 Targeting of lysosomal hydrolases

The mannose 6-phosphate receptors mediate the movement of lysosomal enzymes from the *trans*-Golgi network to lysosomes. In addition, the CI receptor can mediate internalization of extracellular hydrolases from the medium (Figure 3) (9). In both cases, recognition is based on the presence of a unique marker on the hydrolases. This marker consists of a high mannose oligosaccharide to which phosphate groups are esterified. The position and number of the phosphate groups varies, but the presence of two phosphate groups per oligosaccharide has been shown to generate an optimal ligand (11). Attachment of phosphate specifically to lysosomal hydrolases results from the action of an *N*-acetylglucosaminyl phosphotransferase followed by removal of the *N*-acetylglucosamine residue. Multiple features on the surface of lysosomal enzymes, including a critical lysine residue, are recognized by the phosphotransferase and are used to distinguish enzymes destined for lysosomes from other proteins in the common secretory pathway (12, 13).

3.2 Structural features

Both mannose 6-phosphate receptors are type I transmembrane proteins, anchored to the lipid bilayer by hydrophobic stop transfer sequences located near their COOH-termini (9). They share a common sequence motif (Figure 1) in their extracytoplasmic domains, which is present once in the CD receptor and is repeated 15 times in the CI receptor (Figure 2). Mannose 6-phosphate-binding activity has been shown to reside in the single P-type CRD of the CD receptor, and in two

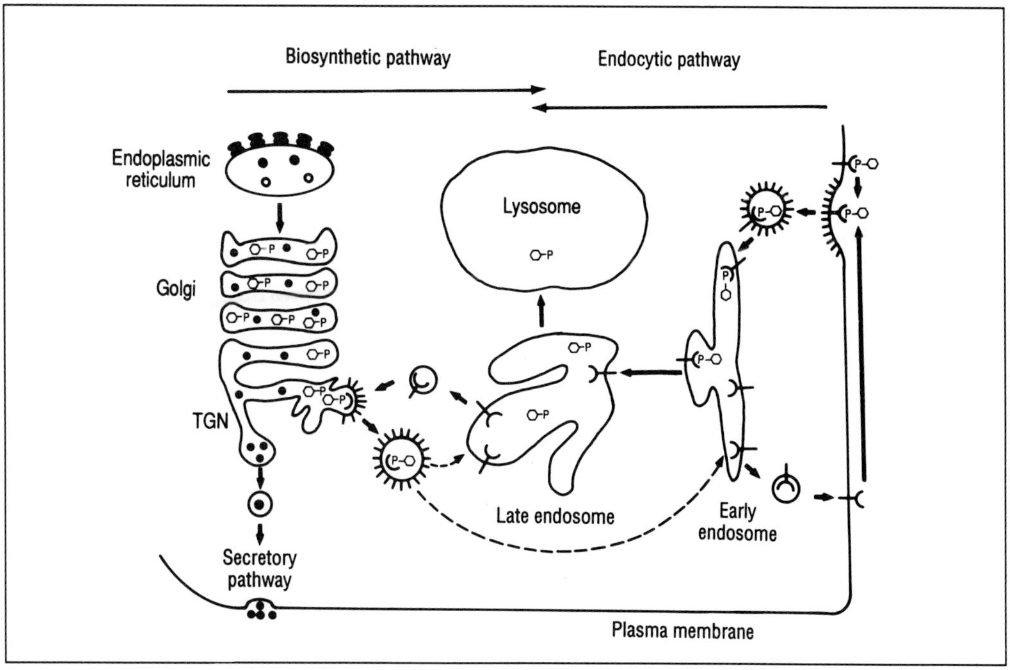

Fig. 3 Intracellular traffic mediated by mannose 6-phosphate receptors. The CD receptor directs ligands from both the *trans*-Golgi network (*TGN*) and the cell surface to lysosomes, while the CI receptor appears to function only in the first process. Adapted from reference 9

domains of the CI receptor (14, 15). Since the CD receptor is a dimer, each native receptor contains a pair of binding sites, which probably accounts for the affinity of the receptors for doubly phosphorylated oligosaccharides.

The role of the 13 CRDs of the CI receptor which are apparently not involved in oligosaccharide binding is unknown. One of the domains is interrupted by a fibronectin type II repeat (16), but except for this, the domains do not fall into clearly defined subsets based on sequence similarity. The finding that the CI mannose 6-phosphate receptor is also the receptor for insulin-like growth factor II suggests that some of the domains may be active in binding other ligands (17). However, it must also be noted that the physiological consequences of insulin-like growth factor binding to the mannose 6-phosphate receptor remain unclear.

While recognition of phosphorylated oligosaccharides is achieved by the CRDs in the extracellular portions of the mannose 6-phosphate receptors, the intracellular trafficking of the receptors shown in Figure 3 is mediated by other portions of the receptors, most notably their COOH-terminal cytoplasmic tails (18, 19). Particular importance has been ascribed to aromatic residues. In the case of the CD receptor, one phenylalanine residue forms part of a short sequence motif common to endocytic receptors. This motif appears to be a marker recognized by the sorting machinery of the *trans*-Golgi network and endosomes (9). The association of sugar binding activity to one portion of the receptor polypeptide and other activities with distinct regions of the protein is a common feature of the animal lectins.

4. Diversity of C-type animal lectins

The proteins which contain C-type CRDs form the most diverse class of animal lectins. As summarized in Table 3, the various groups of C-type animal lectins are found in serum, extracellular matrix, and membrane proteins. All of these proteins share the property of binding ligand in a calcium ion-dependent manner, but they fall into a number of distinct groups, in which the C-type CRD is combined with other protein segments (see, for example, Figure 4). As might be expected from this variety of structures, these lectins perform diverse biological functions. Some

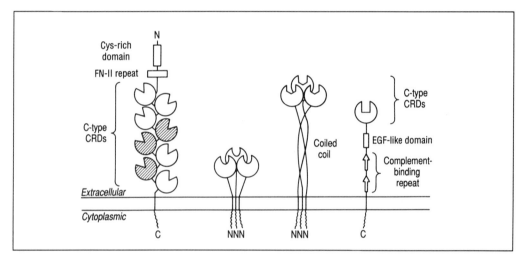

Fig. 4 Organization of membrane-bound C-type animal lectins. The domain organization of three types of endocytic receptors and the selectin cell adhesion molecules is shown from the left, along with the oligomeric structure where it is known.

of these functions are not entirely clear, but tentative suggestions are listed in Table 3. These are discussed in more detail in the next section, after which the three-dimensional structure and evolution of the C-type CRDs will be considered.

4.1 Biological functions of C-type animal lectins

A large number of C-type animal lectins are membrane-bound. Their overall organization is summarized in Figure 4. Following a discussion of the role of the membrane lectins in endocytosis and cell adhesion, the functions of the soluble C-type lectins will be considered.

Table 3 Subgroups of C-type animal lectins

Functions	Evolutionary group	Location	Organization	Ligands	Examples
Endocytic receptors	II	Plasma membrane	Type II transmembrane	Endogenous Endogenous? Endogenous Endogenous Exogenous?	Asialoglycoprotein receptor Kupffer cell receptor Chicken hepatic lectin Lymphocyte IgE F_c receptor Natural killer cell receptors
	V	Plasma membrane	Type I transmembrane	Exogenous	Mannose receptor
Adhesion molecules	IV	Plasma membrane	Type I transmembrane	Endogenous	Selectins
Humoral defence	III	Extracellular	Soluble, collagenous	Exogenous	Mannose-binding proteins Pulmonary surfactant apoproteins
Proteoglycans	I	Matrix	Extended	Endogenous	Aggrecan Versican

4.1.1 Endocytic receptors

Mammalian asialoglycoprotein receptors

The first described C-type animal lectins were the mammalian asialoglycoprotein receptors (20). Most serum glycoproteins bear one or more complex oligosaccharides which terminate in sialic acid residues attached to penultimate galactose residues. Removal of sialic acid by treatment with neuraminidase results in rapid clearance of the glycoproteins from the circulation. They are found to be taken up by parenchymal cells in the liver, and degraded in lysosomes. The proposed biological pathway is summarized in Figure 5. The site of neuraminidase action remains hypothetical. It has been proposed that the half-life of serum glycoproteins may be controlled in part by the number of complex oligosaccharides and their accessibility to neuraminidase (21).

Isolation of the asialoglycoprotein receptor, followed by sequencing and cloning of its constituent polypeptides, and expression of these peptides in fibroblasts have demonstrated the functional receptor consists of two distinct types of polypeptides, both of which have the overall structure shown in Figure 4, but which are only 55% identical in sequence (20, 22, 23). In all species examined to date, including man, rabbit, rat, and mouse, one of the constituent polypeptides is predominant. The purified receptor appears to be a hexamer, although the exact stoichiometry of subunits in the heteroligomer is unclear (23, 24). An important consequence of the formation of oligomers is an increased affinity of the receptor for multivalent ligands (25). This has been most clearly demonstrated with synthetic ligands containing defined clusters of terminal galactose or *N*-acetylgalactosamine residues. Affinity labelling with derivatives of natural oligosaccharides has revealed that a triantennary asialoglycopeptide binds in a specific orientation with

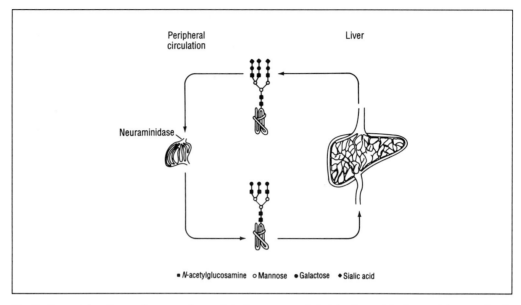

Fig. 5 Proposed pathway of serum glycoprotein turnover mediated by the asialoglycoprotein receptor in mammals. Glycoproteins with complex oligosaccharides are secreted by a number of tissues in addition to the liver. The site of neuraminidase action remains hypothetical

respect to the heteroligomer: the β1,4 branch interacts with the minor subunit, while the other branches interact with the major subunit (26).

The asialoglycoprotein receptor is a prototype for the group II C-type animal lectins shown in Figure 4. All of these proteins consist of type II transmembrane polypeptides, with N-terminal cytoplasmic domains followed by hydrophobic segments which serve as internal, uncleaved signal sequences. The CRDs in these polypeptides are separated from the membrane by necks of varying length.

Chicken hepatic lectin

The chicken hepatic lectin is the avian homologue of the mammalian asialoglycoprotein receptor. The receptor is a homoligomer of three or six subunits, depending on its state of solubilization (27; Verrey and Drickamer, unpublished observations). Each subunit is similar to the mammalian proteins in overall organization, although the neck region between the membrane and the CRD is somewhat shorter. The chicken receptor binds oligosaccharides and glycoproteins terminating in N-acetylglucosamine rather than galactose or N-acetylgalactosamine (28). The chicken protein mediates clearance of serum glycoproteins, although the mechanism by which N-acetylglucosamine would become exposed is not clear. Assuming that the chicken glycoproteins are similar in overall structural to mammalian glycoproteins, action of a β-galactosidase would be required.

Like the mammalian receptor, the intact chicken hepatic lectin binds multivalent ligands containing clusters of terminal N-acetylglucosamine residues with much higher affinity than it binds monovalent ligands (29). This is apparently due in part to the presence of multiple CRDs in the native oligomer, since a proteolytic fragment containing a single CRD shows reduced affinity for multimeric ligands (30).

The glycoprotein receptors must release their ligands in the low pH environment of the endosomes so that the receptors can recycle to the cell surface while the ligands are routed to lysosomes for degradation. The pH-dependent loss of ligand binding activity by the chicken receptor has been traced to a conformational change resulting in reduced affinity for Ca^{2+} (30). Since the concentration of Ca^{2+} in the endosomes is below the dissociation constant of the receptor at the endosomal pH, the ligand is released. Thus, the Ca^{2+}-dependence of saccharide binding to the C-type CRD is utilized as a switch to cycle the receptor between ligand-binding and inactive states.

Kupffer cell fucose receptor

In addition to the asialoglycoprotein receptor of hepatocytes, there are several glycoprotein receptors found on macrophages. For example, peritoneal macrophages express a galactose-specific receptor which is very similar in structure to the hepatocyte asialoglycoprotein receptor, although it is an oligomer composed of a single type of polypeptide (31). Uptake of glycoproteins with additional types of terminal sugars is mediated by the mannose receptor (discussed below) and by a fucose-specific receptor found in the Kupffer cells of the liver.

The fucose receptor is a type II transmembrane protein with a COOH-terminal C-type CRD (32). However, the neck portion between the membrane and the CRD is greatly extended compared to the other group II C-type lectins. The repeated heptad sequences suggest that this portion of the protein is likely to form a 2- or 3-fold coiled coil, as suggested in Figure 4 (33).

Other potential group II C-type animal lectins

Recent cloning work has defined an increasing number of proteins which are structurally similar to the group II lectins, but which have not been shown to possess saccharide-binding activity. Many of these are found on the surface of lymphocytes. Like the group II lectins, they have NH_2-terminal cytoplasmic domains, internal, uncleaved signal sequences, and COOH-terminal domains containing many of the residues characteristic of the C-type CRD motif. The lymphocyte proteins fall into two subgroups. The low affinity lymphocyte IgE F_C receptor, also designated CD23, contains a COOH-terminal extension which appears to be disulphide-bonded to the sequence just NH_2-terminal to the CRD (34, 35). Curiously, the interaction with IgE does not seem to be saccharide-mediated, as it can be inhibited by specific peptides from the F_C region, but not by sugars (36, 37).

Additional potential C-type lectins are found on the natural killer subpopulation of T cells. These proteins are apparently encoded by a cluster of genes (38, 39). The possibility that they might represent the natural killer cell antigen receptor has been disputed based on evidence assigning this function to another family of proteins (40). However, their potential role in ligand binding or cell–cell interactions, as well as their possible saccharide-binding activity are being intensively investigated.

Mannose receptor of macrophages and endothelium

The mannose receptor is found on a variety of macrophage subtypes, as well as on hepatic endothelial cells (41–43). It is responsible for internalization of glycoproteins containing N-linked high mannose structures. Ligands include β-glucuronidase, yeast invertase, and mannosylated serum albumin (41). The observation that a lysosomal hydrolase such as β-glucuronidase can be taken up by this route led to the suggestion that one function of the receptor may be to scavenge for misrouted lysosomal enzymes which have been secreted (41, 43). These enzymes could have been underphosphorylated in the cell in which they were synthesized or they could have become dephosphorylated extracellularly.

The fact that very large, mannose-containing oligosaccharides, found in yeast mannan and invertase for example, are excellent ligands for the receptor suggested that recognition of potential pathogens might be an alternative function of the receptor. In this case, the primary function of the receptor might be to bind exogenous rather than endogenous ligands. Evidence has been presented that the mannose receptor mediates phagocytosis of yeast, although it has also been suggested that another receptor, specific for β-glucans, is responsible for this process (44–46).

The structure of the mannose receptor is strikingly different from the structure of the other endocytic receptors discussed so far (Figure 4) (47). The mannose receptor is a type I transmembrane protein, synthesized with a transient, NH_2-terminal signal sequence and anchored to the membrane by a stop transfer sequence near the COOH terminus. Most strikingly, the receptor contains eight repeats of the C-type CRD motif. There are also two additional cysteine-rich domains at the NH_2 terminus of the mature protein, one of which is a fibronectin type II repeat.

Expression of truncated forms of the receptor in mammalian, insect, and bacterial cells and in an *in vitro* translation system has been used to demonstrate that several of the CRDs are involved in binding a complex ligand such as yeast invertase, although only one shows detectable binding to mannose in isolation (48). This suggests that, like the asialoglycoprotein receptor, the chicken hepatic lectin, and the collectins (see below), the mannose receptor achieves high affinity binding to multivalent oligosaccharides by clustering of multiple domains, each with relatively weak affinity for monosaccharides. However, in the mannose receptor the clustering of CRDs is achieved within a single polypeptide rather than by forming an oligomer.

In addition to achieving high affinity for multivalent ligands, the clustering of multiple CRDs allows the receptor to achieve selectivity for complex oligosaccharides with specific geometrical properties. While high affinity binding to a synthetic ligand such as heavily mannosylated serum albumin can be achieved with just two of the CRDs, four or more of the domains are required to reproduce the affinity of the intact receptor for yeast mannose (48; Taylor and Drickamer, unpublished observations). This suggests that the exact arrangement of the CRDs in the receptor may be a critical determinant of specificity in ligand binding. These results emphasize that the binding of ligands by C-type lectins is a function both of the interaction of individual CRDs with individual saccharides and the arrangement of the multiple CRDs within each lectin.

4.1.2 C-type lectins as cell adhesion molecules: the selectins

The variety of oligosaccharides found at the surface of cells has long been taken as an indication that these structures might be involved in interactions between cells. Adhesion between leukocytes and endothelia has provided a system in which the molecular basis for such saccharide-mediated adhesion can be convincingly demonstrated. The molecules which are involved in this type of recognition share overall structural features, and are grouped together under the term selectin. The specific interactions mediated by each of the three known selectins are summarized in Table 4. L-selectin on T cells targets them to peripheral lymph nodes (49), while E- and P-selectins on endothelium interact with neutrophils and monocytes (50). In addition, P-selectin on platelets mediates their binding to endothelium (51).

Like the mannose receptor, the selectins are type I transmembrane proteins, with NH_2-terminal signal sequences and COOH-terminal membrane anchors and cytoplasmic domains (52–55). The extracellular domain of each, shown diagrammatically in Figure 4, consists of an NH_2-terminal C-type CRD, an epidermal growth factor-like domain, and variable numbers of complement binding repeats (Table 4). The oligomeric structure of the selectins has not yet been established.

Even before the primary structures of the selectins were known, evidence from competition studies with ligands such as mannose 6-phosphate and fucoidin as well as the effects of neuraminidase suggested that L-selectin-mediated binding of T cells to the high endothelium of peripheral lymph nodes involves negatively charged carbohydrate structures on the surface of the endothelium (49). The presence of a C-type CRD in L-selectin suggested a molecular basis for this obser-

Table 4 Members of the selectin gene family

Selectin type	Other designations	Number of CH domains	Location of selectin	Location of target	Known ligands
L	LECAM-1 Mel-14 Homing receptor	2	T lymphocytes	High endothelium of periferal lymph nodes	Sialyl-Le^x Phosphomannan
E	LECAM-2 ELAM-1	6	Endothelium	Neutrophils Monocytes	Sialyl-Le^x (Sialyl-Le^a)
P	LECAM-3 GMP-140 PADGEM CD62	9	Platelets Endothelium	Endothelium Neutrophils Monocytes	Sialyl-Le^x (Sialyl-Le^a) Sulphatides

vation. Interestingly, truncation studies indicate that the epidermal growth factor-like domain is required in addition to the CRD for full ligand binding activity of L-selectin (56).

The range of ligands which can be bound by the selectins remains to be fully established, and the natural ligands on the surface of target cells is still being investigated. However, it is clear that all three of the selectins can bind structures related to the sialyl-Le^x antigen, in which both terminal sialic acid and fucose are present (57). Sulphatides also appear to be candidate cell surface ligands for at least P-selectin (58, 59). Evidence that the major ligand for L-selectin is a cell surface glycoprotein has been presented (60). The selectins are considered in more detail in Chapter 4.

4.1.3 C-type CRDs associated with collagenous domains: collectins

Of the several groups of soluble animal lectins which contain C-type CRDs, the most extensively studied are those consisting of higher oligomers (9–27 subunits) of polypeptides in which COOH-terminal CRDs are linked to NH$_2$-terminal collagen-like segments. The overall structure of these lectins is highlighted in Figure 6. The term collectin is used here to refer to this group of proteins. Although the supporting data are incomplete, it can be proposed that all of the collectins are involved in primative or antibody-independent immunity. Because this idea has evolved in part from observations on the structure and saccharide-binding activities of the lectins, it is useful to consider these topic before summarizing evidence relating to their biological functions.

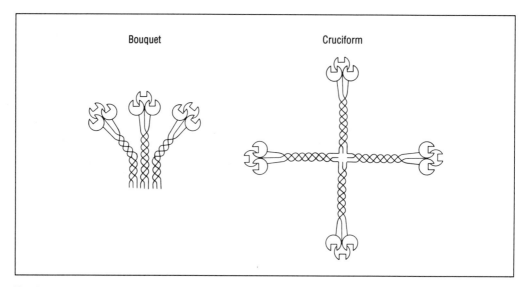

Fig. 6 Summary of subunit and oligomeric structure of collagen-containing lectins. The native molecular weights of some of the lectins in each subgroup indicate that the oligomers shown here must associate into larger structures

Structural features of collectin subgroups

The collagen-like regions of the collectins are characterized by repeats of the tripeptide Gly-X-Y (61–65). The X position is often occupied by proline, and hydroxyproline is found in many of the Y positions. In addition, in some cases hydroxylysine residues conjugated to galactose or glucosyl galactose are present (66). Thus, it appears that the NH_2-terminal portions of these proteins are very much like the triple helical regions of true collagens. The collectins are similar in design to complement protein C1q, which also contains an NH_2-terminal collagen-like domain linked to COOH-terminal, globular domains. The terminal domains of C1q bind immunoglobulin (67).

Based on the detailed structure of the collagen-like regions, the collectins can be categorized into two subgroups (Table 5). The mannose-binding proteins of liver and serum, as well as pulmonary surfactant apoprotein SP-A contain 18–24 Gly-X-Y repeats, interrupted somewhere near the midpoint by an irregularity in the repeat pattern (61–63). Such interruptions are also found in the polypeptide subunits of complement protein C1q, and lead to the formation of bends in the collagen stalks (68). The interruptions have a similar effect in this subgroup of collectins, which appear in the electron microscope as bouquet-like structures, in which the NH_2-terminal portions of the collagenous stalks are held together in a bundle, while the COOH-terminal CRDs are spread apart (Figure 6, left) (69).

In contrast to the bouquet-like collectins, bovine conglutinin and pulmonary surfactant apoprotein have longer collagenous domains (57–59, Gly-X-Y repeats) which lack central interruptions (64, 65). This leads to long, unbent stalks which associate via their NH_2 termini to form cross-like structures when visualized by rotary shadowing (Figure 6) (69). The significance of these structural differences between subgroups 1 and 2 remains to be established.

Saccharide binding by collectins

The mannose-binding proteins were initially isolated based on their affinity for mannose-rich structures such as yeast mannan (70), so their activity as lectins was

Table 5 Subgroups of collectins

Subgroup morphology	Examples	Gly-X-Y repeats	Central interruption	M_r	Possible oligomer	Ligands
Bouquet	Mannose-binding protein-A (1)	18	+	650–700 000	$(\alpha_3)_6$	Man, GlcNAc, Fuc
	Mannose-binding protein-C (2)	20	+	200–220 000	$(\alpha_3)_3$	Man, Fuc
	Surfactant SP-A	24	+	>400 000	$(\alpha_3)_6$	Man, Glc, Fuc, Gal
Cuciform	Conglutinin	57	–	400 000	$(\alpha_3)_4$	Man, GlcNAc, Fuc
	Surfactant SP-D	59	–		$(\alpha_3)_4$	Glc

realized early and has been extensively studied. In both humans and rats, evidence for two mannose-binding proteins has been presented (61, 71). The properties of human mannose-binding protein 1 and rat mannose-binding protein A seem to be parallel, while the human protein 2 and rat protein C are most similar. Both proteins have been cloned from rat (61), while only human protein 1 has been cloned (62).

In each species, the two proteins show striking differences in ligand-binding properties, both at the level of monosaccharides and oligosaccharides. While all of the proteins bind both mannose and fucose, the rat A protein and human 1 protein show a broader specificity which includes N-acetylglucosamine and glucose (71, 72). Differences in binding to oligosaccharides have been highlighted in studies in which binding of the CRDs from the two rat mannose-binding proteins to neoglycolipids bearing various oligosaccharides released from glycoproteins was analysed using a thin-layer chromatogram overlay assay (72). These studies suggest that mannose-binding protein C binds to core structures of oligosaccharides, while mannose-binding protein A interacts with terminal mannose and N-acetylglucosamine residues. As the natural ligands for the mannose-binding proteins are likely to be saccharides found in lower eukaryotes and bacteria (see below), it will be of interest to compare binding of the two lectins to such ligands to see if they may have distinct functional roles.

In contrast to the mannose-binding proteins, the saccharide-binding activity of the pulmonary surfactant apoproteins was postulated after it was noted that these proteins contain potential CRDs. Evidence for the interaction of these proteins with sugars has initially been derived from affinity chromatography on resins densely substituted with monosaccharides (73). Additional binding studies confirm that SP-A interacts with a broad range of monosaccharides, including mannose, glucose, fucose, and galactose, but not with amino sugars. Certain natural glycoconjugates (both glycoproteins and glycolipids) have been shown to be targets for SP-A binding (74). Binding competition experiments indicate that SP-D interacts with glucose-containing structures (75). Finally, a combination of the techniques of neoglycolipid and glycoprotein blotting, and solid phase competition assays has been used to demonstrate that conglutinin is also a lectin which can interact with a range of sugars similar to mannose-binding protein A (76).

Biological significance of the collectins
There is evidence for two types of collectin-mediated defence, via complement fixation and by opsonization. The close similarity of the bouquet-forming collectins and complement component C1q suggested that these collectins might mediate complement fixation via the classical pathway after binding to saccharide-rich surfaces such as the mannan coat of yeast or the peptidoglycan of bacteria. Such binding would bypass the need for antibody binding and could result in activation of subsequent steps in the complement cascade through direct interaction of the collectin with complement components C1r and C1s. Consistent with this scenario, activation of complement by rat serum mannose-binding protein, in an antibody-

and C1q-independent manner, has been demonstrated (69, 77). On the other hand, opsonization of gram-negative bacteria such *Salmonella montevideo* by the human serum mannose-binding protein has been demonstrated in experiments in which the bacteria are incubated with purified lectin in the absence of serum (78).

Direct binding studies have shown that collectins interact with potentially pathogenic organisms. In addition to binding of mannose-binding protein to *S. montevideo*, pulmonary surfactant SP-A binding to *Pneumocystic carinii*, and surfactant SP-D binding to *Escherichia coli* and other gram-negative bacteria have been demonstrated (79). Most of these interactions have been shown to be inhibitable by appropriate sugars. Thus it appears that these collectins share the ability to bind potentially pathogenic microorganisms through surface carbohydrates without the need for specific antibodies.

Analysis of the structure and expression of the human mannose-binding protein gene is also consistent with a role of this lectin in host defence. The levels of mRNA for mannose-binding protein increase during the acute phase, as would be expected for a protein involved in rapid, relatively nonspecific pathogen neutralization (62). The increase in mRNA probably results from the presence of specific control elements upstream of the mannose-binding protein gene (80). These sequences include segments resembling heat shock response elements and glucocorticoid receptor binding sites.

Important clinical evidence for the role of human mannose-binding protein derives from studies of patients with a propensity for recurrent, severe infections. The sera of some such patients manifest a defect in opsonization of yeast, which can be traced to a genetic defect in the gene for human serum mannose-binding protein (81). The phenotype results from a single Gly→Asp mutation in the collagen-like domain, which presumably interferes with oligomerization. As in the case of similar mutations in collagens, the presence of mutant polypeptides in heterozygotes suppresses the formation of functional oligomers from the normal polypeptides. These genetic studies provide compelling evidence that the collectins function in innate immunity. As would be expected, decreased mannose-binding protein function has its most severe effects in the time between 6 months and 2 years of age, when maternal immunity is waning but the infant's immunoglobulin-producing capacity is still limited.

Although the case for the involvement of collectins in defence appears quite strong, some alternate functions have been proposed. In particular, it has been postulated that surfactant protein SP-A may have roles in uptake and recycling of the surfactant protein-lipid complex into the type II alveolar cells (82). It has also been suggested that the combination of carbohydrate and phospholipid-binding properties of this protein may be critical to formation of the unique tubular myelin structures found in the lung (74). The interaction of SP-A with endogenous glycoconjugates in the lung fluid and at the surface of the alveolar cells is consistent with these types of roles. A clearer definition of whether SP-A and/or SP-D primarily serve defence functions analogous to those of the mannose-binding proteins or if they are predominantly involved in other functions should be forthcoming soon.

4.1.4 C-type CRDs found in proteoglycans

C-type CRDs have been identified in two proteoglycans: aggrecan, the primary proteoglycan of cartilage matrix, and versican, a high molecular weight proteoglycan synthesized by fibroblasts (83–85). The primary structures of the core proteins from these proteoglycans are summarized in Figure 7. Interestingly, the CRDs are juxtaposed with epidermal growth factor-like sequences and complement regulatory repeats as in the selectins. However, the order of the domains is different, and alternative splicing generates forms in which some of these domains are missing. The CRD portion of the aggrecan proteoglycan has been visualized in the electron microscope as a globular domain which is designated G3, to distinguish it from the link protein-like regions found near the NH_2 terminus (86). Interestingly, the G3 domain appears to be absent in older tissue, suggesting that it may serve a transient function during cartilage development.

Aggrecan is another example of a protein in which a potential CRD was first identified based on primary structure and was subsequently shown to have saccharide-binding activity. CRD expressed in an *in vitro* translation system was shown to bind weakly to densely substituted sugar resins (87). It was possible to

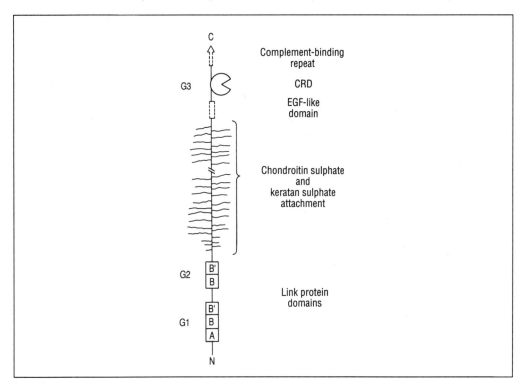

Fig. 7 Domain organization of the proteoglycans which contain C-type CRDs. The G2 domain, found in cartilage aggrecan, is absent from fibroblast aggrecan (83–85). Alternative splicing of aggrecan mRNA generates forms in which either the adjacent epidermal growth factor-like domain or the complement regulatory repeat are not present

use this binding to determine specificity for various saccharides, which revealed that the best ligands are fucose and galactose. A clearer understanding of the possible role of the CRD in organizing the extracellular matrix, presumably by binding to some of the endogenous oligosaccharide structures, awaits further investigation.

4.1.5 Other mosaic proteins containing C-type CRDs

Coagulation factor C from the horseshoe crab *Limulus* seems to lie evolutionarily between the simple proteins which consist of just a C-type CRD (see below) and the complex, multidomain proteins of vertebrates (88). Bacterial lipopolysaccharide activates factor C, initiating a cascade of events which lead to haemolymph coagulation. The protein has features of both vertebrate C-type lectins and clotting factors, since it consists of an epidermal growth factor-like domain, five complement-binding repeats with a C-type CRD between the third and fourth repeats, and a COOH-terminal serine protease domain. The binding of coagulation factors by proteins in the C-type lectin family is a recurrent theme, as demonstrated by the properties of two additional proteins discussed in the next section.

4.1.6 C-type CRDs found as isolated polypeptides

A number of relatively small proteins containing C-type CRDs have been identified in mammals and in invertebrates. The CRD is the dominant feature of each of these proteins, although they sometimes contain short NH_2-terminal or COOH-terminal extensions. The known examples of these proteins are summarized in Table 6.

In higher vertebrates, three small proteins containing C-type CRD sequence motifs have been identified. One of these proteins, tetranectin, was isolated from serum based on its ability to bind kringle domain 4 of plasminogen (89). Two

Table 6 C-type CRDs in isolation

Name	Species	Tissue	Saccharide specificity	Proposed function
Tetranectin	Human	Serum	Unknown	Kringle-domain binding
Stone protein	Human, Rat	Pancreas	Unknown	
Pancreatitis associated protein	Rat	Pancreas	Unknown	
	Rattlesnake	Venom	Gal	Prey capture
	Trimeresurus flavoviridis	Venom	Unknown	Factor IX and X binding
Echinoidin	Sea urchin	Coelomic fluid	Gal	
	Acorn barnacle (2)	Coelomic fluid	Gal	Protection
	Flesh fly	Haemolymph	Gal	Protection
	Cockroach	Haemolymph	Lipopolysaccharide	Protection
	Tunicate		Gal	

protein in the pancreas also contain potential C-type CRDs. One is the major protein component of stones formed in the exocrine pancreas (90). The protein was originally identified as a thread-forming contaminant of trypsin preparations and later, independently, as a gene product specifically expressed in regenerating islets of the endocrine pancreas. The other pancreatic C-type lectin may also be a modulator of islet cell function. The mRNA encoding this protein was found to be expressed at high levels during pancreatic inflammation, so the protein is referred to as pancreatitis-associated proteins (91). Carbohydrate-binding activity has not been demonstrated in any of these proteins.

A galactose-binding, low molecular weight C-type lectin is found in rattlesnake venom (92). In addition, an anticoagulant from *Trimeresurus flavoviridis* venom, which binds to factors IX and X, has the structural features of a C-type CRD but has not been demonstrated to have carbohydrate-binding activity (93).

The invertebrate C-type lectins are alike in that they generally bind galactose as primary monosaccharide ligand (94–98). They are often found in the haemolymph or coelomic fluid. Based on the appearance of the haemolymph lectins in the fly *Sarcophaga perigrina* and the cockroach *Periplanta americana* following traumatization, it has been suggested that these lectins may serve protective functions (96, 97). However, the effector pathway by which protection might be achieved has not been elucidated.

4.2 Evolution of C-type CRDs

One of the striking features of the C-type lectin family is its diversity. Compared with either the P- or S-type lectins, the C-type lectins are more structurally diverse, and show specificity for a much broader range of saccharides. The structural diversity takes the form of juxtaposition of the CRD with a variety of other domains, many of which are common to other nonlectin proteins. This phenomenon has generally been ascribed to shuffling of exons encoding the various domains which can function as largely independent modules, each conferring on the assembled protein particular functional characteristics. The gene structures for a number of C-type CRDs are known, and are generally consistent with this suggestion, since the CRD-coding regions are inevitably separated from the other portions of the genes by introns (99).

Curiously, some of the CRD-coding regions are interrupted by additional introns. It has been argued that the variable positioning of these introns may reflect insertion events relatively late in evolution, but the alternative possibility that this phenomenon represents shifting of primordial introns which linked three separate structural subdomains cannot be ruled out.

Within the C-type lectin family, it is instructive to construct a summary of the relative relatedness of the different CRD sequences. Such a comparison is best undertaken with sequences which have been subjected to optimal multiple alignment, a procedure which is difficult to achieve with total rigour when dealing with such a large number of relatively diverse sequences. However, a reasonable approximation to such a comparison is presented in Figure 8. It is important to emphasize that this dendrogram is based exclusively on the comparison of the CRD portions of the proteins, and thus is independent of the degree to which the flanking sequences are similar. Nevertheless, the results clearly show that the CRDs within each structural group of lectins are generally more similar to each other than they are to the CRDs of lectins in different structural classes.

The observation that CRDs in a particular subclass of lectins are structurally the most closely related allows one to conclude that the divergence of CRDs within each group in the figure followed the shuffling of exons which brought about the overall arrangement of the lectins in that group. An example of this is the chicken hepatic lectin, which, like the asialoglycoprotein receptor, is a type II transmembrane protein, and which contains a CRD relatively closely related in sequence to the CRD of the asialoglycoprotein receptor. This observation indicates that the shuffling event which juxtaposed the CRD with an NH_2-terminal membrane anchor preceded the divergence of the CRDs in these two proteins, although they have completely nonoverlapping saccharide-binding specificities.

Since the saccharide-binding properties of the CRDs within each group can have diverged relatively recently, it can be further concluded that the binding site must have evolved over a relatively short time span to accommodate diverse sugars. This suggests, in turn, that the sugar-binding activity probably results from contact with relatively few amino acid side chains in the lectin. The structural results discussed in the next section are consistent with this hypothesis.

4.3 Structure of a prototype C-type CRD

As indicated in the preceding sections, the fundamental basis for complex oligosaccharide recognition by C-type animal lectins is interaction of an individual component of the oligosaccharide with a CRD. Combination of multiple CRDs in appropriate spatial arrangement then endows the lectins with specificity for the higher order structure of the oligosaccharide. Progress in understanding the first part of this process at the molecular level has resulted from the ability to produce large quantities of a C-type CRD, from rat mannose binding protein A (section 4.1.3), making possible analysis of its three-dimensional structure.

4.3.1 Determinants of C-type CRD structure

Definition of the minimum fragment of mannose-binding protein A which displays full carbohydrate-binding activity has been achieved by molecular biological and protein chemical approaches. Fragments of a cDNA for the lectin expressed by *in vitro* transcription and translation bind immobilized saccharide as long as they contain at least the COOH-terminal 112 amino acid residues of the protein, indicating that this segment of polypeptide contains the CRD (100). Proteolysis studies reveal that the COOH-terminal 115 residues of the protein form a protease-resistant domain with saccharide-binding activity (101).

The structure of the protease-resistant, CRD-containing fragment was elucidated

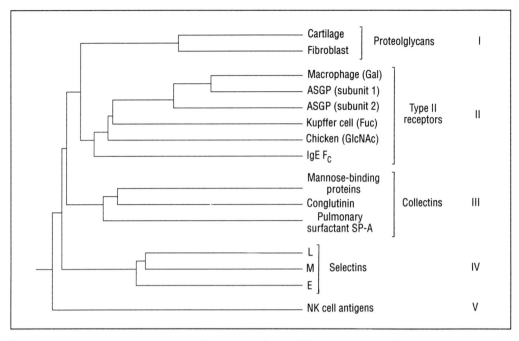

Fig. 8 Dendrogram showing sequence similarity of the C-type CRDs and their classification into groups I to V

by X-ray crystallography. Experimental phases were determined utilizing the anomalous diffraction properties of holmium ions substituted for calcium ions in the CRD. The structure is summarized in Figure 9. The backbone folding of the polypeptide is distinct from any other known protein structures. The regular secondary structure, consisting of two α-helices and two reasonably well-defined β-sheets, is confined to the lower two-thirds of the molecule. This portion of the molecule is stabilized by extensive close packing of aliphatic and aromatic residues in two hydrophobic cores and by two disulphide bonds. The upper third of the molecule consists of several loops which are arranged to ligate two holmium (calcium) ions.

The structure provides striking rationalization for the pattern of conservation of amino acid residues in the C-type CRD motif (Figure 1). Conserved residues of the motif fall into four categories: (1) the four cysteines which form the two disulphide bonds, (2) seven oxygen-containing residues, most of which are ligands for the calcium ions, (3) four proline or glycine residues which form critical turns, and (4) 19 aliphatic and aromatic residues which form the two hydrophobic cores.

4.3.2 Modulation of C-type CRD structure

Further understanding of how the CRD functions can be derived from the results of various modifications in the CRD induced by biochemical manipulations and by

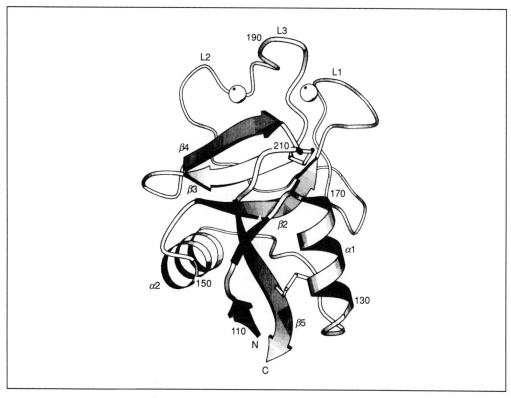

Fig. 9 Ribbon diagram showing the structure of the C-type CRD from a rat mannose-binding protein. The two spheres indicate the positions of the calcium ions. α-Helices are denoted by coils, and β-sheets are shown as flat arrows. Adapted from reference 102

mutagenesis. At the biochemical level, it has been shown that the mannose-binding protein CRD is resistant to proteolysis only in the presence of calcium ions (101). The dependence of protease-resistance on calcium-ion concentration exactly parallels the calcium-ion dependence of ligand binding. As expected, both processes are second order in calcium-ion concentration, indicating that both calcium ions must be bound in order for a saccharide-binding site to be formed. These results indicate that the bound calcium ions stabilize the CRD, and the crystallographic results summarized in Figure 9 suggest that the loops of the upper third of the domain are probably the portion of the structure which is stabilized. In the absence of calcium ions, these loops are apparently accessible to proteolytic digestion. The fact that very similar proteolysis results have been obtained for the CRD of the chicken hepatic lectin (section 4.1.1) indicates that the behaviour of the mannose-binding protein CRD is probably typical of the C-type CRDs in general (30).

The close correlation of the structural change upon loss of calcium ions and the ability to bind ligands indicates that the structure stabilized by the bound calcium ions is critical for formation of a saccharide-binding site. One way to explain this finding is to postulate that the saccharide-binding site is formed by the loops when they are held in the appropriate conformation by the calcium ions.

The structure of the mannose-binding proteins has also been modified by extensive random mutagenesis (103). The results of these studies confirm the structural and evolutionary analyses. For instance, changes in any of the residues which ligate calcium ions result in complete loss of activity, with the exception of one aspartic acid to glutamic acid change which causes a 50-fold increase in the calcium-ion dissociation constant. Changes in residues found on the surface are generally tolerated, while changes in residues in the interior of the molecule are usually deleterious. These studies confirm the importance of the evolutionarily conserved residues in forming the basic CRD fold.

A large and particularly interesting group of changes in the conserved hydrophobic residues of the core have a rather surprising phenotype. These mutants show no saccharide binding under physiological conditions, but bind sugars normally when the calcium ion concentration is raised. This observation leads to two important conclusions. (1) The role of the lower two-thirds of the molecule is to position the loops of the upper third exactly correctly so that they can ligate calcium ions. Any slight perturbation in the structure of the lower two-thirds results in loops which cannot form calcium ion-binding sites with optimal affinity. (2) The finding that the saccharide-binding site, once formed at high enough calcium-ion concentration, is completely normal in these mutants, is most easily explained by postulating that the binding site is formed as the loops are brought into their normal conformation by the calcium ions. The finding that residues on a large portion of the surface can be modified without affecting saccharide binding also suggests that the total area of interaction with the sugars is likely to be small. This is consistent with recent structural data on the saccharide-CRD complex (Weis, W. I., Drickamer, K., and Hendrickson, H. M. K. (1992) *Nature*, **360**, 127).

5. Soluble β-galactoside-binding (S-type) lectins

5.1 Structural features of S-type lectins

The family of S-type lectins has been investigated in a wide range of species (Table 7) (104). All of the proteins which share the S-type CRD motif also share the ability to bind β-galactosides, leading to the alternative designation S-Lac lectins. While three size classes of S-type lectins (13–16 kDa, 18–22 kDa, and 28–35 kDa) have been identified in many species (105), the most thoroughly characterized have been the 11–16 kDa (L-14) and 28–35 kDa (L-30) proteins (104). The relative levels of these lectins vary from tissue to tissue, but both are widely distributed (Table 8).

As summarized in Figure 10, the L-14 group consists of CRDs without any other flanking domains, while lectins in the L-30 group contain an NH_2-terminal extension. Expression of L-14 lectins truncated from either end reveals that the functional CRD corresponds to nearly the entire lectin, thus providing an experimental

Table 7 Groups of S-type lectins

Group	Species	Tissues/cells
L-14	Human	Placenta
	Cow	Spleen
	Rat	Intestine
	Mouse	Muscle
	Chicken (2)	Liver
		Heart
		Lung
		Skin
		Brain
		Thymus
		Bone marrow
		Fibroblasts
L-30	Human	Lung
	Rat	Brain
	Mouse	Fibroblasts
		Macrophages
		Lymphocytes

Table 8 Other types of lectins

Type	Examples	Location	Ca^{2+}-dependence
Pentraxins	C-reactive protein	Serum	+
	Serum amyloid protein	Serum	+
Growth factors	Interleukin 1	Serum	±
	Interleukin 2	Serum	?
	Tumour necrosis factor	Serum	?
Sulphated saccharide binding	Thrombospondin	Serum	−
	Antistasin	Saliva	−
	Circumsporozoite coat protein	Plasma membrane	−
Sialic acid binding	Frog oocyte lectins	Cytoplasm	−

definition of the CRD analogous to that for the C-type lectins (106). In chickens, two L-14 group lectins have been characterized. The L-14 proteins are found as homodimers, while dimers of the L-30 proteins may only form under certain conditions (107).

The L-14 lectins from various mammalian and avian species share some 19 invariant and 36 conserved residues with the COOH-terminal portion of the L-30 lectins (Figure 10) (104, 106, 108). Homologous L-14 lectins from amphibians and fish share many of the same residues (109). The pattern of conserved amino acids bears no discernible resemblance to the pattern seen in P- and C-type CRDs. Although one of the invariant positions is occupied by a cysteine residue, other cysteine residues are not conserved, and all appear to be present as sulphhydryl groups rather than as disulphides. This is consistent with the activity of these proteins in the cytoplasmic compartment of cells. Indeed, the designation 'S-type'

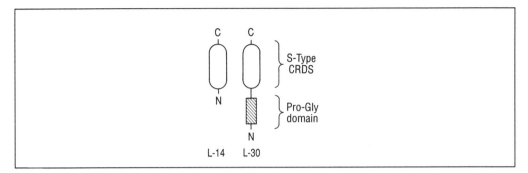

Fig. 10 Structure of two groups of S-type animal lectins. The S-type CRDs are denoted by ovals

was originally based on the fact that many of these lectins are stabilized by the presence of thiols. The chemical basis for this thiol effect remains unclear. It seems unlikely that sulphhydryl groups *per se* are required for activity, because alkylation does not inactivate the proteins, and replacement of each cysteine with serine in one of the chicken S-type lectins does not impair saccharide binding (108).

5.2 Biological functions of S-type lectins

Although all S-type lectins bind β-galactosides, different oligosaccharide chains bind with different affinities (110) and the specificity of each of the different S-type lectins is unique (111). Particular emphasis has been placed on the ability of these lectins to bind blood group and polylactosamine structures (110, 112). Affinity for polylactosamine may be the basis for binding of these lectins to laminin (113, 114).

The localization of the S-type lectins is interesting in two respects. First, the same lectins are synthesized by many different cell types (104). Second, they appear to have a dual localization, both inside the cell in the cytoplasm and nucleus and outside the cell, associated with the cell surface and extracellular matrix (104, 114). In some cases, it is possible to observe a change in localization with changes in the state of the cell. For example, in myoblast cell culture, the L-14 lectin shifts from the cytoplasm to the medium during differentiation (115). The mechanism of post-translational secretion of the proteins is unclear, although mechanisms involving protein pumps or intracellular vesiculation have been proposed.

5.2.1 L-14 lectins

The properties of the L-14 lectins have led to various proposals about their functions, but at the moment none of these ideas has been convincingly proven. The change in localization of the lectins during myoblast differentiation formed the basis for the suggestion that the lectin may have a role in cell-matrix interactions during muscle development (115). The high affinity of the L-14 lectins for poly-lactosamine structures such as those found on laminin indicates that a significant fraction of the extracellular L-14 lectin is likely to be bound to the matrix (113, 114).

Additionally, the bivalent nature of these lectins, resulting from their dimerization, suggests that they may link together glycoconjugates.

A difficulty in assigning functions to the L-14 lectins is their relatively high concentration in a variety of tissues. Thus they may be found as contaminants in preparations that have biological activities resulting from lower abundance proteins. Unless this problem is dealt with carefully, association of activities such as growth promotion or suppression with the L-14 group of lectins must be considered tentative (116, 117). Similarly, correlation between L-14 lectin expression and transformation (118, 119) cannot yet be taken as proof that the lectin itself is a determinant of the transformed phenotype.

As noted above, expression of the L-14 lectin is developmentally regulated, as its synthesis and secretion are altered at specific times during cellular differentiation. Particularly striking is the appearance of L-14 in the trophectoderm of the pre-implantation blastocyst, later in the myotomes, and on specific subsets of neurones (120, 121).

5.2.2 L-30 lectins

Two entirely different sets of functions have been associated with the L-30 group of lectins. In its extracellular form, the rat L-30 lectin has been independently identified as an IgE-binding protein found in basophilic leukaemia cells (122). IgE-binding activity has also been demonstrated in human L-30, although in neither case has the binding activity been demonstrated to be saccharide-dependent (123). Mouse L-30 lectin has also been identified as an antigen found on the surface of macrophages, designated Mac-2 (124). Thus, there are hints that extracellular L-30 lectin may be involved in modulation of immune function, but no specific model for what it does has been proposed.

The NH_2-terminal extension of the L-30 lectins bears similarity to sequences found in ribonuclear protein components, and the L-30 lectin in mouse 3T3 cells has been found to be part of the ribonuclear protein fraction (125). Movement of the lectin from the cytoplasm to the nucleus as the cell changes from the proliferating to the quiescent state seems to result from movement of the entire RNA-protein complex (126). The role of this type of particle in transcriptional control and RNA transport suggests that the lectin may be involved in these control processes. One intriguing possibility is that the lectin may interact with O-linked sugars found in transcription complexes and nuclear pore complexes (127). The primary sugar found associated with these structures is N-acetylglucosamine, which is not a ligand for the L-30 lectins, but the possibility that other sugars are present needs to be investigated.

6. Other groups of animal lectins

Not all sugar-binding proteins found in animals fall into the three groups which have been considered in detail here. Some of additional animal lectins are listed in

Table 8. They have been grouped based on a combination of shared structural and functional properties.

The pentraxins, C-reactive protein and serum amyloid P protein, are homologous serum proteins which share the property of forming oligomers based on an annular pentamer (128). Although the sequences are not related to the C-type CRD motif, these proteins show calcium ion-dependent binding properties, including binding to bacterial cell surfaces. Serum amyloid P shows affinity for specific phosphorylated and sulphated saccharides (129). C-reactive protein, which is an acute phase reactant, has been identified as a galactose-specific particle receptor on liver macrophages (130). The properties of these proteins are in some respects parallel to those of the mannose-binding proteins.

The urinary protein uromodulin has been used as a test ligand to demonstrate saccharide-binding activity for tumour necrosis factor, interleukin-1 and interleukin-2 (131–133). The saccharide-binding activity of these soluble factors is most easily detected with immobilized ligands or factors. The activity of interleukin-2 shows unusual pH and divalent cation requirements, since dependence on cations is observed only at certain pH values. Competition studies suggest that tumour necrosis factor and interleukin-2 bind to the mannose core of N-linked oligosaccharides as well as to oligomers of N-acetylglucosamine. Since uromodulin is primarily found in the urine, its role in immunosuppression resulting from binding of serum factors remains unclear.

A common sequence motif has been described in proteins which bind sulphated glycoconjugates (Table 8) (134). It has been suggested that this motif may reflect a common structure which is responsible for the related saccharide-binding specificities of these proteins. While the similarity between P, C, and S-type CRDs extends over 100 or more amino acid residues, the sulphated oligosaccharide-binding proteins share a relatively small region of sequence similarity.

Finally, sialic acid-binding lectins from oocytes of two species of frog have been cloned and sequenced (135, 136). The sequences are distinct from any of the groups so far discussed, suggesting that additional classes of animal carbohydrate-binding proteins remain to be described.

7. Conclusions

In spite of the great diversity of animal lectins, some common themes are emerging from the studies described here. First, the ability to bind saccharides often resides in discrete domains (CRDs) which function somewhat independently of the rest of the molecule. Interestingly, although the P, C, and S-type CRDs are not apparently related to each other in sequence, they are of a similar size. This probably reflects the need for about 100 amino acid residues to form a self-contained sugar-binding site. The division of the lectins into carbohydrate-binding and effector domains is of great experimental value, and provides a useful way to think about the structure of the lectins. It may also reflect an evolutionary coming together of the effector functions with sugar-binding domains able to target a desired activity with greater precision. The existence of domains which are clearly homologous with the P or C-type CRDs, but which do not have demonstrable saccharide-binding activity, suggests that in some cases the CRDs may have evolved to recognize structures other than sugars.

A second theme is the importance of combining multiple CRDs in a native lectin molecule in order to increase affinity for multivalent ligands. The balance between the dual effects of individual CRD binding specificity and the geometry with which the CRDs are combined can be different. For example, the affinity of individual CRDs of the mannose-phosphate receptors for the mannose 6-phosphate haptene is much higher than the affinity of the mannose receptor for mannose, although the affinity of each receptor for its cognate oligosaccharide is similar.

The field of animal lectins is evolving rapidly. Our molecular understanding of some of the lectins is becoming quite advanced, although our knowledge of the biological functions of these proteins is not always keeping pace. One challenge will be to show how the unique biological properties of the lectins derive from the molecular details which are not being deciphered. On the other hand, many proteins with sugar-binding activity have been described but have not yet been characterized at the primary structure level. A second major challenge will be to pursue the characterization of these additional lectins.

Acknowledgements

I thank Maureen Taylor for critical reading of the manuscript. Work in the author's laboratory is supported by grant GM42628 from the National Institutes of Health and a Faculty Salary Award from the American Cancer Society.

References

1. Heinegård, D. and Oldberg, Å. (1989) Structure and biology of cartilage and bone matrix noncollagenous macromolecules. *FASEB J.*, **3**, 2042.
2. Rademacher, T. W., Parekh, R. B., and Dwek, R. A. (1988) Glycobiology. *Annu. Rev. Biochem.*, **57**, 785.
3. Parekh, R. B. (1991) Effects of glycosylation on protein function. *Curr. Opinion Struct. Biol.*, **1**, 750.
4. Wittwer, A. and Howard, S. C. (1990) Glycosylation at Asn-184 inhibits the conversion of single-chain tissue-type plasminogen activator by plasmin. *Biochemistry*, **29**, 4175.
5. Sharon, N. and Lis, H. (1989) Lectins as recognition molecules. *Science*, **246**, 227.
6. Facy, P., Seve, A.-P., Hubert, M., Monsigny, M., and Hubert, J. (1990) Analysis of nuclear sugar binding components in undifferentiated and *in vitro* differentiated human promyelocytic leukemia cells (HL60). *Exp. Cell Res.*, **190**, 151.
7. Barondes, S. H. (1981) Lectins: their multiple endogenous cellular functions. *Annu. Rev. Biochem.*, **50**, 207.
8. Drickamer, K. (1988) Two distinct classes of carbohydrate-recognition domains in animal lectins. *J. Biol. Chem.*, **263**, 9557.

9. Kornfeld, S. and Mellmann, I. (1989) The biogenesis of lysosomes. *Annu. Rev. Cell Biol.*, **5**, 483.
10. Ma, Z., Grubb, J. H., and Sly, W. S. (1991) Cloning, sequencing, and function characterization of the murine 46-kDa mannose 6-phosphate receptor. *J. Biol. Chem.*, **266**, 10 589.
11. Hoflack, B., Fujimoto, K., and Kornfeld, S. (1987) The interaction of phosphorylated oligosaccharides and lysosomal enzymes with bovine liver cation-dependent mannose 6-phosphate receptor. *J. Biol. Chem.*, **262**, 123.
12. Baranski, T. J., Faust, P. L., and Kornfeld, S. (1990) Generation of a lysosomal enzyme targeting signal in the secretory protein pepsinogen. *Cell*, **63**, 281.
13. Baranski, T. J., Koelsch, G., Hartsuck, J. A., and Kornfeld, S. (1991) Mapping and molecular modeling of a recognition domain for lysosomal enzyme targeting. *J. Biol. Chem.*, **266**, 23 365.
14. Wendland, M., Hille, A., Nagel, G., Waheed, A., von Figura, K., and Pohlman, R. (1989) Synthesis of a truncated Mr 46000 mannose 6-phosphate receptor that is secreted and retains ligand binding. *Biochem. J.*, **260**, 201.
15. Westlund, B., Dahms, N. M., and Kornfeld, S. (1991) The bovine mannose 6-phosphate/insulin-like growth factor II receptor: localization of mannose 6-phosphate binding sites to domains 1–3 and 7–11 of the extracytoplasmic region. *J. Biol. Chem.*, **266**, 23 233.
16. Lobel, P., Dahms, N. M., and Kornfeld, S. (1988) Cloning and sequence analysis of the cation-independent mannose 6-phosphate receptor. *J. Biol. Chem.*, **263**, 2563.
17. Tong, P. Y., Tollefsen, S. E., and Kornfeld, S. (1988) The cation-independent mannose 6-phosphate receptor binds insulin-like growth factor II. *J. Biol. Chem.*, **263**, 2585.
18. Canfield, W. M., Johnson, K. F., Richard, D. Y., Gregory, W., and Kornfeld, S. (1991) Localization of the signal for rapid internalization of the bovine cation-independent mannose 6-phosphate/insulin-like growth factor-II receptor to amino acids 24–29 of the cytoplasmic tail. *J. Biol. Chem.*, **266**, 5682.
19. Johnson, K. F., Chan, W., and Kornfeld, S. (1990) Cation-dependent mannose 6-phosphate receptor contains two internalization signals in its cytoplasmic domain. *Proc. Natl Acad. Sci. USA*, **87**, 10 010.
20. Spiess, M. (1990) The asialoglycoprotein receptor: a model for endocytic transport receptors. *Biochemistry*, **29**, 10 008.
21. Drickamer, K. (1991) Clearing up glycoprotein hormones. *Cell*, **67**, 1029.
22. McPhaul, M. and Berg, P. (1986) Formation of functional asialoglycoprotein receptor after transfection with cDNAs encoding the receptor proteins. *Proc. Natl Acad. Sci. USA*, **83**, 8863.
23. Halberg, D. F., Wager, R. E., Farrell, D. C., Hildreth, J., Quesenberry, M. S., Loeb, J. A. *et al.* (1987) Major and minor forms of the rat liver asialoglycoprotein receptor are independent galactose-binding proteins: primary structure and glycosylation heterogeneity of minor receptor forms. *J. Biol. Chem.*, **262**, 9828.
24. Herzig, M. C. S. and Weigel, P. H. (1990) Surface and internal galactosyl receptors are heterooligomers and retain this structure after ligand internalization or receptor modulation. *Biochemistry*, **29**, 6437.
25. Lee, R. T., Lin, P., and Lee, Y. C. (1984) New synthetic cluster ligands for galactose/N-acetylgalactosamine-specific lectin of mammalian liver. *Biochemistry*, **23**, 4255.
26. Rice, K. G., Weisz, O. A., Barthel, T., Lee, R. T., and Lee, Y. C. (1990) Defined stoichiometry of binding between triantennary glycopeptide and the asialoglycoprotein receptor of rat hepatocytes. *J. Biol. Chem.*, **265**, 18 429.

27. Loeb, J. A. and Drickamer, K. (1987) The chicken receptor for endocytosis of glycoprotein contains a cluster of N-acetylglucosamine-binding sites. *J. Biol. Chem.*, **262**, 3022.
28. Kawasaki, T. and Ashwell, G. (1977) Isolation and characterization of an avian hepatic binding protein specific for N-acetylglucosamine-terminated glycoproteins. *J. Biol. Chem.*, **252**, 6536.
29. Lee, R. T., Rice, K. G., Rao, N. B. N., Ichikawa, Y., Barthel, T., Piskarev, V. et al. (1989) Binding characteristics for N-acetylglucosamine-specific lectin of the isolated chicken hepatocytes: similarities to mammalian hepatic galactose/N-acetylgalactosamine-specific lectin. *Biochemistry*, **28**, 8351.
30. Loeb, J. A. and Drickamer, K. (1988) Conformational changes in the chicken receptor for endocytosis of glycoproteins: modulation of binding activity by Ca^{2+} and pH. *J. Biol. Chem.*, **263**, 9752.
31. Ii, M., Kurata, H., Itoh, N., Yamashina, I., and Kawasaki, T. (1990) Molecular cloning and sequence analysis of cDNA encoding the macrophage lectin specific for galactose and N-acetylgalactosamine. *J. Biol. Chem.*, **265**, 11 295.
32. Hoyle, G. W. and Hill, R. L. (1988) Molecular cloning and sequencing of a cDNA for a carbohydrate binding receptor unique to rat Kupffer cells. *J. Biol. Chem.*, **263**, 7487.
33. Beavil, A. J., Edmeades, R. L., Gould, H. J., and Sutton, B. J. (1992) α-Helical coiled-coil stalks in the low-affinity receptor for IgE (FcεRII/CD23) and related C-type lectins. *Proc. Natl Acad. Sci. USA*, **89**, 753.
34. Bettler, B., Maier, R., Rüegg, D., and Hoffstetter, H. (1989) Binding site for IgE of the human lymphocyte low-affinity Fcε receptor (Fcε/CD23) is confined to the domain homologous with animal lectins. *Proc. Natl Acad. Sci. USA*, **86**, 7118.
35. Bettler, B., Texido, G., Raggini, S., Rüegg, D., and Hofstetter, H. (1992) Immunoglobulin E-binding site in Fcε receptor (FcεRII/CD23) identified by homolog-scanning mutagenesis. *J. Biol. Chem.*, **267**, 185.
36. Vercelli, D., Helm, B., Marsh, P., Padlan, E., Geha, R. S., and Gould, H. (1989) The B-cell binding site on human immunoglobulin E. *Nature*, **338**, 649.
37. Richards, M. L. and Katz, D. H. (1990) The binding of IgE to murine Fc RII is calcium-dependent but not inhibited by carbohydrate. *J. Immunol.*, **144**, 2638.
38. Houchins, J. P., Yabe, T., McSherry, C., and Bach, F. H. (1991) DNA sequence analysis of NKG2, a family of related cDNA clones encoding type II integral membrane proteins on human natural killer cells. *J. Exp. Med.*, **173**, 1017.
39. Wong, S., Freeman, J. D., Kelleher, C., Mager, D., and Takei, F. (1991) Ly-49 multigene family: new members of a superfamily of type II membrane proteins with lectin-like domains. *J. Immunol.*, **147**, 1417.
40. Roder, J. (1991) Killing comes naturally. *Curr. Biol.*, **1**, 242.
41. Stahl, P. D. and Schlesinger, P. H. (1980) Receptor-mediated pinocytosis of mannose/N-acetylglucosamine-terminated glycoproteins and lysosomal enzymes by macrophages. *Trends Biochem. Sci.*, **5**, 194.
42. Taylor, M. E., Leaning, M. S., and Summerfield, J. A. (1987) Uptake and processing of glycoproteins by rat hepatic mannose receptor. *Am. J. Physiol.*, **252**, E690.
43. Stahl, P. D. (1990) The macrophage mannose receptor: current status. *Am. J. Respir. Cell Mol. Biol.*, **2**, 317.
44. Warr, S. A. (1980) A macrophage receptor for (mannose/glucosamine)—glycoproteins of potential importance in phagocytic activity. *Biochem. Biophys. Res. Commun.*, **93**, 737.
45. Sung, S. J., Nelson, R. S., and Silverstein, S. C. (1983) Yeast mannans inhibit binding and phagocytosis of zymosan by mouse peritoneal macrophages. *J. Cell Biol.*, **96**, 160.

46. Czop, J. K. and Kay, J. (1991) Isolation and characterization of β-glucan receptors on human mononuclear phagocytes. *J. Exp. Med.*, **173**, 1511.
47. Taylor, M. E., Conary, J. T., Lennartz, M. R., Stahl, P. D., and Drickamer, K. (1990) Primary structure of the mannose receptor contains multiple motifs resembling carbohydrate-recognition domains. *J. Biol. Chem.*, **265**, 12 156.
48. Taylor, M. E., Bezouska, K., and Drickamer, K. (1992) Contribution to ligand binding by multiple carbohydrate-recognition domains in the macrophage mannose receptor. *J. Biol. Chem.*, **267**, 1719.
49. Yednock, T. A. and Rosen, S. D. (1989) Lymphocyte homing. *Adv. Immunol.*, **44**, 313.
50. Osborn, L. (1980) Leukocyte adhesion to endothelium in inflammation. *Cell*, **62**, 3.
51. Larsen, E., Celi, A., Gilbert, G. E., Furie, B. C., Erban, J. K., Bonfanti, R. et al. (1989) PADGEM protein: a receptor that mediates the interaction of activated platelets with neutrophils and monocytes. *Cell*, **59**, 305.
52. Lasky, L. A., Singer, M. S., Yednock, T. A., Dowbenko, D., Fennie, C. et al. (1989) Cloning of a lymphocyte homing receptor reveals a lectin domain. *Cell*, **56**, 1045.
53. Johnston, G. I., Cook, R. G., and McEver, R. P. (1989) Cloning of GMP-140, a granule membrane protein of platelets and endothelium: sequence similarity to proteins involved in cell adhesion and inflammation. *Cell*, **56**, 1033.
54. Bevilacqua, M. P., Stengelin, S., Gimbrone, M. A., Jr, and Seed, B. (1989) Endothelial leukocyte adhesion molecule 1: an inducible receptor for neutrophils related to complement regulatory proteins and lectins. *Science*, **243**, 1160.
55. Siegelman, M. H., van de Rijn, M., and Weissman, I. L. (1989) Mouse lymph node receptor cDNA encodes a glycoprotein revealing tandem interaction domains. *Science*, **243**, 1165.
56. Kansas, G. S., Spertini, O., Stoolman, L. M., and Tedder, T. F. (1991) Molecular mapping of functional domains of the leukocyte receptor for endothelium, LAM-1. *J. Cell. Biol.*, **114**, 351.
57. Feizi, T. (1991) Cell–cell adhesion and membrane glycosylation. *Curr. Opin. Struct. Biol.*, **1**, 766.
58. Skinner, M. P., Lucas, C. M., Burns, G. F., Chesterman, C. N., and Berndt, M. C. (1991) GMP-140 binding to neutrophils is inhibited by sulfated glycans. *J. Biol. Chem.*, **266**, 5371.
59. Arufo, A., Kolanus, W., Walz, G., Fredman, P., and Seed, B. (1991) CD62/P-selectin recognition of myeloid and tumor cell sulfatides. *Cell*, **67**, 35.
60. Baumhueter, S., Singer, M. S., Henzel, W., Hemmerich, S., Renz, M., Rosen, S. D., and Lasky, L. A. (1993). Binding of L-selectin to the vascular sialomucin CD34. *Science*, **262**, 436.
61. Drickamer K., Dordal, M. S., and Reynolds, L. (1986) Mannose-binding proteins isolated from rat liver contain carbohydrate-recognition domains linked to collagenous tails. *J. Biol. Chem.*, **261**, 6878.
62. Ezekowitz, R. A. B., Day, L. E., and Herman, G. A. (1988) A human mannose-binding protein is an acute phase reactant that shares sequence homology with other vertebrate lectins. *J. Exp. Med.*, **167**, 1034.
63. Benson, B., Hawgood, S., Schilling, J., Clements, J., Damm, D., Cordell, B. et al. (1985) Structure of canine pulmonary surfactant apoprotein: cDNA and complete amino acid sequence. *Proc. Natl Acad. Sci. USA*, **82**, 6379.
64. Shimizu, H., Fisher, J. H., Papst, P., Benson, B., Lau, K., Mason, R. J. et al. (1992) Primary structure of rat pulmonary surfactant protein D: cDNA and deduced amino acid sequence. *J. Biol. Chem.*, **267**, 1853.

65. Lee, Y.-M., Leiby, K. R., Allar, J., Paris, K., Lerch, B., and Okarma, T. B. (1991) Primary structure of bovine conglutinin, a member of the C-type animal lectin family. *J. Biol. Chem.*, **266**, 2715.
66. Colley, K. J. and Baenziger, J. U. (1987) Identification of the post-translational modifications of the core-specific lectin: the core-specific lectin contains hydroxyproline, hydroxylysine, and glucosylgalactosylhydroxylysine. *J. Biol. Chem.*, **262**, 10 290.
67. Reid, K. B. M. (1983) Proteins involved in the activation and control of the two pathways of human complement. *Biochem. Soc. Trans.*, **11**, 1.
68. Perkins, S. J. (1985) Molecular modeling of human complement subcomponent C1q and its complex with $C1r_2C1s_2$ derived from neutron-scattering curves and hydrodynamic properties. *Biochem. J.*, **228**, 13.
69. Lu, J., Thiel, S., Wiedemann, H., Timpl, R., and Reid, K. B. M. (1990) Binding of the pentamer/hexamer forms of a mannan-binding protein to zymosan activates the proenzyme $C1r_2C1s_2$ complex of the classical pathway of complement, without involvement of C1q. *J. Immunol.*, **144**, 2287.
70. Wild, J., Robinson, D., and Winchester, B. (1983) Isolation of mannose-binding proteins from human and rat liver. *Biochem J.*, **210**, 167.
71. Taylor, M. E. and Summerfield, J. A. (1987) Carbohydrate-binding proteins of human serum: isolation of two mannose/fucose-specific lectins. *Biochim. Biophys. Acta*, **915**, 60.
72. Childs, R. A., Feizi, T., Yuen, C. T., Drickamer, K., and Quesenberry, M. S. (1990) Differential recognition of core and terminal portions of oligosaccharide ligands by carbohydrate-recognition domains of two mannose binding proteins. *J. Biol. Chem.*, **265**, 20 770.
73. Haagsman, H. P., Hawgood, S., Sargeant, T., Buckley, D., White, R. T., Drickamer, K. *et al.* (1987) The major lung surfactant protein, SP 28–36, is a calcium-dependent, carbohydrate-binding protein. *J. Biol. Chem.*, **262**, 13 877.
74. Childs, R. A., Wright, J. R., Ross, G. F., Yuen, C.-T., Lawson, A. M., Chai, W. *et al.* (1992) Specificity of lung surfactant protein SP-A for both the protein and the lipid moieties of certain neutral glycolipids. *J. Biol. Chem.*, **267**, 9972.
75. Persson, A., Chang, D., and Crouch, E. (1990) Surfactant protein D is a divalent cation-dependent carbohydrate-binding protein. *J. Biol. Chem.*, **265**, 5755.
76. Loveless, R. W., Feizi, T., Childs, R. A., Mizouchi, T., Stoll, M. S., Oldroyd, R. G. *et al.* (1989) Bovine serum conglutinin is a lectin which binds non-reducing terminal N-acetylglucosamine, mannose and fucose residues. *Biochem. J.*, **258**, 109.
77. Ikeda, K., Sannoh, T., Kawasaki, N., Kawasaki, T., and Yamashina, I. (1987) Serum lectin with known structure activates complement through the classical pathway. *J. Biol. Chem.*, **262**, 7451.
78. Kuhlman, M., Joiner, K., and Ezekowitz, R. A. B. (1989) The human mannose-binding protein functions as an opsonin. *J. Exp. Med.*, **169**, 1733.
79. Ezekowitz, R. A. B., Williams, D. J., Koziel, H., Armstrong, M. Y. K., Warner, A., Richards, F. F. *et al.* (1991) Uptake of *Pneumocystis carinii* mediated by the macrophage mannose receptor. *Nature*, **351**, 155.
80. Taylor, M. E., Brickell, P. M., Craig, R. K., and Summerfield, J. A. (1989) Structure and evolutionary origin of the gene encoding a human serum mannose-binding protein. *Biochem. J.*, **262**, 761.
81. Sumiya, M., Super, M., Tabona, P., Levinsky, R. J., Takayuki, A., Turner, M. W. *et al.* (1991) Molecular basis of opsonic defect in immunodeficient children. *Lancet*, **337**, 1569.

82. Wright, J. R., Borchelt, J. D., and Hawgood, S. (1989) Lung surfactant apoprotein SP-A (26–36 kDa) binds with high affinity to isolated type II cells. *Proc. Natl Acad. Sci. USA*, **86**, 5410.
83. Doege, K., Sasaki, M., Horigan, E., Hassell, J. R., and Yamada, Y. (1987) Complete primary structure of the rat cartilage proteoglycan core protein deduced from cDNA clones. *J. Biol. Chem.*, **262**, 17 757.
84. Zimmermann, D. R. and Ruoslahti, E. (1989) Multiple domains of the large fibroblast proteoglycan, versican. *EMBO J.*, **8**, 2975.
85. Doege, K. J., Sasaki, M., Kimura, T., and Yamada, Y. (1991) Complete coding sequence and deduced primary structure of the human cartilage large aggregating proteoglycan, aggrecan: human-specific repeats, and additional alternatively spliced forms. *J. Biol. Chem.*, **266**, 894.
86. Paulsson, M., Mörgelin, M., Wiedmann, H., Beardmore-Gray, M., Dunham, D., Hardingham, T. *et al.* (1987) Extended and globular protein domains in cartilage proteoglycans. *Biochem. J.*, **245**, 763.
87. Halberg, D. H., Proulx, G., Doege, K., Yamada. Y., and Drickamer, K. (1988) A segment of cartilage proteoglycan core protein has lectin-like activity. *J. Biol. Chem.*, **263**, 9486.
88. Muta, T., Miyata, T., Misumi, Y., Tokunaga, F., Nakamura, T., Toh, Y. *et al.* (1991) Limulus factor C: an endotoxin-sensitive serine protease zymogen with a mosaic structure of complement-like, epidermal growth factor-like, and lectin-like domains. *J. Biol. Chem.*, **266**, 6554.
89. Fuhlendorff, J., Clemmensen, I., and Magnusson, S. (1987) Primary structure of tetranectin, a plasminogen kringle 4 binding plasma protein: homology with asialoglycoprotein receptor and cartilage proteoglycan core protein. *Biochemistry*, **26**, 6757.
90. Watanabe, T., Yonekura, H., Terazono, K., Yamamoto, H., and Okamoto, H. (1990) Complete nucleotide sequence of human *reg* gene and its expression in normal and tumoral tissues: the *reg* protein, pancreatic stone protein, and pancreatic thread protein are one and the same product of the gene *J. Biol. Chem.*, **265**, 7432.
91. Iovanna, J., Orelle, B., Keim, V., and Dagorn, J.-C. (1991) Messenger RNA sequence and expression of rat pancreatitis-associated protein, a lectin-related protein overexpressed during acute experimental pancreatitis. *J. Biol. Chem.*, **266**, 24 664.
92. Hirabayashi, J., Kusunoki, T., and Kasai, K. (1991) Complete primary structure of a galactose-specific lectin from the venom of the rattlesnake *Crotalus atrox*: homologies with Ca^{2+}-dependent-type lectins. *J. Biol. Chem.*, **266**, 2320.
93. Atoda, H., Hyuga, M., and Morita, T. (1991) The primary structure of coagulation factor IX/factor X-binding protein isolated from the venom of *Trimeresurus flavoviridis*: homology with asialoglycoprotein receptors, proteoglycan core protein, tetranectin, and lymphocte Fcε receptor for immunoglobulin E. *J. Biol. Chem.*, **266**, 14 903.
94. Giga, Y., Ikai, A., and Takahashi, K. (1987) The complete amino acid sequence of echinoidin, a lectin from the coelomic fluid of the sea urchin *Anthocidaris crassispina*: homologies with mammalian and insect lectins. *J. Biol. Chem.*, **262**, 6197.
95. Muramoto, K. and Kamiya, H. (1990) The amino-acid sequence of multiple lectins from the acorn barnacle *Megabalanus rosa* and its homology with animal lectins. *Biochim. Biophys. Acta*, **1039**, 42.
96. Takahashi, H., Komano, H., Kawaguchi, N., Kitamura, N., Nakanishi, S., and Natori, S. (1985) Cloning and sequencing of cDNA of *Sarcophaga peregrina* humoral lectin induced on injury of the body wall. *J. Biol. Chem.*, **260**, 12 228.

97. Jomori, T. and Natori, S. (1991) Molecular cloning of cDNA for lipopolysaccharide-binding protein from the hemolymph of the American cockroach, *Periplanta americana*: similarity of the protein with animal lectins and its acute phase expression. *J. Biol. Chem.*, **266**, 13 318.
98. Suzuki, T., Takagi, T., Furukohri, T., Kawamure, K., and Nakauchi, M. (1990) A calcium-dependent galactose-binding lectin from the tunicate *Polyandtocarpa misakiensis*: isolation, characterization, and amino acid sequence. *J. Biol. Chem.*, **265**, 1274.
99. Bezouska, K., Crichlow, G. V., Rose, J. M., Taylor, M. E., and Drickamer, K. (1991) Evolutionary conservation of intron position in a subfamily of genes encoding carbohydrate-recognition domains. *J. Biol. Chem.*, **266**, 11 604.
100. Quesenberry, M. S. and Drickamer, K. (1991) Determination of the minimum carbohydrate-recognition domain in two C-type animal lectins. *Glycobiology*, **1**, 615.
101. Weis, W. I., Crichlow, G. V., Murthy, H. M. K., Hendrickson, W. A., and Drickamer, K. (1991) Physical characterization and crystallization of the carbohydrate-recognition domain of a mannose-binding protein from rat. *J. Biol. Chem.*, **266**, 20 678.
102. Weis, W. I., Kahn, R., Fourme, R., Drickamer, K., and Hendrickson, W. A. (1991) Structure of the calcium-dependent lectin domain from a rat mannose-binding protein determined by MAD phasing. *Science*, **254**, 1608.
103. Quesenberry, M. S. and Drickamer, K. (1992) Role of conserved and nonconserved residues in the Ca^{2+}-dependent carbohydrate-recognition domain of a rat mannose-binding protein: analysis by random cassette mutagenesis. *J. Biol. Chem.*, **267**, 10831.
104. Wang, J. L., Laing, J. G., and Anderson, R. L. (1991) Lectins in the cell nucleus. *Glycobiology*, **1**, 243.
105. Cerra, R. F., Gitt, M. A., and Barondes, S. H. (1985) Three soluble rat β-galactoside-binding lectins. *J. Biol. Chem.*, **260**, 10 474.
106. Abbott, W. M. and Feizi, T. (1991) Soluble 14-kDa β-galactoside-specific bovine lectin: evidence from mutagenesis and proteolysis that almost the complete chain is necessary for integrity of the carbohydrate-recognition domain. *J. Biol. Chem.*, **266**, 5552.
107. Woo, H.-J., Lotz, M., Jung, J. U., and Mercurio, A. M. (1991) Carbohydrate-binding protein 35 (Mac-2), a laminin-binding lectin, forms functional dimers using cysteine 186. *J. Biol. Chem.*, **266**, 18 419.
108. Hirabayashi, J. and Kasai, K. (1991) Effect of amino acid substitution by site-directed mutagenesis on the carbohydrate-recognition and stability of human 14-kDa β-galactoside-binding lectin. *J. Biol. Chem.*, **266**, 23 648.
109. Marschal, P., Herrmann, J., Leffler, H., Barondes, S. H., and Cooper, D. N. W. (1992) Sequence and specificity of a soluble lactose-binding lectin from *Xenopus laevis* skin. *J. Biol. Chem.*, **267**, 12942.
110. Abbott, M. W., Hounsell, E. F., and Feizi, T. (1988) Further studies of oligosaccharide recognition by the soluble 13 kDa lectin of bovine heart muscle: ability to accommodate the blood-group-H and -B-related sequences. *Biochem. J.*, **252**, 283.
111. Leffler, H. and Barondes, S. H. (1986) Specificity of binding of three soluble rat lung lectins to substituted and unsubstituted mammalian β-galactosides. *J. Biol. Chem.*, **261**, 10 119.
112. Sato, S. and Hughes, R. C. (1992) Bind specificity of a baby hamster kidney lectin for H type I and II chains, polylactosamine glycans, and appropriately glycosylated forms of laminin and fibronectin. *J. Biol. Chem.*, **267**, 6983.
113. Zhou, Q. and Cummings, R. D. (1990) The S-type lectin from calf heart tissue binds selectively to the carbohydrate chains of laminin. *Arch. Biochem. Biophys.*, **281**, 27.

114. Woo, H.-J., Shaw, L. M., Messier, J. M., and Mercurio, A. M. (1990) The major non-integrin laminin binding protein of macrophages is identical to carbohydrate binding protein 35 (Mac-2). *J. Biol. Chem.*, **265**, 7097.
115. Cooper, D. N. and Barondes, S. H. (1990) Evidence of export of a muscle lectin from cytosol to extracellular matrix and for a novel secretory mechanism. *J. Biol. Chem.*, **110**, 1681.
116. Sanford, G. L. and Harris-Hooker, S. (1990) Stimulation of vascular cell proliferation by β-galactoside specific lectins. *FASEB J.*, **4**, 2912.
117. Wells, V. and Mallucci, L. (1991) Identification of an autocrine negative growth factor: mouse β-galactoside-binding protein is a cytostatic factor and cell growth regulator. *Cell*, **64**, 91.
118. Raz, A., Zhu, D., Hogan, V., Shah, N., Raz, T., Karkash, R. et al. (1990) Evidence for the role of 34 kDa galactoside-binding lectin in transformation and metastasis. *Int. J. Cancer*, **46**, 871.
119. Yamaoka, K., Ohno, S., Kawasaki, H., and Suzuki, K. (1991) Overexpression of a β-galactoside binding protein causes transformation of BALB3T3 fibroblast cells. *Biochem. Biophys. Res. Commun.*, **179**, 272.
120. Hynes, M. A., Gitt, M., Barondes, S. H., Jessell, T. M., and Buck, L. B. (1990) Selective expression of an endogenous lactose-binding lectin gene in subsets of central and periferal neurons. *J. Neurosci.*, **10**, 1004.
121. Poirier, F., Timmons, P. M., Chan, C.-T. J., Guénet, J.-L., and Rigby, P. W. J. (1992) Expression of the L-14 soluble lectin during mouse embryogenesis suggests multiple roles during pre- and post-implantation development. *Development*, **115**, 143.
122. Albrandt, K., Orida, N. K., and Liu, F. T. (1987) An IgE-binding protein with a distinctive repetitive sequence and homology with an IgG receptor. *Proc. Natl Acad. Sci. USA*, **84**, 6859.
123. Robertson, M. W., Albrandt, K., Keller, D., and Liu, F. T. (1990) Human IgE-binding protein: a soluble lectin exhibiting a highly conserved interspecies sequence and differential recognition of IgE glycoforms. *Biochemistry*, **29**, 8093.
124. Cherayil, B. J., Weiner, S. J., and Pillai, S. (1989) The Mac-2 antigen is a galactose-specific lectin that binds IgE. *J. Exp. Med.*, **170**, 1959.
125. Jia, S. and Wang, J. L. (1988) Carbohydrate binding protein 35. Complementary DNA sequence reveals homology with proteins of the heterogeneous nuclear RNP. *J. Biol. Chem.*, **263**, 6009.
126. Agrwal, N., Wang, J. L., and Voss, P. G. (1989) Carbohydrate-binding protein 35. Levels of transcription and mRNA accumulation in quiescent and proliferating cells. *J. Biol. Chem.*, **264**, 17 236.
127. Hart, G. W., Haltiwanger, R. S., Holt, G. D., and Kelly, W. G. (1989) Glycosylation in the nucleus and cytoplasm. *Annu. Rev. Biochem.*, **58**, 841.
128. Pepys, M. B. and Baltz, M. L. (1983) Acute phase proteins with special reference to C-reactive protein and related proteins (pentraxins) and serum amyloid A protein. *Adv. Immunol.*, **34**, 141.
129. Loveless, R. W., Floyd-O'Sullivan, G., Raynes, J. G., Yuen, C.-T., and Feizi, T. (1992) Human serum amyloid P is a multispecific adhesive protein whose ligands include 6-phosphorylated mannose and the 3-sulphated saccharides galactose, N-acetylgalactosamine and glucuronic acid. *EMBO J.*, **11**, 813.
130. Kempka, G., Roos, P. H., and Kolb-Bachoven, V. (1990) A membrane-associated form of C-reactive protein is the galactose-specific particle receptor on rat liver macrophages. *J. Immunol.*, **144**, 1004.

131. Muchmore, A. V. and Decker, J. M. (1987) Evidence that recombinant IL 1 α exhibits lectin-like specificity and binds homogeneous uromodulin via N-linked oligosaccharides. *J. Immunol.*, **138**, 2541.
132. Sherblom, A. P., Decker, J. M., and Muchmore, A. V. (1988) The lectin-like interaction between recombinant tumor necrosis factor and uromodulin. *J. Biol. Chem.*, **263**, 5418.
133. Sherblom, A. P., Amoorthy, N., Decker, J. M., and Muchmore, A. V. (1989) IL-2, a lectin with specificity for high mannose glycopeptides. *J. Immunol.*, **143**, 939.
134. Holt, G. D., Krivan, H. C., Gasic, G. J., and Ginsburg (1989) Antistasin, an Inhibitor of coagulation and metastasis, binds to sulfatide (Gal(3-SO$_4$)β1-1Cer) and has a sequence homology with other proteins that bind sulfated glycoconjugates. *J. Biol. Chem.*, **264**, 12 138.
135. Titani, K., Takio, K., Kuwada, M., Nitta, K., Sakakibara, F., Kawauchi, H. *et al.* (1987) Amino acid sequence of sialic acid binding lectin from frog (*Rana catesbeina*) eggs. *Biochemistry*, **26**, 2189.
136. Kamiya, Y., Oyama, F., Oyama, R., Sakakibara, F., Nitta, K., Kawauchi, H. *et al.* (1990) Amino acid sequence of a lectin from Japanese frog (*Rana japonica*) eggs. *J. Biochem. (Tokyo)*, **108**, 139.

3 | Molecular cloning of glycosyltransferase genes

HARRY SCHACHTER

1. Introduction

Oligosaccharides differ from proteins and nucleic acids, the other two classes of large biological polymers, in two important characteristics—they are usually highly branched molecules and their monomeric units are connected to one another by many different linkage types. Proteins and nucleic acids are linear and there is only a single type of linkage between the monomers of these molecules (amide bonds for proteins and 3'-5'-phosphodiester bonds for nucleic acids). Although all three classes of macromolecule share an almost infinite capacity for information storage, branching and multiple linkage types bestow on oligosaccharides an ability to store more information per unit of molecular size—more information 'bang' for the molecular weight 'buck'. Perhaps it is not surprising that a major site for information exchange mediated by oligosaccharides is at membrane surfaces, for example interactions between cells and their fluid or cellular environments. The cell surface and its environment are busy places and the spatial efficiency of oligosaccharides may make them uniquely suitable for these interactions.

The complex nature of oligosaccharide structure has important biosynthetic implications. Proteins and nucleic acids are synthesized by well-established template mechanisms. Only linear molecules can be assembled in a factory in which a new molecule is simply copied from a pre-existing template molecule. Complex structures like automobiles and oligosaccharides must be manufactured on an assembly line. The assembly line for protein- and lipid-bound oligosaccharides is the endomembrane system—the endoplasmic reticulum and Golgi apparatus (1). A series of membrane-bound glycosidases and glycosyltransferases act sequentially on the growing oligosaccharide as it moves along the lumen of the endomembrane system.

A large number of individual glycosidase and glycosyltransferase reactions have been described and these reactions have been organized into pathways leading to the assembly of lipid-bound oligosaccharides (2, 3) and protein-bound *N*-glycans (Asn-GlcNAc *N*-glycosidic linkage) (1, 4–6) and *O*-glycans (Ser/Thr-GalNAc *O*-glycosidic linkage) (7–10). Unfortunately, there is an appreciable void in our understanding of how these glycan synthetic pathways are controlled. The question of control is intimately linked to that of function. As other chapters in this book describe in detail, the number of different glycan structures that have been isolated and characterized has reached awesome proportions. Further, these structures vary with the time and place, for example during development, differentiation, oncogenic transformation, etc. Why do proteins and lipids carry so many different oligosaccharides? Why does the organism make such a vast panorama of structures? How does it decide what to make and where and when to make it?

These questions are frequently raised, especially by nonspecialists in the field, and as frequently the answers provided by the specialists are not definitive. The elaborate intracellular glycosylation machinery must play a major role in these processes (4, 11, 12) but the factors which control expression of the glycosyltransferases in the assembly line remain to be elucidated. This chapter will report on recent advances in the isolation, characterization, and expression of the genes which encode mammalian glycosyltransferases.

2. Glycosyltransferases

Glycosyltransferases are enzymes which carry out the biosynthesis of oligosaccharides. The most common reaction catalysed by these enzymes is:

$$\text{nucleotide—sugar} + \text{R—OH} \longrightarrow \text{R—O—sugar} + \text{nucleotide}$$

where R is a free saccharide, or a saccharide linked to an aglycone (for example a protein or lipid), or a protein or lipid. The 'one linkage-one glycosyltransferase' rule states that there is one distinct glycosyltransferase for every glycosidic linkage. The Lewis blood group-dependent α3/4-fucosyltransferase can make more than one linkage and is an exception to this rule. Conversely, there are examples where a particular linkage can be made by more than a single glycosyltransferase, at least *in vitro*, for example there are at least four distinct human α3-fucosyltransferases which can synthesize the Galβ1-4 (Fucα1-3)GlcNAc moiety (13–16). The glycosyltransferases are usually membrane-bound and detergent treatment of the membrane preparations is required for solubilization and full expression of enzymatic activity. Most of the transferase assays require addition of exogenous divalent cation (usually Mn^{2+}). In some cases, Mn^{2+} can be replaced by other divalent cations. Exogenous cation is not required for the assay of sialyltransferases and several β-6-GlcNAc-transferases.

Glycosyltransferases have traditionally been endowed with a very precise specificity for both the acceptor and nucleotide-sugar donor substrate. Whereas this substrate specificity is displayed at physiological concentrations in the *in vivo* situation, it has been shown that promiscuous reactions can occur *in vitro* at relatively slow rates. An interesting example of this phenomenon is the ability of the human blood group B α3-Gal-transferase to transfer GalNAc instead of Gal to its acceptor substrate *in vitro* and thereby make the human blood group A epitope (17). The consequences of such a reaction *in vivo* would clearly be disastrous. Other examples of promiscuous glycosylations have been reported (18–23). These reactions

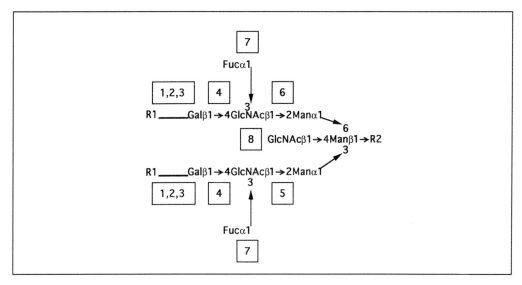

Fig. 1 Glycosyltransferases involved in the synthesis of N-glycans. (1) R1 = Galα1-3; UDP-Gal:Galβ1-4GlcNAc-R α3-Gal-transferase. (2) R1 = sialylα2-6; CMP-sialic acid:Galβ1-4GlcNAc-R α2,6-sialyltransferase. (3) R1 = sialylα2-3; CMP-sialic acid:Galβ1-3/4GlcNAc-R α2,3-sialyltransferase. (4) UDP-Gal:GlcNAc-R β4-Gal-transferase. (5) UDP-GlcNAc:Manα1-3R (GlcNAc to Manα1-3) β2-GlcNAc-transferase I. (6) UDP-GlcNAc:Manα1-6R (GlcNAc to Manα1-6) β2-GlcNAc-transferase II. (7) GDP-Fuc:Galβ1-4GlcNAc (Fuc to GlcNAc) α3-Fuc-transferase. (8) UDP-GlcNAc:R$_1$-Manα1-6(GlcNAcβ1-2Manα1-3)R$_2$ (GlcNAc to Manβ1-4) β4-GlcNAc-transferase III

have proved useful as adjuncts to classic organic chemistry for the synthesis of interesting oligosaccharides.

The large number of complex oligosaccharide structures that have been described indicates the existence of well over 100 different glycosyltransferases. The genes of only a relatively small number of mammalian glycosyltransferases have been cloned to date. These fall into three groups (12, 24) — enzymes involved in the biosynthesis of N-glycan antennae (Figure 1), O-glycan-specific linkages (Figure 2), and human blood group carbohydrate epitopes (Figure 3). The biosynthetic roles of these three classes of glycosyltransferase are discussed below; the remainder of the chapter is devoted to the molecular biology of these enzymes. Table 1 summarizes the methods used to clone glycosyltransferase genes and Table 2 lists some of the properties of the genes.

2.1 Biosynthesis of N-glycan antennae

The biosynthesis of N-glycans begins in the rough endoplasmic reticulum with the transfer of a large oligosaccharide Glc$_3$Man$_9$GlcNAc$_2$ from dolichol pyrophosphate oligosaccharide to an Asn residue in the polypeptide (1, 25, 26). This is followed by oligosaccharide processing within the lumen of the endoplasmic reticulum and Golgi apparatus, i.e. removal of glucose and mannose residues to form high-mannose N-glycans. The latter are converted to hybrid and complex N-glycans

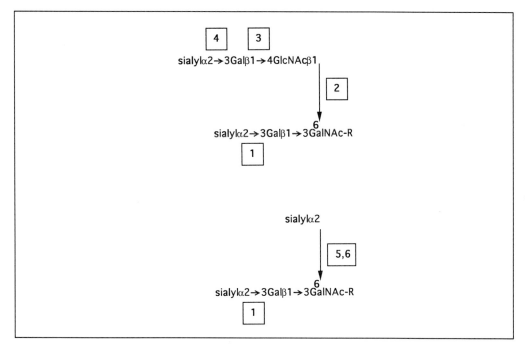

Fig. 2 Glycosyltransferases involved in the synthesis of O-glycans with core types 1 and 2. (1) CMP-NAN:Galβ1-3GalNAc-R α2,3-sialyltransferase. (2) O-glycan core 2 UDP-GlcNAc:Galβ1-3GalNAc-R (GlcNAc to GalNAc) β6-GlcNAc transferase (3) UDP-Gal:GlcNAc-R β4-Gal-transferase. (4) CMP-sialic acid:Galβ1-3/4GlcNAc-R α2,3-sialyltransferase. (5) CMP-sialic acid: R_1-GalNAc-R_2 (sialic acid to GalNAc) α2,6-sialyltransferase I. (6) CMP-sialic acid: sialylα2-3Galβ1-3GalNAc-R (sialic acid to GalNAc) α2,6-sialyltransferase II

when antennae are initiated by addition of GlcNAc residues to the N-glycan core (Figure 4). The antennae are usually elongated and terminated by addition of Gal, Fuc and sialyl residues (Figure 1).

The key enzyme for the conversion of high-mannose to complex and hybrid N-glycans is UDP-GlcNAc:Manα1-3R (GlcNAc to Manα1-3) β2-GlcNAc-transferase I (EC 2.4.1.101, Figure 4). The presence of a β2-linked GlcNAc residue at the nonreducing terminus of the Manα1-3Manβ1-4GlcNAc arm is essential for subsequent action of several enzymes in the processing pathway, for example α3/6-mannosidase II and UDP-GlcNAc:Manα1-6R (GlcNAc to Manα1-6) β2-GlcNAc-transferase II (EC 2.4.1.143, Figure 4), as well as other enzymes involved in complex N-glycan synthesis (GlcNAc-transferases III, IV, and V, and the α6-fucosyltransferase which adds fucose in α1-6 linkage to the Asn-linked GlcNAc).

The addition of GlcNAc by GlcNAc-transferase I allows removal of 2 mannose residues from the GlcNAcMan$_5$GlcNAc$_2$-Asn-X intermediate by α3/6-mannosidase II and sets the stage for the action of GlcNAc-transferase II (Figure 4). The GlcNAcMan$_3$GlcNAc$_2$-Asn-X intermediate (Figure 4) is the only known physiological substrate for GlcNAc-transferase II; the enzyme cannot act on substrates lacking the GlcNAcβ1-2Manα1-3-moiety.

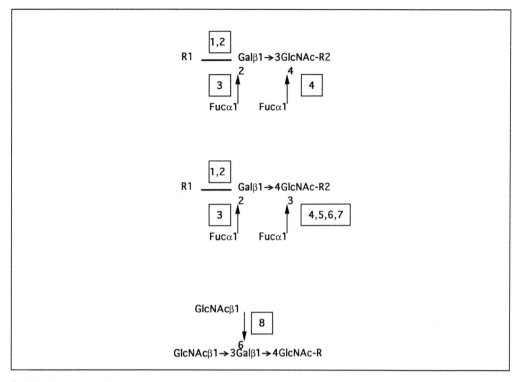

Fig. 3 Glycosyltransferases involved in the synthesis of human blood group antigens. (1) R1 = Galα1-3; human blood group B UDP-Gal:Fucα1-2Galβ1-3/4GlcNAc-R α3-Gal-transferase. (2) R1 = GalNAcα1-3; human blood group A UDP-GalNAc:Fucα1-2Galβ1-3/4GlcNAc-R α3-GalNAc-transferase. (3) Blood group H GDP-Fuc:Galβ1-3/4GlcNAc-R α2-Fuc-transferase. (4) Blood group Lewis GDP-Fuc:Galβ1-3/4GlcNAc-R (Fuc to GlcNAc) α4/3-Fuc-transferase III. (5, 6 and 7) GDP-Fuc:Galβ1-4GlcNAc-R (Fuc to GlcNAc) α3-Fuc-transferases IV, V, and VI. (8) Blood group I UDP-GlcNAc:GlcNAcβ1-3Galβ1-4GlcNAc-R (GlcNAc to Gal) β6-GlcNAc-transferase

Table 1 Cloning of glycosyltransferases

(A) Screening of libraries

Vector	Library	Enzyme	Reference
Antibody probe			
λgt11	Bovine kidney cDNA	β4-Gal-T	39
λgt11	Human fetal liver cDNA	β4-Gal-T	201
λgt11	Calf thymus cDNA	α3-Gal-T	90
λgt11	Rat liver cDNA	α2,6-sialyl-T (Galβ1-4GlcNAc)	107
Oligonucleotide probe (based on amino acid sequences)			
pCD	Lactating bovine mammary gland cDNA	β4-Gal-T	37
λgt10	Human liver cDNA	β4-Gal-T	49

Table 1 (*Continued*)

Vector	Library	Enzyme	Reference
λgt10	Porcine submaxillary gland cDNA	α2,3-sialyl-T (Galβ1-3GalNAc)	135
λgt11	Rat liver cDNA	β2-GlcNAc-T II	188

Homology screening
(based on previously isolated clones)

Vector	Library	Enzyme	Reference
λgt10	Murine mammary cDNA	β4-Gal-T	52
λgt10	Bovine kidney cDNA	β4-Gal-T	41
λgt11	Human placenta cDNA	β4-Gal-T	51
λgt11	Murine F9 embryonal carcinoma cDNA	β4-Gal-T	53
λgt11	Bovine liver cDNA	β4-Gal-T	38
λgt10	Murine C127 cDNA	α3-Gal-T	95
λGEM-11	Murine genomic DNA	α3-Gal-T	95
λgt10	Human colon adenocarcinoma cell line cDNA	α3-Gal-T (blood group B)	104
λgt11	Human submaxillary gland cDNA	α2,6-sialyl-T (Galβ1-4GlcNAc)	110
λgt10	Human placenta cDNA	α2,6-sialyl-T (Galβ1-4GlcNAc)	109
pCDM7	Human A431 cDNA	α2-Fuc-T (blood group H)	149
cosmid	Human genomic DNA	α3-Fuc-T III	165
λFIX	Human genomic DNA	α3-Fuc-T IV	31
λFIX	Human genomic DNA	α3-Fuc-T V	150
λFIX	Human genomic DNA	α3-Fuc-T VI	151
cosmid	Human genomic DNA	α3-Fuc-T VI	165
λEMBL3	Human genomic DNA	β2-GlcNAc-T I	176, 177
λEMBL3	Murine genomic DNA	β2-GlcNAc-T I	179
pCDM7	Murine F9 teratocarcinoma cell cDNA	β2-GlcNAc-T I	180
λEMBL3	Human genomic DNA	β2-GlcNAc-T II	190
λCharon	Rat genomic DNA	β2-GlcNAc-T II	189

PCR-based methods
(based on amino acid or nucleotide sequences)

Vector	Library	Enzyme	Reference
None	A431 and HL60 cDNA	α3-Fuc-T VI	164
λgt10	Rat liver cDNA	α2,3-sialyl-T (Galβ1-3/4GlcNAc)	137
λgt10	Human stomach endothelial cell line cDNA	α3-GalNAc-T (blood group A_1)	170
λgt10	Rabbit liver cDNA	β2-GlcNAc-T I	173, 174
λgt10	Rat kidney cDNA	β4-GlcNAc-T III	191

(B) Expression screening
transfection of cDNA library into COS or CHO cells

Vector	Library	Enzyme	Reference
pCDM7	Murine F9 teratocarcinoma cell cDNA	α3-Gal-T	91
pCDM8	Human Daudi cell cDNA (B lymphoblastoid cell line)	α2,6-sialyl-T (Galβ1-4GlcNAc)	111

Table 1 (Continued)

Vector	Library	Enzyme	Reference
pCDM8	Human spleen cDNA	α2,6-sialyl-T (Galβ1-4GlcNAc)	112
pCDM7	Human A431 cDNA	α3/4-Fuc-T III (blood group Lewis)	156
pCDM8	HL-60 cell cDNA	α3-FucT IV	161
pCDM8	Human NK-like cell line YT cDNA	β4-GalNAc-T	171
pCDNAI	HL-60 cell cDNA	O-glycan core 2 β6-GlcNAc-T	196
pCDNA1	PA1 human teratocarcinoma cDNA	Blood Group I β6-GlcNAc-T	199

(C) Expression screening transfection of total genomic DNA into mammalian cells

Source of genomic DNA	Host cells	Enzyme	Reference
Human A431 cell line	Mouse L cells	α2-Fuc-T (blood group H)	147, 148
Human HL-60 cells	Chinese hamster ovary cells	α3-FucT IV	34, 162
Human HL-60 cells	Lec 1 Chinese hamster ovary mutant cells	β2-GlcNAc-T I	178

Table 2 Domain structures of cloned mammalian glycosyltransferases

Glycosyltransferase	Species	Domains (number of amino acids)			Total number of amino acids	3'-Untranslated sequence (kb)	Chromosome localization
		N-terminal cytoplasmic	Signal/ anchor domain	C-terminal lumenal/ catalytic			
UDP-Gal:GlcNAc-R β4-Gal-transferase (EC 2.4.1.38; EC 2.4.1.90)	Human	23	20	354	397	2.9	9p13-p21
	Murine	11/24	20	355	386/399	2.6	4
	Rat						5
	Bovine	11/24	20	358	389/402	2.6–2.8	U18
Blood group B UDP-Gal:Fucα1-2Gal-R α1,3-Gal-transferase (EC 2.4.1.37)	Human	16	21	316	353		9q34
UDP-Gal:Galβ1-4GlcNAc-R α1,3-Gal-transferase (EC 2.4.1.151)	Bovine	6	16	346	368	2.4	—
	Murine	41	16	337	394	2.1	2
	Murine	6	16	337	359		
	Murine	6	16	315	337		
	Murine	6	16	327	349		
	Murine	6	16	349	371		
CMP-sialic acid: Galβ1-4GlcNAc-R α2,6-sialyltransferase (EC 2.4.99.1)	Rat	9	17	377	403	2.8	—
	Human	9	17	380	406		3q27-q28

Table 2 (Continued)

Glycosyltransferase	Species	Domains (number of amino acids)			Total number of amino acids	3'-Untranslated sequence (kb)	Chromosome localization
		N-terminal cytoplasmic	Signal/ anchor domain	C-terminal lumenal/ catalytic			
CMP-sialic acid: Galβ1-3GalNAc-R α2,3-sialyltransferase (EC 2.4.99.4)	Porcine	11	16	316	343	~1.0	—
CMP-sialic acid: Galβ1-3(4)GlcNAc-R α2,3-sialyltransferase (EC 2.4.99.6)	Rat	8	20	346	374	~1.0	—
GDP-Fuc:β-galactoside α2-Fuc-transferase (blood group H) (EC 2.4.1.69)	Human	8	17	340	365	2.2	19
Lewis GDP-Fuc: Galβ1-3/4GlcNAc-R α4/3-Fuc-transferase III (EC 2.4.1.65)	Human	15	19	327	361	0.9	19
GDP-Fuc: Galβ1-4GlcNAc-R α3-Fuc-transferase V (plasma type)	Human	15	19	340	374		19
GDP-Fuc: Galβ1-4GlcNAc-R α3-Fuc-transferase VI	Human	14	20	324	358		19
GDP-Fuc: Galβ1-4GlcNAc-R α3-Fuc-transferase IV (myeloid type)	Human	22	34	349	405		11q
Blood Group A UDP-GalNAc:Fucα1-2Gal-R α3-GalNAc-transferase (EC 2.4.1.40)	Human	16	21	316	353		9q34
UDP-GalNAc:G_{M3}/G_{D3} β4-GalNAc-transferase (EC 2.4.1.92)	Human	7	18	536	561	>0.8	
UDP-GlcNAc:Manα1-3R β2-GlcNAc-transferase I (EC 2.4.1.101)	Human Rabbit Murine	4 4 4	25 25 25	416 418 418	445 447 447	1.09 1.06 0.99	5q35 — 11
UDP-GlcNAc:Manα1-6R β2-GlcNAc-transferase II (EC 2.4.1.143)	Human Rat	9	22	416	447 430		14q21
UDP-GlcNAc:R_1-Manα1-6-[GlcNAcβ1-2Manα1-3]-Manβ1-4R_2 (GlcNAc to Manβ) β4-GlcNAc-transferase III (EC 2.4.1.144)	Rat	5	16	515	536		

Table 2 (Continued)

Glycosyltransferase	Species	Domains (number of amino acids)			Total number of amino acids	3'-Untranslated sequence (kb)	Chromosome localization
		N-terminal cytoplasmic	Signal/ anchor domain	C-terminal lumenal/ catalytic			
O-glycan core 2 UDP-GlcNAc: Galβ1-3GalNAc-R (GlcNAc to GalNAc) β6-GlcNAc-transferase (EC 2.4.1.102)	Human	9	23	396	428	0.6	9q21
Blood group I UDP-GlcNAc: GlcNAcβ1-3Galβ1-4GlcNAc-R (GlcNAc to Gal) β6-GlcNAc-transferase	Human	6	19	375	400		9q21

N-Glycan antennae can be elongated and terminated in a variety of ways (Figure 1). The most common N-glycan antenna is sialylα2-6Galβ-1-4GlcNAc- formed by the action of UDP-Gal:GlcNAc-R β4-Gal-transferase (EC 2.4.1.38/90) and CMP-sialic acid:Galβ1-4GlcNAc-R α2,6-sialyltransferase (ST6N) (EC 2.4.99.1). Sialic acid can also be incorporated in α2-3 linkage to Gal by CMP-sialic acid:Galβ1-3/4GlcNAc-R α2,3-sialyltransferase (ST3N) (EC 2.4.99.6); the latter enzyme cannot act on fucosylated substrates but there may be an α2,3-sialyltransferase in neutrophils which can convert Galβ1-4(Fucα1-3)GlcNAc to sialylα2-3Galβ1-4(Fucα1-3)GlcNAc (see below). The Galβ1-4 residue may be replaced by a Galβ1-3 residue or by a GalNAcβ1-4 residue. In some glycoproteins, the terminal sialic acid residue is replaced by a Galα1-3 residue due to the action of UDP-Gal:Galβ1-4GlcNAc-R α3-Gal-transferase (EC 2.4.1.124/151). The genes for GlcNAc-transferases I, II, and III, α3/6-mannosidase II, β4-Gal-transferase, the N-glycan α2,6- and α2,3-sialyltransferases and α3-Gal-transferase have been cloned (Tables 1 and 2).

2.2 Biosynthesis of O-glycans

Whereas N-glycans all share a common core (Figure 1), O-glycans have at least four major core types (9, 10). The synthetic pathway for O-glycans differs from the N-glycan pathway in that there is no transfer of a large oligosaccharide from lipid carrier to protein nor is there processing by the action of glycosidases. The O-glycans are built up by the successive addition of individual hexose or hexosamine units. Sialic acid termination of O-glycan core types 1 and 2 is effected by four sialyltransferases (Figure 2), three of which are distinct from the two N-glycan sialyltransferases (Figure 1). The O-glycan-specific glycosyltransferase genes cloned to date are CMP-sialic acid:Galβ1-3GalNAc-R α2,3-sialyltransferase (ST3O) (EC 2.4.99.4, Figure 2) and core 2 UDP-GlcNAc:Galβ1-3GalNAc-R (GlcNAc to GalNAc) β6-GlcNAc-transferase (EC 2.4.1.102, Figure 2).

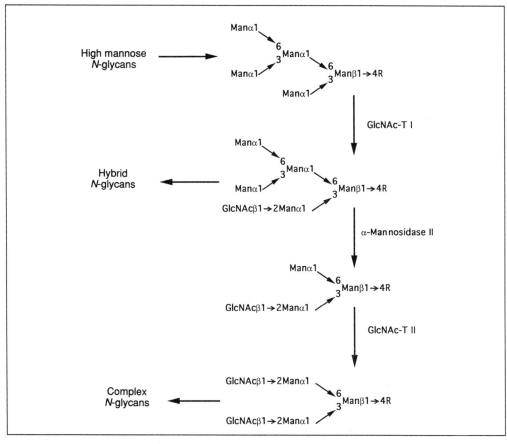

Fig. 4 Conversion of high mannose *N*-glycans to hybrid and complex *N*-glycans. GlcNAc-T I, UDP-GlcNAc:Manα1-3R (GlcNAc to Manα1-3) β2-GlcNAc-transferase I. GlcNAc-T II, UDP-GlcNAc:Manα1-6R (GlcNAc to Manα1-6) β2-GlcNAc-transferase II

2.3 Biosynthesis of human blood group carbohydrate epitopes

Human blood group carbohydrate epitopes are found at the nonreducing termini of both protein- and lipid-bound oligosaccharides. The genes for eight enzymes which synthesize human blood group antigen determinants have been cloned (Figure 3, Tables 1 and 2). These enzymes act on the Galβ1-3/4GlcNAc moieties of *N*- and *O*-glycans and glycolipids. The human blood group A, B, H, Le (Lewis), and I epitopes (Figure 3) are synthesized, respectively, by UDP-GalNAc:Fucα1-2Gal-R (GalNAc to Gal) α3-GalNAc-transferase (EC 2.4.1.40), UDP-Gal:Fucα1-2Gal-R (Gal to Gal) α3-Gal-transferase (EC 2.4.1.37), GDP-Fuc:Galβ-R α2-Fuc-transferase (EC 2.4.1.69), GDP-Fuc:Galβ1-4/3GlcNAc (Fuc to GlcNAc) α3/4-Fuc-transferase III (EC 2.4.1.65) and UDP-GlcNAc:GlcNAcβ1-3Galβ1-4GlcNAc-R (GlcNAc to Gal) β6-GlcNAc-transferase.

There are at least four distinct human α3-Fuc-transferases (13–16), including the Le-dependent GDP-Fuc:Galβ1-4/3GlcNAc (Fuc to GlcNAc) α3/4-Fuc-transferase III. These enzymes catalyse the reaction:

$$\text{GDP-Fuc} + \text{Gal}\beta\text{1-4GlcNAc-R} \longrightarrow \text{Gal}\beta\text{1-4(Fuc}\alpha\text{1-3)GlcNAc-R} + \text{GDP}$$

Three of these enzymes (Fuc-transferases IV, V, and VI) differ from the Le-dependent α3/4-Fuc-transferase III in that they cannot form the Fucα1-4GlcNAc linkage. These four enzymes have been differentiated from one another by their reactivities towards sialylated substrates (NeuAcα2-3Galβ1-4GlcNAc-R), kinetic properties and inhibition by N-ethylmaleimide. Analysis of Chinese hamster ovary cell mutants has identified four distinct α3-Fuc-transferase activities, none of which are able to add Fuc in α1-4 linkage (27).

The Galβ1-4(Fucα1-3)GlcNAc moiety (termed blood group Lewis X or Le^X or X or CD15) has been called stage-specific embryonic antigen 1 (SSEA-1) because of its presence during embryonic development (28). Interest in the Le^X antigen and its biosynthesis has been heightened recently because of the demonstration that the sialylα2-3Galβ1-4(Fucα1-3)GlcNAc moiety (sialyl-Le^X) is the most likely ligand for an endothelial cell receptor called endothelial-leukocyte adhesion molecule-1 (ELAM-1) (13, 29–35). ELAM-1 (more recently termed E-selectin) is involved in inflammation-activated homing of leukocytes to endothelial cells and is therefore a potential target for anti-inflammatory drugs. The importance of sialyl-Le^X in this process has become apparent from a recent report of two children with a congenital defect in the synthesis of sialyl-Le^X which results in an inability of the leukocytes to adhere properly to endothelial walls (36). The new disease has been named leukocyte adhesion deficiency type 2. The affected children suffer recurrent severe infections and other problems.

3. Cloning of glycosyltransferase genes

The glycosyltransferases which have been cloned to date are indicated in Tables 1 and 2. The following sections discuss these studies in detail.

3.1 Galactosyltransferases
3.1.1 UDP-Gal:GlcNAc-R β4-Gal-transferase (EC 2.4.1.38/90)

Bovine β4-Gal-transferase

Bovine UDP-Gal:GlcNAc-R β4-Gal-transferase was the first mammalian glycosyltransferase to be cloned. Narimatsu *et al.* (37) isolated a partial cDNA clone (3.7 kb) by screening a cDNA library from lactating bovine mammary gland with degenerate synthetic oligonucleotides based on partial amino acid sequences. The clone encoded a 984 bp coding region and a 2.7 kb 3'-untranslated region. Hybrid selection was used to purify a human mRNA and *in vitro* translation of this message followed by immunoprecipitation yielded several bands, the largest of which was about M_r 65 kDa. The predicted molecular mass of the glycosylated membrane-bound form of the enzyme is about 60 kDa (38). Northern blots on mRNA from bovine, human, mouse and rat tissues revealed 2–3 bands, the largest being about 4.2–4.5 kb.

Shaper et al. (39) detected 10 positive clones by screening a λgt11 bovine kidney cDNA expression library with antibody to bovine β4-Gal-transferase; only one of these clones hybridized with a mixture of three degenerate oligonucleotide probes based on amino acid sequence data. This clone contained an insert of 1.3 kb which encoded 334 amino acids.

The full-length membrane-bound form of the bovine enzyme was shown by D'Agostaro et al. (38) to have 402 amino acids (Table 2). The cDNA encoding the full-length protein was obtained by screening a 5'-extended bovine liver cDNA library in λgt11; 5'-extension was carried out with a primer based on the β4-Gal-transferase sequence.

Full-length bovine β4-Gal-transferase clones have also been reported by Masibay and Qasba (40) and Russo et al. (41). Comparison of the bovine sequences reported by different laboratories (37–39) show several base substitutions and an in-frame three nucleotide deletion. These sequence differences do not alter the reading frame and cause conservative amino acid substitutions; they have been attributed to genetic polymorphism.

Domain structure of glycosyltransferases

Full-length bovine β4-Gal-transferase is a type II integral membrane protein (N_{in}/C_{out} orientation) with a 24 amino acid amino-terminal cytoplasmic domain, a 20 amino acid noncleavable signal/anchor hydrophobic transmembrane domain and a 358 amino acid carboxy-terminal catalytic domain within the lumen of the endoplasmic reticulum and Golgi apparatus. This domain structure has been found for all the glycosyltransferase genes cloned to date (Table 2, Figure 5) (12, 24). The full-length 4.5 kb mRNA has a 5'-untranslated region of about 600 bp, a coding region of 1206 bp and a 3'-untranslated region of 2.7 kb. There is a 13-nucleotide poly(A) sequence in the 3'-untranslated region (37) which resulted in fortuitous mispriming of oligo(dT)-primed cDNA synthesis during the preparation of cDNA libraries and allowed the isolation of partial coding regions (38, 39).

Size heterogeneity has been observed for several glycosyltransferases purified from tissues, mammalian secretions such as milk and colostrum, and body fluids such as blood (42–48). D'Agostaro et al. (38) obtained N-terminal amino acid sequences for two soluble forms of β4-Gal-transferase (M_r 51 and 42 kDa) purified from bovine milk. These N-termini were located 35 and 62 amino acids downstream from the C-terminal end of the hydrophobic transmembrane domain. Since the mixture of soluble proteins is catalytically active, the data indicate that the membrane-bound enzyme can be cleaved at a so-called 'neck' or 'stem' region between the transmembrane domain and the catalytic domain to yield active soluble enzyme (Figure 5). It is likely that this cleavage is due to proteolysis during

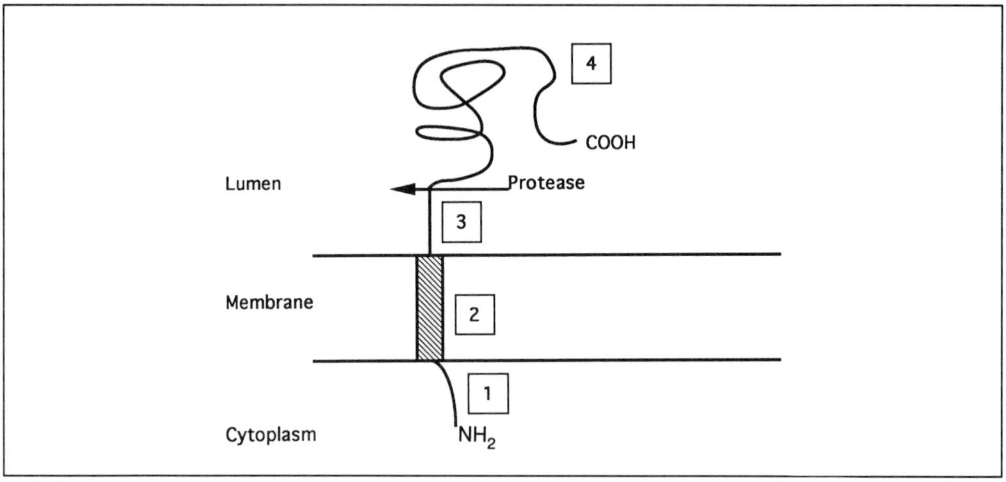

Fig. 5 Domain structure of the glycosyltransferases. All the mammalian glycosyltransferases which have been cloned to date are type II transmembrane glycoproteins (N_{in}/C_{out}) (Table 2). (1) Short N-terminal cytoplasmic segment. (2) Transmembrane noncleavable signal/anchor sequence. (3) 'Stem' or 'neck' region which can be cleaved by proteases to release a catalytically active soluble enzyme. (4) The globular C-terminal catalytic domain

purification. Although the 'stem' region shows relatively low amino acid sequence conservation between species, there are some common features which indicate that this domain serves as a flexible tether to hold the catalytic domain within the Golgi lumen, i.e. the presence of Pro and Gly residues and of consensus sites for N-glycosylation.

Similar studies carried out on other glycosyltransferases indicate that the transmembrane domain is not required for the catalytic activity of these enzymes; as will be discussed below, the transmembrane domain appears to be essential for targeting to the Golgi apparatus and anchoring to the Golgi membrane. It also appears that the truncated form of the enzyme that has been cleaved in the 'stem' region is more amenable to purification than is the full-length protein.

Human β4-Gal-transferase
Human milk β4-Gal-transferase was purified and partial amino acid sequence data was used to design a 60 mer 'optimal' probe (49, 50). Screening of a human liver cDNA library yielded a 982 bp fragment encoding the C-terminal 261 amino acids of the enzyme. This 982 bp fragment was used to probe a human placenta cDNA library (51); two overlapping cDNA clones were isolated which together yielded a 1.4 kb sequence containing the full coding region (Table 2). Amino-terminal sequence analysis of the purified milk enzyme (50) indicated proteolytic cleavage had occurred in the 'neck' region (Figure 5) 35 residues downstream from the end of the transmembrane segment; a similar cleavage was reported for the bovine milk enzyme (see above).

Murine β4-Gal-transferase

Shaper et al. (52) used the 1.3 kb bovine β4-Gal-transferase clone (see above) to probe a murine cDNA library. Since all the resulting clones were missing the 5'-terminus, a primer extension cDNA library was probed to obtain the upstream sequence. A series of overlapping clones were obtained which yielded a full-length cDNA spanning 4038 bp and containing a 1197 bp coding region, a 2582 bp 3'-untranslated region and an 88 bp poly(A) tract. The 3'-untranslated region contains two distinct families of repetitive sequences, one of which is the murine equivalent of the human Alu consensus sequence.

Nakazawa et al. (53) used a bovine β4-Gal-transferase cDNA probe to obtain a full-length murine 3.9 kb cDNA.

The predicted primary structure of murine β4-Gal-transferase shows the domain pattern typical of all cloned glycosyltransferases (Table 2, Figure 5).

Alternate transcription initiation sites

Murine β4-Gal-transferase cDNA has two in-frame ATG codons at its 5'-end and a termination codon 18 bp upstream of the first ATG (Figure 6). S1 nuclease protection and primer extension analyses of RNA from a murine mammary cell line (52, 54, 55) showed the presence of two sets of transcripts differing in length by about 200 bp. The first transcription initiation site was 145–190 bp upstream of the first ATG codon and the second site was between the two ATG codons 15–25 bp downstream of the first ATG (Figure 6), i.e. there are at least two promoters controlling the gene (Figure 7). Northern analysis confirmed the presence of two transcripts at 3.9 and 4.1 kb. Translation from the two in-frame AUG codons predicts proteins of 399 and 386 amino acids respectively. The long 4.1 kb transcript should produce only the 399 amino acid long form of the protein but could also produce the short form by a process called 'leaky scanning' in which an internal Met codon is used as the translation initiation point. The short 3.9 kb

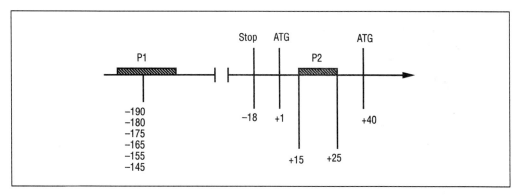

Fig. 6 Transcription start sites for murine UDP-Gal:GlcNAc-R β4-Gal-transferase (see text). There are at least two promoters (P1 and P2, shaded boxes). The transcription start sites are indicated by vertical lines below the promoters. Numbering is based on the first ATG translation start site at +1. Data from references 52, 54, and 55

Fig. 7 Genomic structure of murine UDP-Gal:GlcNAc-R β4-Gal-transferase (see text). The numbered boxes indicate the six exons. The filled boxes represent the coding region. The diagram is not drawn to scale. Data from references 52, 54, and 55

transcript can only result in the short protein species. The two Met codons are conserved in the bovine, human and murine sequences (Figure 8).

The roles of the two transcripts were investigated by Russo *et al.* (41) using the bovine β4-Gal-transferase gene. S1 nuclease protection and primer extension analyses of RNA from Madin–Darby bovine kidney cells showed two sets of transcripts very similar to those previously described for murine mammary cells (Figure 6). The two transcripts encode proteins of 389 and 402 amino acids (Table 2). Plasmids were constructed with an SP6 RNA polymerase promoter and cDNA inserts encoding truncated (TGT), short (SGT) and long (LGT) forms of bovine β4-Gal-transferase (Figure 9). SGT is the short 389 amino acid form of the enzyme, LGT is the long 402 amino acid form, and TGT is a truncated form lacking both initiation Met residues as well as the transmembrane segment. All three forms retain their two N-glycosylation sites (Figure 9). *In vitro* transcription was carried out with these three plasmids using SP6 RNA polymerase; the RNA was used as a template for *in vitro* translation in the absence of microsome membranes. The protein products were analysed by sodium dodecyl sulphate-polyacrylamide gel electrophoresis. Only a single protein product of the predicted size was obtained from each of the plasmids. This proved that the long transcript initiated translation only at the up-stream Met codon and that 'leaky scanning' did not occur, at least under the conditions tested.

When *in vitro* translation was carried out in the presence of dog pancreas microsomes, TGT did not increase in size whereas the apparent M_r of both SGT and LGT increased by about 3000. Endoglycosidase H treatment removed most of this extra material indicating that SGT and LGT had both been glycosylated. It was concluded that both SGT and LGT contain a noncleavable transmembrane segment and are oriented as type II integral membrane proteins (Figure 5) with the large C-terminal catalytic domain within the lumen of the microsomes. Protease protection experiments confirmed this orientation. Evidence has been obtained (see below) for tissue-specific expression of SGT and LGT.

Sequence similarities
Figure 8 compares the amino acid sequences of bovine, human, and murine β4-Gal-transferases. The conservation of a four Pro sequence in the 'stem' region and of all seven Cys residues is indicated in the figure. There is a single conserved

```
                                    *      *
Cow     MKFREPLLGGSAAMPGASLQRACRLLVAVCALHLGVTLVYYLAGRDLRRLPQLVGVHPPL
Human   -RL-----S-.-------------------------------------S--------ST--
Mouse   -R---QF---------T-----------------------S----S--------SST-

                                +                                    +
Cow     QGSSHGAAAIGQPSGELRLRGVAPPPPLQNSSKPRSRAPSNLDAYSHPGPGPGPGSNLTS
Human   --G-NS------S-----TG-AR-----GA--Q--P....GG-SSPVVDS----A-----
Mouse   --GTN----SK--P--Q-P--AR-----GV-P---P....G--SSPGAAS---LK---S-

                    *                                             *
Cow     APVPSTTTR.SLTACPEESPLLVGPMLIEFNIPVDLKLIEQQNPKVKLGGRYTPMDCISP
Human   V---H--AL.--P-----------------M----E-VAK---N--M----A-R--V--
Mouse   L---T--GL1--P-----------------D---A--E-LAKK--EI-T----S-K--V--

Cow     HKVAIIILFRNRQEHLKYWLYYLHPMVQRQQLDYGIYVINQAGESMFNRAKLLNVGFKEA
Human   ------P-----------------VL----------------DTI----------Q--
Mouse   ------P-----------------IL----------------DT---------I--Q--

              *                     *
Cow     LKDYDYNCFVFSDVDLIPMNDHNTYRCFSQPRHISVAMDKFGFSLPYVQYFGGVSALSKQ
Human   ------T-----------------A-----------------------------------
Mouse   ------------------D-R-A-------------------------------------

                                                  *
Cow     QFLSINGFPNNYWGWGGEDDDIYNRLAFRGMSVSRPNAVIGKCRMIRHSRDKKNEPNPQR
Human   ---T-----------------F---V-----I------V-R------------------
Mouse   ---A-----------------F---VHK---I------V-R------------------

Cow     FDRIAHTKETMLSDGLNSLTYMVLEVQRYPLYTKITVDIGTPS
Human   --------------------Q--D--------Q---------
Mouse   -----------RF--------K--D--------Q-------R
```

Fig. 8 Comparison of the amino acid sequences of cow, human, and mouse UDP-Gal:GlcNAc-R β4-Gal-transferases using the 1-letter code (see Table 1 for references). The transmembrane segment is under- and over-lined. The 7 Cys residues are indicated by an asterisk. The two possible N-glycan sequons are indicated by a plus sign. The four conserved Pro residues are underlined

potential N-glycosylation site. The greatest variability occurs in the 'stem' region (Figure 8); the sequence similarity is over 90% in regions where functional domains have been located, i.e. the transmembrane anchor and C-terminal catalytic domains. No significant sequence similarities have as yet been found between β4-Gal-transferase and other proteins; a short hexapeptide region has been suggested as a potential UDP-Gal binding site on the basis of similarity to UDP-Gal:Galβ1-4GlcNAc-R (Gal to Gal) α3-Gal-transferase (see below).

Genomic structure

A murine genomic DNA library was screened with a murine β4-Gal-transferase cDNA probe and several clones were isolated and sequenced (54). The gene contains six exons (Figure 7). A clone overlapping exons 1 and 2 has not as yet been

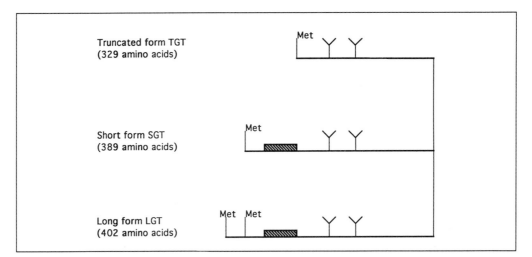

Fig. 9 Protein products formed by expression of various bovine UDP-Gal:GlcNAc-R β4-Gal-transferase DNA constructs (see text). The shaded rectangle represents the transmembrane segment. The Y structures represent N-glycans. Met = methionine residues which initiate translation of the short and long forms of the enzyme. Data from reference 41

isolated but the exons are at least 10 kb apart indicating that the gene is at least 50 kb in size. Exon 1 encodes the 5'-untranslated region, the N-terminal cytoplasmic domain, the transmembrane segment and the first 94 amino acids of the C-terminal domain. Exon 6 encodes the C-terminal segment of the protein and the entire 3'-untranslated region. The region 5' to both promoters lacks both classic CAAT and TATA sequences but has a high G/C content and several regions similar to the 'GC box' consensus sequence required for interaction with the cellular transcription initiation factor Sp1. The control of gene expression is discussed below.

A human genomic library was screened with a murine cDNA probe and several clones were isolated and analysed (56). The human gene has a structure similar to the murine gene containing six exons spanning over 50 kb of genomic DNA. A clone overlapping exons 1 and 2 was not isolated. Southern analysis indicated the presence of only a single gene.

The chromosomal localization of the β4-Gal-transferase gene has been determined in four species (Table 2) (57–60).

Expression of β4-Gal-transferase

Masibay and Qasba (40) inserted a full-length bovine β4-Gal-transferase cDNA into an Okayama–Berg plasmid expression vector under the control of a simian virus 40 promoter. After transient transfection of this vector into monkey kidney COS-7 cells, the authors observed a 12-fold increase in β4-Gal-transferase activity compared to extracts of COS-7 cells which were mock-transfected or transfected with various control plasmids. Mixing experiments showed that the cDNA did not encode an activator protein. The transfected COS-7 cells showed a 10-fold increase over controls in lactose synthetase activity measured in the presence of α-lactalbumin. Treatment of the transfected cell extracts with polyclonal antibody to bovine milk β4-Gal-transferase resulted in an antibody-antigen complex with β4-Gal-transferase activity; a monoclonal antibody to a synthetic peptide based on the bovine β4-Gal-transferase sequence showed the same results.

An 8-fold increase in β4-Gal-transferase activity over appropriate controls was obtained by transient transfection into COS-1 cells of a plasmid constructed by insertion of full-length murine β4-Gal-transferase cDNA into the mammalian expression vector pCMV4 (61); pCMV4 is driven by a human cytomegalovirus promoter and has a translational enhancer from alfalfa mosaic virus 4. Free Gal was released from the product of the enzyme reaction by streptococcal β-galactosidase which cleaves Galβ1-4GlcNAc but not Galβ1-3GlcNAc. The acceptor specificity of the plasmid-directed enzyme was changed from GlcNAc to Glc by addition of α-lactalbumin.

Although human β4-Gal-transferase is a glycoprotein (62), Aoki et al. (63) were able to express an unglycosylated but enzymatically active protein in E. coli by using the pIN-III-ompA$_2$ vector. This vector has the ompA signal peptide driven by combined lpp and lac promoters. The vector also has the lac operator so that fusion protein synthesis is induced by isopropyl-β-D-thiogalactoside. The ompA signal peptide directs the fusion protein to the outer membrane where the signal is cleaved by bacterial signal peptidase with resulting secretion of the expressed protein into the periplasmic space and culture medium. Aoki et al. (63) inserted into this vector cDNA encoding the catalytic region of human β4-Gal-transferase which lacks the N-terminal and transmembrane domains (Figure 5). After transformation of E. coli by the vector, β4-Gal-transferase was detected in the periplasmic space and culture medium by both immunological and enzymatic assays. The soluble enzyme in the culture medium was purified to homogeneity in a single step using a GlcNAc-Sepharose affinity column. These findings indicate that the E. coli periplasm can provide an environment which allows proper folding of the protein even in the absence of glycosylation. Further, the data provide additional proof that the catalytic activity does not require either the N-terminal cytoplasmic segment nor the transmembrane segment (Figure 5).

The authors used site-directed mutagenesis of the recombinant enzyme to carry out preliminary studies on the catalytic site of the enzyme. They first identified a putative UDP-binding site by reacting purified human milk β4-Gal-transferase with [^{14}C]4-azido-2-nitrophenyluridylpyrophosphate in the presence of ultraviolet light. A labelled 53-amino acid peptide was purified and identified after cyanogen bromide cleavage of the protein. Site-directed mutagenesis in this region of the protein suggested three Tyr and two Trp residues as potential sites for UDP-Gal and/or acceptor binding.

The most definitive proof that a cloned fragment of DNA encodes a particular protein is the demonstration that expression of the DNA leads to synthesis of the protein. The above experiments firmly establish this point for the β4-Gal-transferase genes.

The functional significance of the long and short transcripts of β4-Gal-transferase
The β4-Gal-transferase gene has at least two promoters that initiate two transcripts; translation of each transcript results in two proteins, a long and a short form that differ in the length of their amino-terminal cytoplasmic domains (Figures 6 and 9, see above). The functional significance of these two transcripts has been the subject of controversy. A series of papers from the laboratory of Shur (64–68) have suggested that the long form may be preferentially targeted to the cell surface. However, other workers have concluded that both the long and short forms are targeted to the Golgi apparatus (69–75); Golgi targeting is discussed in the following section.

The enzyme has been implicated in three different cellular functions: (a) biosynthesis of the Galβ1-4GlcNAc moiety in *N*-glycans, *O*-glycans, and glycolipids; (b) when complexed with α-lactalbumin, the enzyme catalyses the biosynthesis of lactose in the lactating mammary gland; (c) although most of the intracellular β4-Gal-transferase is located in the Golgi apparatus, there is general agreement that a small proportion of the total intracellular enzyme is on the cell surface of some cells (67); this cell surface β4-Gal-transferase may play a role in cell–cell recognition (67, 76, 77).

Harduin-Lepers *et al.* (74) have recently shown that the long and short forms of β4-Gal-transferase are expressed in a tissue-specific manner in the mouse and provide a mechanism for regulation of enzyme levels. Northern blot analysis and S1 nuclease protection experiments were carried out on several mouse tissues (74). Tissues which express relatively low levels of β4-Gal-transferase (such as brain) are under the control of the upstream promoter (P1 in Figure 6) and contain only the longer 4.1 kb transcript. Most somatic mouse tissues express intermediate levels of enzyme and are under the control of both P1 and P2 (Figure 6); the ratio of the 4.1 kb to 3.9 kb transcripts is about 5:1. The short 3.9 kb transcript is the predominant mRNA species in tissues which express β4-Gal-transferase at very high levels such as parietal yoke sac cells and lactating mammary gland (65, 66, 68, 74); lactating mammary gland has a ratio of the 4.1 to 3.9 kb transcripts of 1:10 indicating that the enzyme in this tissue is primarily under the control of the P2 promoter (74).

Both promoters are weak and are unable to effect expression of chloramphenicol acetyltransferase (CAT) in transient expression experiments unless an enhancer element is incorporated into the construct (74). The P1 promoter appears to be a typical constitutive 'housekeeping' promoter. It lacks classical CAAT and TATA boxes and has six upstream GC boxes (*cis*-acting positive regulatory elements) under the control of the Sp1 transcription factor. Upstream of the P2 promoter are three more GC boxes, at least two mammary gland-specific positive regulatory elements and a negative *cis*-acting regulatory element. It is suggested that in the mammary gland, mammary gland specific transcription factors bind to their regulatory elements and deactivate the negative control thereby causing a marked increase in transcription of the 3.9 kb mRNA.

As mentioned above, it has been suggested that the short form of murine β4-Gal-transferase is destined primarily for the Golgi apparatus whereas the long form is preferentially targeted to the cell surface (65–67). S1 nuclease protection studies on RNA from a variety of tissues indicated that the relative abundance of the short and long forms of β4-Gal-transferase mRNA correlates with β4-Gal-transferase specific activities in the Golgi and plasma membrane respectively. Both stable and transient transfections have been carried out with DNA constructs encoding either the short or long forms of β4-Gal-transferase; some reports indicated no preferential targeting (69, 70, 72, 73) whereas Shur's group reported that the long form is targeted to the plasma membrane fraction at a level 2.4-fold higher than the short form (66, 67). These discrepancies may be due to the different systems used by the various groups and to the difficulty of establishing by immunohistochemical localization that a small fraction of the total intracellular enzyme is or is not present on the cell surface. However, Masibay et al. (72) used subcellular fractionation methods to separate Golgi and plasma membrane fractions and found that the long and short isoforms of β4-Gal-transferase gave identical distribution patterns for these fractions.

Evans et al. (68) created a dominant negative mutation that interferes with cell surface β4-Gal-transferase activity. Murine F9 embryonal carcinoma cells and Swiss 3T3 fibroblasts were stably transfected with expression plasmid constructs encoding a truncated version of the long form (LGT) of β4-Gal-transferase (Figure 9) which contains both Met start codons and the transmembrane domain but lacks the catalytic domain. Cells expressing this truncated protein became nonadhesive to a laminin substratum. Cells transfected with control plasmids lacking inserts or with plasmids encoding the equivalent truncated short form (SGT) of β4-Gal-transferase (Figure 9) remained adhesive. The only difference between truncated LGT and SGT is the presence of an extra 13-amino acid amino-terminal peptide in LGT. The transfection with truncated LGT resulted in an increase in both intracellular and cell surface β4-Gal-transferase (primarily the long form) but caused a reduction in the amount of cell surface β4-Gal-transferase associated with the actin cytoskeleton. The data suggest that the long form of β4-Gal-transferase is a cell surface adhesion molecule. Overexpression of truncated LGT displaces β4-Gal-transferase from the cytoskeleton suggesting that the 13-amino acid N-terminal extension mediates binding of β4-Gal-transferase to the cytoskeleton. The data also indicate that binding of the transferase to the cytoskeleton is essential for its function as a cell adhesion molecule.

Targeting to the Golgi apparatus
All Golgi-resident proteins identified to date belong to the type II integral membrane protein category (Table 2) (12, 24, 78). β4-Gal-transferase has been localized by immunoelectron microscopy to the *trans*-cisternae of the Golgi apparatus (62). Proteolytic cleavage of β4-Gal-transferase in the 'stem' region (Figure 5) produces a catalytically active enzyme which is not retained within the Golgi apparatus *in vivo* (62) suggesting that the information responsible for Golgi localization probably resides in the N-terminal cytoplasmic and transmembrane domains. If there is

differential targeting of the short and long forms of β4-Gal-transferase to Golgi apparatus and plasma membrane respectively (see above), the extra 13 amino acid segment at the cytoplasmic N-terminus of the long form may also play a role in targeting.

Chimeric cDNAs have been constructed to generate hybrid proteins in which the putative Golgi anchoring sequences of β4-Gal-transferase have been connected to reporter proteins that would normally not be retained in the Golgi apparatus (69, 70). Following transfection and transient expression of these hybrid constructs in mammalian cells, the intracellular destination of the reporter protein was determined by immunofluorescence and immuno-electron microscopy.

Nilsson et al. (69) used human β4-Gal-transferase cDNA as a donor for Golgi targeting sequences, two forms of the human invariant chain as reporter proteins, the eukaryotic expression plasmid pCMUIV as vector and HeLa cells as hosts. They found that the β4-Gal-transferase cytoplasmic domain did not cause retention in the Golgi. The transmembrane domain without the cytoplasmic domain did result in Golgi retention but there was also some reporter protein detected at the cell surface. It was found that the cytoplasmic domain from either β4-Gal-transferase or from human invariant chain in combination with the β4-Gal-transferase transmembrane domain could result in complete Golgi retention. The transmembrane domain is clearly the signal for Golgi retention while the cytoplasmic domain plays a nonspecific accessory role. As few as 10 amino acids on the lumenal half of the transmembrane domain were sufficient to localize the reporter protein to the Golgi. Immunoelectron microscopy using gold particles localized the reporter protein to the *trans*-Golgi. Metabolic labelling of transfected cells showed that the Golgi-localized reporter protein was N-glycosylated but was only partially resistant to endoglycosidase H; resistance to endoglycosidase H indicates the synthesis of complex N-glycans, a process that occurs primarily in the *trans*-Golgi.

Teasdale et al. (70) carried out similar experiments with cDNA hybrids constructed from bovine β4-Gal-transferase, influenza hemagglutinin, human transferrin receptor and hen ovalbumin cDNAs. Removal of most of the cytoplasmic tail from β4-Gal-transferase did not alter the Golgi localization of the enzyme in either transiently transfected COS-1 cells or stably transfected murine L cells. Replacement of the cytoplasmic tail and transmembrane domain of β4-Gal-transferase with the cleavable signal peptide from influenza hemagglutinin led to rapid secretion of the enzyme from transfected COS-1 cells. Replacement of the transmembrane domain of β4-Gal-transferase with the signal/anchor domain of the plasma membrane-localized human transferrin receptor resulted in transport of β4-Gal-transferase to the cell surface of transfected COS-1 cells. Finally, replacement of the internal signal sequence of ovalbumin, which is normally secreted, with the transmembrane domain of β4-Gal-transferase resulted in retention of ovalbumin in the Golgi of transfected COS-1 cells.

Aoki et al. (71) used transient expression with the vector pcDNAI to express various hybrid proteins in COS cells. They used the C-terminal portion of human chorionic gonadotropin as a reporter molecule for localization of Gal-transferase;

this technique allowed them to distinguish Gal-transferase produced by their constructs from the endogenous Gal-transferase in COS cells. Immunofluorescence localized the recombinant protein primarily to the Golgi apparatus although overexpression led to some fluorescence throughout the cytoplasm. Deletions of large portions of the cytoplasmic tail or stem region of Gal-transferase did not alter its Golgi localization. Site-directed mutagenesis of the transmembrane segment identified a 5-amino acid region required for Golgi retention; in particular residues Cys^{29} and His^{32} in this region of the signal-anchor domain were shown to be essential. It was suggested that these two amino acids may be involved in a metal-binding site involved in Golgi retention.

Russo et al. (73) found that both the long and short forms of β4-Gal-transferase targeted to the Golgi apparatus after stable transfection into Chinese hamster ovary (CHO) cells. Immunolocalization at the electron microscope level showed both forms of the enzyme to be located in the transregion of the Golgi. No immunoreactivity was detected at the cell surface thereby contradicting the evidence that the long and short forms are differentially targeted (see above); these contradictory data may be due to differences between cell types. Chimeric proteins were stably expressed in CHO cells such that the cytoplasmic tail, the transmembrane domain and 8 amino acids from the stem region of Gal-transferase were upstream of pyruvate kinase as a reporter molecule. Pyruvate kinase is normally a soluble cytoplasmic protein. Both the short and long fusion proteins targeted to the Golgi apparatus with no evidence of immunoreactivity at the cell surface.

It has been shown (63) that a truncated form of β4-Gal-transferase lacking the transmembrane segment can be expressed in E. coli in an enzymatically active nonglycosylated form. It is therefore surprising that attempted transient expression of a similar construct in COS-7 cells (79) resulted in mRNA synthesis but no enzyme protein. Possible explanations are (i) improper translation of the message or (ii) the lack of a transmembrane segment prevents targeting of the enzyme to the Golgi apparatus with consequent degradation of the protein in the cytoplasm. The lack of expression of enzyme activity after transient transfection of truncated forms of the bovine β4-Gal-transferase gene into COS-7 cells was used (72) to study the requirements for Golgi targeting. It was found that deletion of the cytoplasmic domain had no effect but that removal of even the first five amino acids at the amino-terminal side of the transmembrane domain resulted in loss of enzyme expression. These authors also showed that the cytoplasmic and transmembrane domains of bovine UDP-Gal:Galβ1-4GlcNAc-R α3-Gal-transferase could replace the equivalent regions of the bovine β4-Gal-transferase without any effect on expression of enzyme activity. However, when the equivalent regions of CMP-sialic acid:Galβ1-4GlcNAc-R α2,6-sialyltransferase were used, no activity was obtained unless flanking sequences encoding part of the neck region were also included in the construct. The role of the neck region in sialyltransferase targeting has also been reported by Munro (80) as is discussed in section 3.2.1.

Masibay et al. (72) point out that the transmembrane domains of plasma membrane-targeted proteins are broader and more hydrophobic than those of

Golgi-targeted proteins. Replacement of amino acids in the transmembrane domain of β4-Gal-transferase with isoleucine residues or insertion of extra isoleucine residues caused a significant shift of targeting from the Golgi to the cell surface. The authors suggest that a posttranslational modification of the transmembrane segment to a more hydrophobic state may play a role in targeting some of the transferase to the cell surface.

The above series of experiments strongly support the hypothesis that the transmembrane domain of β4-Gal-transferase is the Golgi retention signal. The studies of Nilsson *et al.* (69) and the suggestion that the extra 13 amino acids at the N terminus of the long form of β4-Gal-transferase may target the enzyme to the plasma membrane (66, 67) (see above) indicate that the cytoplasmic domain also plays a role in targeting. Sequence comparisons of the transmembrane domains of different glycosyltransferases show no similarities other than the presence of charged residues flanking the transmembrane domain (75). This suggests the following possible mechanisms of Golgi retention: (a) there may be a unique retention signal and a unique receptor for every glycosyltransferase; (b) the retention signal may be dependent on a three-dimensional signal patch, (c) the transmembrane domain causes oligomerization of the transferase forming a complex which is unable to leave the Golgi; or (d) the Golgi membrane lipid bilayer controls retention.

Galactosyltransferase-associated p58 protein kinase

In 1986, a report appeared on the cloning of the cDNA for human β4-Gal-transferase from an expression library using an antibody to the transferase (81). The authors did not express their gene. The sequence was later shown to be totally different from the proven β4-Gal-transferase sequences (Figure 8) and its expression did not correlate with β4-Gal-transferase activity (64). This work indicates the danger of identifying cloned genes in the absence of expression studies.

Subsequent work on this gene (82, 83) has shown that it encodes a novel calmodulin-regulated serine/threonine protein kinase, named $p58^{GTA}$, with homology to yeast and human $p34^{cdc2}$ kinases (involved in the control of the cell division cycle). $p58^{GTA}$ co-purifies with β4-Gal-transferase (GTA = galactosyltransferase associated) thereby explaining why an antibody to β4-Gal-transferase resulted in the expression cloning of the $p58^{GTA}$ gene. Elevated expression of $p58^{GTA}$ kinase in Chinese hamster ovary cells results in cell cycle defects while decreased expression enhances DNA replication. Increased and decreased expression of $p58^{GTA}$ kinase in Chinese hamster ovary cells also correlates with increased and decreased β4-Gal-transferase activity. This effect on β4-Gal-transferase activity was not mediated at either the transcriptional or translational levels. Preliminary data suggests that $p58^{GTA}$ kinase may activate β4-Gal-transferase by protein phosphorylation (82). If correct, this work indicates that one of the mechanisms controlling tissue and time-dependent variations in oligosaccharide structure may be post-translational phosphorylation of glycosyltransferases.

Control of galactosyltransferase expression during spermatogenesis

As described above, Northern and S1 nuclease analyses have shown that both the 3.9 and 4.1 kb galactosyltransferase transcripts are present in a variety of murine somatic tissues, cells and cell lines under the respective control of promoters P2 and P1 (Figure 6). However, murine male germ cells (spermatogonia, pachytene spermatocytes, and round spermatids) do not express the 3.9 kb transcript (84). Spermatogonia show a strong 4.1 kb signal, pachytene spermatocytes show very faint bands at 2.9, 3.1, and 4.1 kb, and round spermatids show strong bands at 2.9 and 3.1 kb. The two truncated transcripts and the 4.1 kb transcript all produce the same 399-amino acid long form of β4-Gal-transferase protein. Truncation is due to the use of alternate polyadenylation signals.

The 2.9 kb truncated transcript (and probably also the 3.1 kb transcript) has an additional 5'-terminal untranslated 559 bp sequence not present in the 4.1 kb transcript (85). This extra sequence is immediately upstream and contiguous to the 4.1 kb transcriptional start site on genomic DNA (P1, Figure 6). The data suggest the presence of a germ cell-specific promoter upstream to the first somatic cell promoter P1; it is suggested that this promoter regulates expression of β4-Gal-transferase in haploid round spermatids.

3.1.2 UDP-Gal:Galβ1-4GlcNAc-R (Gal to Gal) α3-Gal-transferase (EC 2.4.1.124/151)

Galα1–3 (Figure 1) occupies the terminal nonreducing position on *N*-glycan antennae and is an uncharged alternative to sialic acid. Galili (86–89) has carried out extensive studies on the evolution of the Galα1-3Galβ1-4GlcNAc epitope. The epitope is absent from fish, amphibian, reptile, and bird fibroblasts but has a wide distribution in nonprimate mammals, lemurs and New World monkeys. The epitope is absent from Old World monkeys, apes and humans. The hypothesis is that the epitope appeared early in the evolution of mammals. Ancestral primates in the Old World may have been exposed to an endemic detrimental infectious agent which expressed the epitope. Such a pathogen could have exerted selective pressure for the evolution of primates which produced anti-α-Gal antibodies as a protective response. In order to prevent autoimmune disease, the ability to produce these antibodies was accompanied by loss of expression of α3-Gal-transferase, the enzyme responsible for synthesis of the Galα1-3 terminus. Anti-α-Gal antibodies are natural antibodies produced in all humans and constitute 1% of circulating IgG. The appearance of aberrant α3-Gal-transferase activity in human cells is a possible trigger for some forms of autoimmune disease (87, 89).

Bovine α3-Gal-transferase

Joziasse *et al.* (90) isolated a 1.8 kb cDNA encoding bovine α3-Gal-transferase. The domain structure of the enzyme is similar to that described above for β4-Gal-transferase (Table 2, Figure 5). The only amino acid sequence similarity noted between the α3- and β4-Gal-transferases is a Cys(X)$_{5-6}$Lys(or Arg)AspLysLys-AsnAsp(or Glu) sequence just downstream from the sequence identified by Aoki *et al.* (63) as a possible UDP-binding site (Figure 8). S1 nuclease protection, primer extension and Northern blot analyses on calf thymus poly(A)-RNA showed at least two sets of transcripts (3.6–3.9 kb) indicating the presence of at least two promoters; however, unlike the situation with β4-Gal-transferase (Figures 6 and 7), both transcripts initiate translation at the same Met residue and only a single protein is encoded. As expected, Northern blot analysis showed no α3-Gal-transferase message in poly(A)-RNA from human and Old World monkey tissues. However, sequences homologous to bovine α3-Gal-transferase were detected in human genomic DNA by Southern blot analysis.

Murine α3-Gal-transferase

Larsen *et al.* (91) used an approach to cloning murine α3-Gal-transferase cDNA that required neither the purification of the enzyme nor the availability of antibodies to the enzyme. They used a transient expression screening approach involving gene transfer of a murine cDNA library ligated into the mammalian expression vector pCDM7 developed by Seed and Aruffo (92, 93). The vector has an SV40 origin of replication allowing efficient replication in mammalian cells such as COS-1 African green monkey kidney cells which express the SV40 large T antigen. The human cytomegalovirus promoter of the vector allows highly efficient transcription of the inserted gene. The vector is thus capable of high level transient expression of DNA.

The screening procedure used to detect expressed α3-Gal-transferase activity was based on identifying cells with the product of α3-Gal-transferase action on their cell surfaces. COS-1 cells were selected because they do not express α3-Gal-transferase and therefore would have a low background. Further, they are known to express cell surface polylactosamine molecules that are required as substrate for α3-Gal-transferase.

The pCDM7 cDNA library was transfected into COS-1 cells and 72 hours were allowed for expression. The cells were then panned on dishes coated with *Griffonia simplicifolia* isolectin-I-B$_4$ which recognizes and binds to the Galα1-3Gal epitope on the cell surface. Cells adhering to the lectin were collected and episomal plasmid DNA was isolated by the Hirt procedure (94). The pCDM7 vector carries a suppressor tRNA gene (supF) to allow selection in *E. coli* by antibiotic resistance. *E. coli* strain MC1061(p3) transformed with pCDM7 DNA will show tetracycline and ampicillin resistance due to suppression of amber mutations in the genes for tetracycline resistance and β-lactamase carried on the p3 plasmid. Plasmid DNA was prepared from the *E. coli* transformants and transfected once again into COS-1 cells to repeat the screening procedure.

Sib selection was applied to *E. coli* transformants containing plasmid DNA rescued from the second screening. The transformants were plated to yield 16 pools containing between 100 and 5000 bacterial colonies each. Plasmid DNA prepared from replica plates was transfected separately into COS-1 cells and the transfectants were screened for 'active' DNA by panning to detect lectin-adherent cells. An *E. coli* pool showing the presence of at least a single colony containing 'active' DNA was subdivided into several smaller pools and the process was repeated. Four rounds of sib selection were required to purify a plasmid (pCDM7-αGT) capable of directing expression of *Griffonia simplicifolia* isolectin-I-B$_4$ binding activity in COS-1 cells.

The cDNA insert of pCDM7-αGT was 1.5 kb in size and encoded a 394-amino acid protein with a domain structure typical of glycosyltransferases (Table 2, Figure 5). Northern analysis showed a message of 3.6 kb in murine F9 teratocarcinoma cells.

Joziasse *et al.* (95) screened a murine cDNA library with a bovine probe for α3-Gal-transferase and isolated a cDNA encoding a 337-amino acid protein (Table 2). The deviation from the previously described murine sequence (91) is due to (i) assignment of a downstream ATG as the translation initiation codon thereby giving a cytoplasmic tail of 6 amino acids rather than 41, and (ii) a 22-amino acid deletion in the stem region. Analysis of mouse mRNA by a polymerase chain reaction (PCR) approach identified four distinct messages (all about 3.7 kb in size) encoding four different protein products which differ from one another in the stem region (Table 2). The 359-amino acid protein (Table 2) corresponds to the cDNA previously isolated by Larsen *et al.* (91). Although there may be more than one bovine α3-Gal-transferase transcript (see above), PCR analysis of bovine mRNA (95) showed only a single 368-amino acid protein product (Table 2).

Sequence analysis

The 368-amino acid bovine protein shows a strong similarity (about 80%) to the 371-amino acid murine protein (Table 2); the main difference is a 3-amino acid deletion in the stem region of the bovine protein.

Although α3-Gal-transferase has a domain structure typical of the glycosyltransferases (Table 2, Figure 5), the only sequence similarity is an approximately 55% identity of the human α3-Gal-transferase pseudogene to the human blood group A and B transferases (96). The Cys(X)$_{5-6}$Lys(or Arg)AspLysLysAsnAsp(or Glu) consensus sequence suggested as a UDP-Gal-binding site (see above) appears as Cys(X)$_5$GlnAspLysLysHisAsp in murine α3-Gal-transferase.

Genomic structure

The α3-Gal-transferase is not normally expressed in human tissues; it has been suggested that aberrant expression of the enzyme in man may be associated with autoimmune disease and other pathological conditions (see above). Northern analyses of various human tissues has not revealed the presence of α3-Gal-transferase transcripts; however, Southern analyses of human genomic DNA have revealed sequences homologous to α3-Gal-transferase using bovine (90) and murine (97) probes.

Joziasse *et al.* (98) probed a human genomic DNA library with a bovine α3-Gal-transferase cDNA probe and detected two homologous genes. Both genes are nonfunctional pseudogenes; clone HGT-2 is an intron-less processed pseudogene located on chromosome 12 and clone HGT-10 is an unprocessed pseudogene on chromosome 9. Murine α3-Gal-transferase maps to a region of chromosome 2 with linkage homology to human chromosome 9q22-ter (99). Fluorescent *in situ* hybridization has localized HGT-2 to human chromosome 12q14-q15, and HGT-10 to chromosome 9q33-q34 (100).

HGT-2 has an uninterrupted sequence 81% similar to the complete bovine coding sequence but contains multiple frame-shifts and nonsense codons in all three reading frames. HGT-2 is flanked by *Alu*-like repeats at both ends but lacks the 3'-poly(A) tail usually found in processed pseudogenes. HGT-10 contains a sequence similar to the N terminal and trans-membrane domains of the bovine enzyme and is flanked by introns at both ends. The C-terminal segment of this gene was not cloned by Joziasse et al. (98) but may be present on an 18 kb EcoRI genomic DNA fragment detected by them on Southern blot analysis; a clone isolated by Larsen et al. (97) may contain the C-terminal part of the gene partially encoded by clone HGT-10.

The C-terminal region of HGT-2 shows about 39% sequence similarity to the analogous regions of human blood group A α3-GalNAc-transferase and human blood group B α3-Gal-transferase. Clone HGT-10 is located on human chromosome 9q33-q34 and the gene locus encoding the allelic blood group A and B transferase genes is on human chromosome 9q34. These findings indicate that all four genes are derived from the same ancestral gene by gene duplication and subsequent divergence.

Larsen et al. (97) used a murine α3-Gal-transferase probe to isolate a homologous gene from a human genomic DNA library. As mentioned above, this sequence and clone HGT-10 of Joziasse et al. (98) probably come from the same nonfunctional unprocessed pseudogene. This clone contains a 703 bp region which is 82% similar to the murine α3-Gal-transferase C-terminal coding region but contains two frameshift mutations and a nonsense codon that disrupt the reading frame.

The nonfunctional pseudogenes probably replaced a functional α3-Gal-transferase gene in Old World monkeys some time after the divergence of New World and Old World monkeys (88, 98).

Joziasse et al. (95) isolated five over-lapping murine genomic clones for α3-Gal-transferase and determined that the gene consists of nine exons spanning at least 35 kb. The genomic structure resembles that of the α2,6-sialyltransferase and β4-Gal-transferase genes in that all are relatively large and are divided into 6 or more exons. In each case, the cytoplasmic domain, the transmembrane domain and part of the 'stem' region (Figure 5) are on a single exon while another exon contains the translation termination codon and all of the relatively long 3'-untranslated region.

Analysis of the mouse gene showed that the four transcripts for α3-Gal-transferase (Table 2) are due to alternative splicing of a message transcribed from a single gene on the centromeric region of chromosome 2. Splicing produces mRNAs which either (i) contain both exons 5 and 6, (ii) contain only exon 5, (iii) contain only exon 6, or (iv) contain neither exons 5 nor 6. The four transcripts were found in all mouse tissues tested except for male germinal cells which had no detectable mRNA.

Expression of α3-Gal-transferase

The cDNA encoding the putative catalytic domain of murine α3-Gal-transferase (lacking the N-terminal and transmembrane segments) was fused in-frame downstream to the cDNA encoding a secretable form of the IgG binding domain of *Staphylococcus aureus* protein A in a mammalian expression vector (91). Transfection of COS-1 cells with this fusion construct resulted in the secretion of α3-Gal-transferase activity into the culture medium. The secreted and soluble fusion protein bound specifically to an IgG-Sepharose column. The IgG-Sepharose-bound material was enzymatically active and was shown to transfer Gal from UDP-Gal to Galβ1-4GlcNAc to make Galα1-3Galβ1-4GlcNAc. These experiments indicate that the signal for retention of α3-Gal-transferase in the Golgi apparatus resides in the N-terminal and transmembrane segments.

High-level expression of enzymatically active α3-Gal-transferase was achieved using an insect cell expression system (101). Bovine cDNA containing the entire coding sequence of α3-Gal-transferase was ligated into the transfer plasmid pVL1393. *Spodoptera frugiperda* (Sf9) pupal ovary tissue insect cells were co-transfected with the recombinant plasmid and DNA from wild-type *Autographa californica* nuclear polyhedrosis baculovirus (AcNPV). Homologous recombination within infected insect cells results in insertion of the bovine α3-Gal-transferase DNA into the genome of the baculovirus at the polyhedrin protein site. Expression of the transferase is therefore under the control of the highly efficient polyhedrin promoter. Infected insect cells were screened visually for recombinant baculovirus which can no longer make the polyhedrin particles. Recombinant virus was plaque-purified. Southern blot analysis showed that uninfected or wild-type baculovirus-infected Sf9 cells did not contain α3-Gal-transferase DNA whereas this DNA was readily detected in recombinant baculovirus-infected cells.

Insect cells infected with purified recombinant virus were lysed with detergent and α3-Gal-transferase was detected in the lysate; there was very little enzyme activity in the growth medium. The lysate enzyme was purified by a single step using UDP-hexanolamine-Sepharose affinity chromatography. The expressed enzyme accounted for about 2% of the total cellular protein and had a specific activity 1000-fold higher (> 0.1 μmoles/min/mg) than that found in calf thymus gland, a rich source of bovine α3-Gal-transferase. The recombinant enzyme was shown to have kinetic properties identical to the calf thymus enzyme. Both enzyme preparations incorporate Gal in α1-3 linkage to the terminal Gal of Galβ1-4Glc(NAc) on oligosaccharides, glycolipid and glycoproteins and do not act on Fucα1-2Gal-terminated compounds thereby distinguishing them from blood group B α3-Gal-transferase (see below).

It has been observed in differentiating F9 murine teratocarcinoma cells that an increase in α3-Gal-transferase activity is accompanied by a decrease in poly-*N*-acetyllactosamine chain length suggesting that addition of Galα1-3 residues to the

Galβ1-4GlcNAc termini of these chains may compete with chain elongation (102). An investigation of this phenomenon was carried out (103) by stable transfection of the murine α3-Gal-transferase gene into Chinese hamster ovary (CHO) cells which make poly-N-acetyllactosamines but do not express α3-Gal-transferase. CHO cells were co-transfected with genomic DNA from F9 murine teratocarcinoma cells and pSV2-neo. Transfectants were obtained by culture in G418. Panning on plates coated with *Griffonia simplicifolia* isolectin-I-B$_4$ was used to isolate a CHO clone which expressed surface Galα1-3Gal epitope. Northern and Southern analyses of this clone with a murine α3-Gal-transferase cDNA probe showed the presence, respectively, of homologous 3.6 kb mRNA and genomic DNA. Enzyme assays showed the presence of α3-Gal-transferase activity in extracts from transfected cells but not wild-type cells. The transfected CHO clone and wild-type CHO cells showed no differences in poly-N-acetyllactosamine chain lengths indicating that the α3-Gal-transferase did not in fact compete with β3-GlcNAc-transferase for Galβ1-4GlcNAc termini. However, terminal sialylation with α2,3-linked sialic acid of poly-N-acetyllactosamine chains on N-glycans was reduced in transfected CHO cells in proportion to the appearance of terminal Galα1-3 residues indicating that α2,3-sialyltransferase and α3-Gal-transferase compete for Galβ1-4GlcNAc termini in these cells. The data suggest that α2,3-sialyltransferase and α3-Gal-transferase reside in the same subcellular compartment which is presumably distal to the compartment carrying the β3-GlcNAc-transferase involved in poly-N-acetyllactosamine chain elongation. This study therefore did not clarify the factors controlling poly-N-acetyllactosamine chain length.

3.1.3 Human blood group B UDP-Gal:Fucα1-2Gal-R (Gal to Gal) α3-Gal-transferase (EC 2.4.1.37)

Individuals with the blood group A phenotype express the blood group A α3-GalNAc-transferase, B individuals the blood group B α3-Gal-transferase, AB individuals express both enzymes, and O individuals express neither enzyme. The A and B genes are alleles at the ABO locus and encode the respective transferases. The transferases synthesize the histo-blood group A and B antigens (see Figure 3).

Cloning of the gene for the human blood group B UDP-Gal:Fucα1-2Gal-R (Gal to Gal) α3-Gal-transferase was preceded by isolation and sequencing of the cDNA encoding the human blood group A UDP-GalNAc:Fucα1-2Gal-R (GalNAc to Gal) α3-GalNAc-transferase (see below). Since immunological evidence had indicated that the allelic A and B transferases were structurally very similar, homology screening (Table 1) was used to isolate the B α3-Gal-transferase gene (104). An A-transferase probe was used to screen cDNA libraries from human cells of phenotype O, AB, and B. Several cDNAs were isolated and sequenced. The cDNAs from A, B, and O cells were highly homologous; the A sequence showed consistent differences from the B sequence at seven base positions and from the O sequence at a single base position.

The B α3-Gal-transferase gene encodes a protein with the type II integral membrane protein domain structure typical for the glycosyltransferases (Figure 5; Table

2). The A α3-GalNAc- and B α3-Gal-transferases have the same length (Table 2) but differ at four amino acid positions. The functional role of these four amino acids was investigated by transient transfection and expression in HeLa cells of a series of chimeric constructs in a plasmid expression vector (96). The transfected cells were analysed for the presence of cell surface A or B antigen. Untransfected HeLa cells were shown to have the OO genotype. Transfection with the A or B cDNA resulted in the appearance, respectively, of the A or B antigen on the cell surface. Sixteen chimeric DNAs were constructed in which each of the four amino acid sites contained the residue present in either the A or B gene. Transfection with these chimeric DNAs showed that only two of the four amino acid sites specify which nucleotide-sugar donor is utilized by the enzyme. Some of the chimera produced enzymes which utilized either UDP-Gal or UDP-GalNAc equally well; normal A and B transferases can also utilize both nucleotide-sugars but the nonspecific nucleotide-sugar is utilized at an appreciably lower rate than the specific nucleotide-sugar.

The cDNA isolated from O cells differs from the A and B sequences by having a single base deletion near the N-terminus of the coding region thereby resulting in a shift of the reading frame. Northern analysis shows identical mRNA patterns for A, B, and O cells. Thus the O phenotype expresses messages very similar to the A and B phenotypes but there is either premature termination of translation or the production of an entirely different noncross-reactive protein. Indeed, antibody to the A-transferase does not detect a cross-reactive protein in O cells.

The molecular basis for ABO polymorphism was determined by analysis of genomic DNA from white cell samples from 14 individuals with known A, B, AB and O phenotypes (104). The approach depended on the existence of allele-specific restriction enzyme sites in the A, B, and O cDNA sequences that were typical of these sequences. The polymerase chain reaction was used to amplify segments of genomic DNA containing these diagnostic restriction sites and the amplified DNA was cut with the relevant restriction enzyme and analysed by electrophoresis. There was a perfect match between the presence of the diagnostic DNA fragments and the phenotype of the sample. A similar approach called allele-specific polymerase chain reaction (ASPCR) was used to analyse genomic DNA from 41 individuals of known ABO phenotype (105); again there was a perfect match between the serology and the ASPCR analysis. The data provide strong evidence that ABO polymorphism is due to the sequence differences described above for the A- and B-transferases and the nonfunctional homologous gene present in O individuals.

There is an approximately 55% identity of the human α3-Gal-transferase pseudogene to the human blood group A and B transferases (96). In an attempt to find other homologous genes, Yamamoto et al. (106) carried out polymerase chain reaction amplification of human genomic DNA with degenerate oligonucleotide primers that are conserved in the A- and B-transferases and in α3-Gal-transferase. A fragment was cloned which was homologous to but different from the A- and B-transferases and α3-Gal-transferase; the function of this gene (named hgt4) has not been determined.

3.2 Sialyltransferases
3.2.1 CMP-sialic acid:Galβ1-4GlcNAc-R α2,6-sialyltransferase (ST6N) (EC 2.4.99.1)

Rat α2,6-sialyltransferase

Screening of a rat liver cDNA λgt11 expression library with antibody to α2,6-sialyltransferase yielded a single positive clone (107). The clone was used to obtain overlapping cDNA clones encoding a 403 amino acid protein with a type II integral membrane protein domain structure typical of the glycosyltransferases (Table 2, Figure 5). There are three potential N-glycan sites; two of these are glycosylated in the native rat liver enzyme.

Northern analysis showed a tissue-specific distribution of mRNAs (108). An abundant mRNA of 4.3 kb was detected only in liver, a 4.7 kb mRNA was seen in many tissues and a 3.6 kb mRNA was seen only in kidney (Figure 10).

Human α2,6-sialyltransferase

The coding region of rat liver α2,6-sialyltransferase was used to isolate the sialyltransferase cDNA from human cDNA libraries (Table 1) (109, 110). Human α2,6-sialyltransferase cDNA has also been isolated by expression cloning (Table 1). Stamenkovic et al. (111) transformed COS cells with a cDNA expression library from Daudi cells (a human B lymphoblastoid cell line) and screened for the appearance of antigen CDw75 on the COS cell surface; Bast et al. (112) used a cDNA expression library from human spleen cells to isolate a cDNA essential for expression of the antigen HB-6 on the host cell surface. Both groups isolated α2,6-sialyltransferase cDNA, an enzyme which is essential for the synthesis of the CDw75 and HB-6 antigens.

The human enzyme is 406 amino acids long and has a domain structure similar to the rat enzyme (Table 2). Northern analyses of human tissues show mRNAs between 4 and 5 kb containing a long 3'-untranslated region (110, 112).

Sequence analysis

The rat and human α2,6-sialyltransferase amino acid sequences show 87.6% similarity (109). The human cDNA encodes three additional amino acids (Glu-Lys-Lys) at the C-terminal end of the transmembrane segment. Comparison of the sequences for human cDNAs isolated from four different tissues (Table 1) show minor differences in both the coding and noncoding regions (112).

Although there is no obvious sequence similarity between the α2,6-sialyltransferases and β4-Gal-transferases, the enzymes have the same domain structure (Table 2, Figure 5) and both have seven Cys residues. One Cys is in the transmembrane domain of both enzymes and four other Cys residues are in equivalent positions in the catalytic domains of both enzymes.

Genomic structure

The organization of the rat α2,6-sialyltransferase gene (Figure 10) resembles that of the UDP-Gal:GlcNAc-R β4-Gal-transferase gene (Figure 7) (113–115) and the UDP-

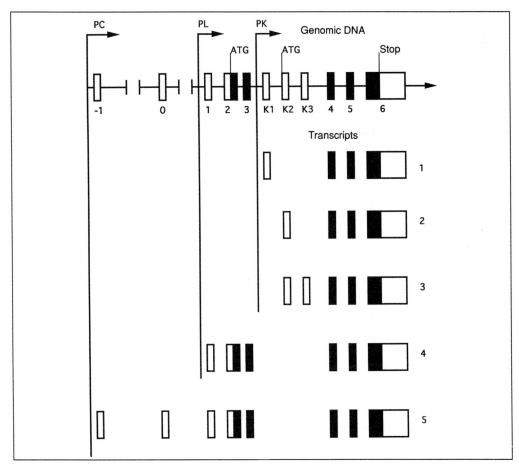

Fig. 10 Genomic map of rat CMP-sialic acid:Galβ1-4GlcNAc-R α2,6-sialyltransferase gene spanning 80 kb (see text). The numbered boxes indicate the eleven exons. The filled boxes represent the coding region of the nonkidney specific sequence. The diagram is not drawn to scale. The top line represents the genomic DNA. P_C, P_L, and P_K represent the constitutive, liver-specific and kidney-specific promoters respectively. Lines 1, 2, and 3 represent the three kidney-specific 3.6 kb mRNAs. Line 4 is the 4.3 kb liver-specific mRNA. Line 5 is the 4.7 kb mRNA that is present in most tissues. Data from references 113–117

Gal:Galβ1-4GlcNAc-R (Gal to Gal) α3-Gal-transferase gene (95) although there is no sequence similarity between the three enzymes. The α2,6-sialyltransferase gene spans at least 80 kb of genomic DNA and contains at least 11 exons. There are at least three promoters responsible for the production of five or more messages from a single gene sequence (Figure 10) (108, 114–116). Promoter P_L produces a liver-specific 4.3 kb mRNA (exons 1 to 6). Promoter P_C is constitutive in several tissues and produces a 4.7 kb mRNA (exons -1 to 6). Promoter P_K is restricted to the kidney and produces several 3.6 kb mRNAs missing the 5'-half of the coding sequence (kidney-specific exons K1, K2, and K3 and exons 4 to 6). Although the 3.6 kb messages produce a protein in kidney that is immunologically cross-reactive with α2,6-sialyltransferase, this protein does not appear to encode an active α2,6-sialyltransferase (117).

Three overlapping rat genomic DNA clones were isolated by screening a cosmid library; these clones contained exons 1-6 spanning 40 kb genomic DNA (Figure 10) (113). The site of transcription initiation in rat liver was determined by primer extension, nuclease S1 analysis and 5'-extension using the polymerase chain reaction. A major initiation site (promoter P_L) responsible for the 4.3 kb transcript was found at the start of exon 1 (Figure 10). However, evidence was also obtained for the presence in liver, spleen, and ovary of transcripts with at least one exon upstream to exon 1 (Figure 10); spleen and ovary did not contain the 4.3 kb initiation site. Promoter P_L was demonstrated to be about 50-fold more active in a hepatoma cell line (HepG2) known to express α2,6-sialyltransferase than in a cell line (Chinese hamster ovary) which does not express this enzyme; promoter activity was assayed using luciferase as a reporter protein. The P_L promoter sequence contains consensus binding sites for liver-restricted transcription factors. These findings explain the high levels of α2,6-sialyltransferase activity in rat liver.

Since rat kidney expresses both 4.7 and 3.6 kb mRNAs, Wen *et al.* (114) screened rat kidney cDNA libraries with probes from the coding region of the 4.3 kb rat liver α2,6-sialyltransferase. Clones corresponding to both the 4.7 and 3.6 kb transcripts were obtained and characterized. The 4.7 kb mRNA differs from the 4.3 kb transcript only in containing a longer 5'-untranslated region; both transcripts encode the same α2,6-sialyltransferase protein (Figure 10).

Analysis of rat α2,6-sialyltransferase genomic DNA clones (114, 115) showed the presence of a functional liver-specific promoter P_L at the start of exon 1 and of a functional kidney-specific promoter P_K in the intron between exons 3 and 4 (Figure 10). This intron contains kidney-specific exons K1, K2, and K3 under the control of promoter P_K. Exons K1, K2, and K3 are spliced out during expression of α2,6-sialyltransferase under the control of promoters P_C and P_L. Alternative splicing of exons is also involved in the generation of the three 3.6 kb mRNAs (Figure 10). The functional significance of these truncated kidney-specific transcripts is not known. Mouse kidney expresses only one 4.7 kb mRNA suggesting that the 3.6 kb transcripts in rat kidney are not essential for kidney function (114). Wang *et al.* (115) found a P_K promoter between K1 and K2 indicating that there may be at least two kidney-specific promoters.

Wang *et al.* (118) have characterized human α2,6-sialyltransferase genomic DNA (Figure 11). Human exons Y, Z, and I to VI are homologous to rat exons -1, 0, and 1 to 6 respectively. The position of human exon X on genomic DNA is not known and no equivalent exon has been found in the rat genome. There is good evidence for the presence of at least three promoters (Figure 11). Promoter P_Y produces a transcript [Y-Z-I to VI] (Figure 11) in human placenta, HL-60 cells and in six human lymphoblastoid cells. Promoter P_X produces appreciably higher levels of transcript [X-I to VI] (Figure 11) in three lymphoblastoid cell lines with a mature B-cell phenotype (Daudi, Jok-1, and Louckes). The [X-I to VI] transcript was not found in the other tissues tested. The presence of a high level of sialyltransferase expression

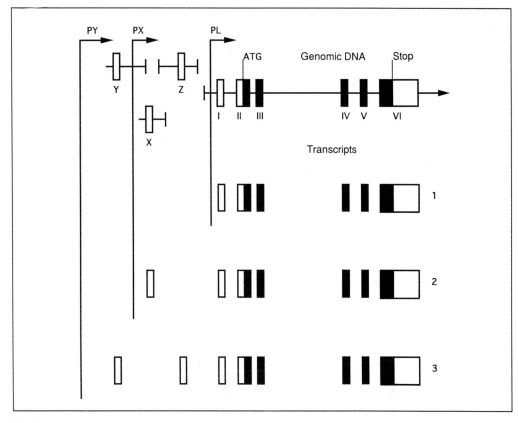

Fig. 11 Genomic map of human CMP-sialic acid:Galβ1-4GlcNAc-R α2,6-sialyltransferase gene (118). The boxes indicate the nine known exons (Roman numerals I to VI and X, Y, and Z). The filled boxes represent the coding region. The diagram is not drawn to scale. The top line represents the genomic DNA. P_X, P_Y, and P_L represent the postulated promoters. Line 1 is the liver-specific mRNA under the control of P_L. Line 2 is the [X-I-IV] transcript that is increased in human lymphoblastoid lines exhibiting the mature B-cell phenotype. Line 3 is the [Y-Z-I-VI] transcript that is present in most tissues and is homologous to the 4.7 kb constitutive transcript found in the rat genome (see Figure 10)

therefore correlates with the [X-I to VI] transcript. It is of interest that a high level of sialyltransferase expression was also found to correlate with the presence of the CDw75 antigen on the lymphocyte cell surface. CDw75 appears during the transition from the pre-B stage of development into mature activated B-cells. It has been shown by others that expression of this antigen requires the presence of the α2,6-sialyltransferase (112, 119, 120). Human hepatoma HepG2 cells express neither the [Y-Z-I to VI] nor the [X-I to VI] transcripts but express high levels of a third transcript which is presumed to be under the control of a liver-specific promoter P_L similar to the one reported for the rat gene.

Comparison of the α2,6-sialyltransferase gene (Figures 10 and 11) with the genes for UDP-Gal:GlcNAc-R β4-Gal-transferase (Figure 7) and UDP-Gal:Galβ1-4GlcNAc-R (Gal to Gal) α3-Gal-transferase indicates that all are relatively large and

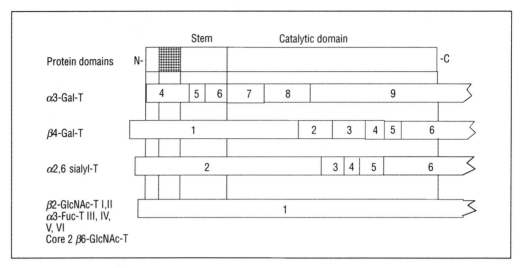

Fig. 12 Comparison of the protein domains of several glycosyltransferases with their exon structures (adapted from reference 200). The hatched region is the trans-membrane domain of the protein. The numbered boxes represent the exons. Some genes (for example UDP-Gal:Galβ1-4GlcNAc-R α3-Gal-transferase, CMP-sialic acid:Galβ1-4GlcNAc-R α2,6-sialyltransferase and UDP-GlcNAc:Manα1-3R (GlcNAc to Manα1-3) β2-GlcNAc-transferase I) have at least one up-stream exon containing only noncoding sequences. See text for details

are divided into six or more exons. In each case, a single exon contains the cytoplasmic domain, the trans-membrane domain and part of the 'stem' region (Figures 5 and 12). The 3'-terminal exon is very large and contains the translation termination codon and all of the relatively long 3'-untranslated region. The catalytic domain is divided between several small exons (Figure 12). Multiple transcripts are produced from all three genes, at least in some species, due either to more than one site of transcription initiation, or alternative exon splicing, or alternative polyadenylation signals, or a combination of these processes.

Expression of α2,6-sialyltransferase

Tissue-specific expression of rat α2,6-sialyltransferase appears to be regulated at the transcriptional level (108). The 4.3 kb mRNA (Figure 10) is expressed only in liver under the control of the liver-specific promoter P_L; the 4.7 kb mRNA is expressed in lesser amounts (10–100-fold less) in liver and several other tissues. This correlates with a 10–100-fold higher activity of α2,6-sialyltransferase activity in liver relative to other rat tissues. The levels of the 4.7 kb mRNA vary 50–100-fold in various rat tissues and these levels also correlate with enzyme activity. The use of alternative promoters also appears to control the tissue-specific expression of human α2,6-sialyltransferase (118).

Cell type-specific transcription of genes is controlled by DNA-binding proteins called transcription factors which themselves have a restricted pattern of expression. The intron between α2,6-sialyltransferase exons 0 and 1 (Figure 10) has a liver-specific promoter P_L (113) which is active in the hepatoma cell line HepG2 but not in Chinese hamster ovary cells presumably due to the presence of liver-specific transcription factors in the hepatoma cells. Svensson et al. (121) have used several techniques to study the transcription factor binding sites on the P_L promoter. Deletion analysis, footprinting, and site-directed mutagenesis identified four such sites within 250 bp of the transcriptional start site for the 4.3 kb mRNA. Transactivation experiments were also carried out in which cells were co-transfected with a plasmid carrying the α2,6-sialyltransferase promoter region driving a luciferase reporter gene and a plasmid capable of expressing a particular transcription factor. Transcription factors HNF-1α (hepatocyte nuclear factor 1α), DBP (D-binding protein) and LAP (liver-enriched transcriptional activator protein) were found to be strong activators of promoter P_L and are most likely responsible for the high level expression of α2,6-sialyltransferase in rat liver. The promoter also has binding sites for general transcription factors that are ubiquitous in their expression such as MLTF (adenovirus major late transcription factor) and the NF-1 (nuclear factor 1) like proteins.

The hepatic acute phase response is accompanied by increased levels of α2,6-sialyltransferase in liver and in the circulation (122–126). Dexamethasone is believed to mediate this response. Dexamethasone was shown to cause 3–4-fold enrichment in α2,6-sialyltransferase mRNA in rat hepatoma cells and a corresponding increase in enzyme activity (127); the induction was mediated by a transcriptional enhancement mechanism (128).

Chinese hamster ovary (CHO) cells produce N-glycans with the terminal sequence NeuAcα2-3Galβ1-4GlcNAc but not NeuAcα2-6Galβ1-4GlcNAc, i.e. they do not express α2,6-sialyltransferase activity. CHO cells were co-transfected with a full-length α2,6-sialyltransferase cDNA in the pECE expression vector and pSV2neo to allow selection of transfected cells with the antibiotic G418 (129). Pure clones of transfected cells carrying the NeuAcα2-6Galβ1-4GlcNAc structure on their surfaces were selected by repeated fluorescence-activated cell sorting using fluorescein isothiocyanate-labelled *Sambucus nigra* agglutinin. This lectin has 50–100-fold higher affinity for NeuAcα2-6Gal- than for NeuAcα2-3Gal-. Clones which bind this lectin express a 2.2 kb mRNA for α2,6-sialyltransferase. Although the level of α2,6-sialyltransferase activity in these cells was relatively low, N-glycans with NeuAcα2-6Galβ1-4GlcNAc termini were detected both by chemical and immunohistochemical methods indicating that the α2,6-sialyltransferase competed effectively with the endogenous α2,3-sialyltransferase for available Galβ1-4GlcNAc termini.

Bast et al. (112) reported that transformation of COS cells with an α2,6-sialyltransferase expression plasmid generated three distinct cell surface sialic-acid dependent antigens (HB-6, CDw75, and CD76). The sialyltransferase was shown to be present in the Golgi apparatus and no enzyme was detected at the cell surface (112, 119); a report suggesting that the CD75 antigen is cell surface sialyltransferase (111) has been withdrawn (120). Although sialyltransferase mRNA is widely distributed in many human lymphocyte cell types, the levels of expression vary widely and there are differences in mRNA size (112). A recent report has shown the presence of at least three promoters controlling the expression of the human α2,6-sialyltransferase (118) (Figure 11).

Targeting to the Golgi apparatus

Rat α2,6-sialyltransferase has been localized to the *trans*-Golgi cisternae and the *trans*-Golgi network of hepatocytes, hepatoma cells, and intestinal goblet cells but may be more diffusely distributed in the Golgi apparatus of other cell types (130). Purified rat liver α2,6-sialyltransferase is a soluble form of the enzyme with a lower molecular weight than the Golgi-bound enzyme; this difference in molecular weights suggested that proteolysis results in solubilization of a catalytically active enzyme (107, 131). N-terminal analysis of the purified soluble enzyme showed the cleavage to be in the 'stem' region (Figure 5) 38 amino acids downstream of the C-terminal end of the transmembrane domain (107). Inflammation has been shown to cause the release of soluble and catalytically active α2,6-sialyltransferase into the serum due to cleavage of the enzyme at the 'stem' region by a cathepsin D-like protein (122–124). The subcellular location of the endogenous protease is probably within the Golgi apparatus (107) and proteolysis may occur both *in vivo* and during enzyme purification. These studies indicate that the amino-terminal cytoplasmic tail, the transmembrane segment and the stem region are not needed for catalysis but may be required for Golgi targeting and anchoring.

The Golgi targeting signal for α2,6-sialyltransferase was investigated (130) by construction of a eukaryote expression plasmid in which the N-terminal 57 amino acids of α2,6-sialyltransferase (containing the cytoplasmic tail, transmembrane domain, and 31 amino acids from the 'stem' region) were replaced with the human γ-interferon signal peptide and signal cleavage site. The plasmid contained an SV40 early gene promoter and a mouse dihydrofolate reductase (DHFR) gene to allow selection of amplified genes with methotrexate. Chinese hamster ovary (CHO) cells were co-transfected with this plasmid and pSV2neo and stable transfectants were obtained by selection with G418 and methotrexate. CHO cells transfected with the fusion protein construct secreted active α2,6-sialyltransferase into the growth medium. The secreted protein was mainly resistant to endo-β-N-acetylglucosaminidase H digestion indicating the formation of complex N-glycans; intracellular enzyme was digested by the endoglycosidase. Control CHO cells transfected with full-length α2,6-sialyltransferase cDNA did not secrete any detectable enzyme.

A detailed investigation of the targeting signal for α2,6-sialyltransferase was carried out by Munro (80). Sequences encoding the cytoplasmic domain, transmembrane domain and 'stem' region of α2,6-sialyltransferase and of dipeptidyl-peptidase IV (DPPIV), a cell surface type II integral membrane protein, were fused to the coding region of the secretory protein chicken lysozyme. The constructs were placed into a eukaryote expression plasmid, transfected into COS monkey cells, and the destination of the lysozyme reporter protein was determined by immunofluorescence. The results are shown in Table 3 and indicate that the transmembrane segment of α2,6-sialyltransferase is necessary and sufficient to

Table 3 Intracellular targeting of α2,6-sialyltransferase. Sequences encoding the cytoplasmic domain, trans-membrane domain and 'stem' region of α2,6-sialyltransferase (S) and of dipeptidylpeptidase IV (D) were fused to the coding region of the secretory protein chicken lysozyme. The constructs were placed into a eukaryote expression plasmid, transfected into COS monkey cells, and the destination of the lysozyme reporter protein was determined by immunofluorescence. DKK, addition of two Lys residues to the dipeptidylpeptidase IV cytoplasmic domain. Fusion protein constructs were also made with either poly(Leu)$_{17}$ or poly(Leu)$_{23}$ trans-membrane domains. The data are from Munro et al. (80)

Cytoplasmic domain	Trans-membrane domain	Stem region	Primary intracellular location
D	D	D	Cell surface
S	S	S	Golgi
D	S	D	Golgi (some cell surface)
D	D	S	Cell surface
S	D	D	Cell surface
S	D	S	Cell surface
S	S	D	Golgi
D	S	S	Golgi
DKK	S	D	Golgi (no cell surface)
DKK	S	S	Golgi
S	Poly(Leu)17	S	Golgi
DKK	Poly(Leu)17	D	Cell surface
S	Poly(Leu)23	S	Cell surface
D	Poly(Leu)23	D	Cell surface
DKK	Poly(Leu)17	S	Golgi
S	Poly(Leu)17	D	Cell surface

prevent movement of lysozyme to the cell surface. However, constructs with either the cytoplasmic or stem region from α2,6-sialyltransferase in addition to the α2,6-sialyltransferase transmembrane region showed less leakage of lysozyme to the plasma membrane than constructs lacking these regions. The cytoplasmic domain of α2,6-sialyltransferase has two more Lys residues than the cytoplasmic domain of DPPIV; addition of two Lys residues to the DPPIV cytoplasmic domain prevented leakage to the cell surface (Table 3) indicating the importance of charged residues flanking the transmembrane domain for retention in the Golgi. The transmembrane domains of α2,6-sialyltransferase and of DPPIV are, respectively, 17 and 23 amino acids long. Fusion protein constructs were therefore made with either poly(Leu)$_{17}$ or poly(Leu)$_{23}$ transmembrane domains (Table 3). The data show that a poly(Leu)$_{17}$ but not a poly(Leu)$_{23}$ transmembrane domain can target to the Golgi provided a part of the α2,6-sialyltransferase stem region is also present. The stem region therefore contains Golgi targeting information. However, the stem and cytoplasmic domains must be spaced apart as in native α2,6-sialyltransferase. Constructs containing a full-length cDNA for α2,6-sialyltransferase in which the transmembrane domain was replaced with the transmembrane domain of DPPIV targeted to the cell surface.

Colley et al. (132) carried out transient transfection of COS cells with cDNA encoding a variety of mutant sialyltransferases and also observed that both the transmembrane domain and the stem region play a role in Golgi targeting.

Wong et al. (133) used stable transfection of MDCK cells to study the problem. They tested expression constructs in which various segments of the amino-terminal 64 amino acids of sialyltransferase replaced equivalent regions in the cDNA of DPPIV. The 17-residue transmembrane domain of sialyltransferase was sufficient for complete retention of the catalytic domain of DDPIV in the Golgi apparatus. Unlike the studies described above, these workers found no evidence to support the presence of Golgi targeting sequences in either the cytoplasmic tail or the stem region of sialyltransferase.

Retention of both α2,6-sialyltransferase and β4-Gal-transferase in the Golgi is dependent on the transmembrane domain. However, there are no obvious homologies between these enzymes. Even a poly(Leu)$_{17}$ sequence can replace the transmembrane domain of α2,6-sialyltransferase. The regions on either side of the transmembrane domain of α2,6-sialyltransferase may play a role in Golgi retention. Although both enzymes have been localized to the *trans*-Golgi cisternae, the differences in their Golgi targeting signals lend support to the immunohistochemical evidence that α2,6-sialyltransferase may be localized more distally in the Golgi than β4-Gal-transferase (134).

3.2.2 CMP-sialic acid:Galβ1-3GalNAc-R α2,3-sialyltransferase (ST30) (EC 2.4.99.4)

Porcine α2,3-sialyltransferase

Two forms of α2,3-sialyltransferase (45 and 48 kDa) were purified from porcine liver extracts. A 53 bp nondegenerate oligonucleotide based on the amino-terminal amino acid sequence of the 48 kDa enzyme was used to screen a cDNA library from porcine submaxillary glands (135). A single 1.6 kb clone was isolated which was then used to obtain a second overlapping clone from the same library.

Sequence analysis

The two cDNA clones for the α2,3-sialyltransferase differ by a 120 bp segment in the coding region. PCR analysis indicated that both cDNA sequences were present in mRNA from porcine submaxillary gland. The shorter message probably represents alternate splicing with deletion of one or more exons and does not appear to encode active enzyme. Northern analysis of several rat tissues showed only a single 5.7 kb mRNA. Sequence was obtained for 2.6 kb of cDNA (600 bp 5'-untranslated region, 1029 bp open reading frame and 1000 bp 3'-untranslated region). The protein has a type II integral membrane protein structure typical of all the glycosyltransferases (Table 2). Purified porcine sialyltransferase exists in two catalytically active forms (45 and 48 kDa). Sequence analysis shows that the 48 kDa form contains the hydrophobic transmembrane segment whereas the 45 kDa form is a truncated version of the same protein which has been cleaved in the stem region 56 amino acids downstream of the initiation Met residue of the full-length form of the enzyme. These 56 residues are not required for enzyme activity.

Comparison of the sequences of CMP-sialic acid:Galβ1-4GlcNAc-R α2,6-sialyltransferase and CMP-sialic acid:Galβ1-3GalNAc-R α2,3-sialyltransferase reveal a short 45 amino acid region of 64% sequence identity in the centre of the coding region. This homologous domain overlaps exons 2 and 3 of the α2,6-sialyltransferase (Figures 10 and 11). The two proteins show no primary sequence similarities to other known glycosyltransferases.

Expression of α2,3-sialyltransferase

The cDNA encoding the catalytic domain of α2,3-sialyltransferase (lacking the cytoplasmic tail and transmembrane segment) was placed in a eukaryote expression vector downstream of the cleavable insulin signal sequence. COS cells transiently transfected with this DNA construct secreted active enzyme into the medium. These experiments prove that the cytoplasmic tail and transmembrane segment are not required for enzyme activity but are necessary for targeting to the Golgi apparatus.

Northern analysis of several rat tissues showed differential expression of the α2,3-sialyltransferase; the mRNA levels in these tissues correlated with enzyme activity. The patterns of tissue-specific expression of the α2,3- and α2,6-sialyltransferases were different.

The maturation of thymocytes is accompanied by a change in cell surface carbohydrate from structures that react with peanut agglutinin (PNA) to structures which are PNA-negative. Transient expression of CMP-sialic acid:Galβ1-3GalNAc-R α2,3-sialyltransferase in a PNA-positive T-lymphoblastoid cell line resulted in 11% conversion to the PNA-negative phenotype (136). The data indicate that the conversion from PNA-positive to PNA-negative cells is due to the masking of Galβ1-3GalNAc-R by α2,3-linked sialic acid.

3.2.3 CMP-sialic acid:Galβ1-3(4)GlcNAc-R α2,3-sialyltransferase (ST3N) (EC 2.4.99.6)

Rat liver α2,3-sialyltransferase

Rat liver CMP-sialic acid:Galβ1-3(4)GlcNAc-R α2,3-sialyltransferase was purified, the protein was reduced and alkylated, tryptic digests were carried out and the resulting peptides were fractionated by reverse phase microbore HPLC (137). In a novel and sensitive sequencing approach, 300 pmol (12 μg) of purified enzyme were used to obtain 25% of the total protein sequence; the tryptic peptides were analysed by liquid secondary ion mass spectrometry (LSIMS) and the 30 most abundant molecular ions (out of a total of about a hundred) were subjected to tandem mass spectrometry on a four-sector machine using high energy collision-induced dissociation analysis (CID).

Degenerate oligonucleotide sense and antisense primers based on the amino acid sequence were used to carry out PCR on rat liver cDNA. Screening a rat liver cDNA library with one of the PCR products yielded two clones with inserts of 2.1 and 1.5 kb respectively.

Sequence analysis

The 2.1 kb clone contained a complete 1122 bp open reading frame encoding a 374 amino acid protein with all the peptide sequences revealed by mass spectrometry. The protein shows the domain structure typical of the glycosyltransferases (Figure 5, Table 2). The amino terminus of the purified enzyme was in the stem region 49 residues downstream from the initiation Met residue showing that this enzyme preparation was due to proteolysis and that the transmembrane domain is not required for enzyme activity.

Comparison of CMP-sialic acid:Galβ1-4GlcNAc-R α2,6-sialyltransferase (ST6N), CMP-sialic acid:Galβ1-3GalNAc-R α2,3-sialyltransferase (ST30) and CMP-sialic acid:Galβ1-3(4)GlcNAc-R α2,3-sialyltransferase (ST3N) reveals a region of 55 amino acids with extensive homology (40% identity; 58% conservation) in the middle of the catalytic domain. There is a second 23 residue region of similarity near the COOH terminus (138). Each similarity region contains a conserved Cys residue suggesting that these two Cys residues may form a disulphide bridge in the native proteins.

PCR based on primers in the conserved region (sialylmotif) was used to clone additional cDNA sequences (139). One of these new cDNAs appears to be specific to the newborn brain.

Expression of α2,3-sialyltransferase

A DNA construct encoding a fusion protein with the catalytic domain of α2,3-sialyltransferase (lacking the cytoplasmic tail and transmembrane segment) and the cleavable insulin signal sequence was transiently expressed in COS cells. Active enzyme was secreted into the medium proving that the cytoplasmic tail and transmembrane segment are not required for enzyme activity but are necessary for targeting to the Golgi apparatus. The recombinant enzyme transferred sialic acid to both Galβ1-3GlcNAc-R and Galβ1-4GlcNAc-R acceptors.

All rat tissues tested showed a 2.5 kb mRNA; rat colon showed an additional 2.0 kb message. These messages were differentially expressed.

3.3 Fucosyltransferases

Fucose is found in many protein- and lipid-bound oligosaccharides where it is attached to Gal, Glc or GlcNAc residues in α-linkage to carbons 2, 3, 4, or 6. Several of these fucosylated structures form important human blood group antigens (A, B, H, Le^a, Le^b, Le^x, and Le^y antigenic epitopes, Figure 3). Fucosylated oligosaccharides are often expressed in a regulated manner in development, differentiation, and progression of metastasis (28, 140–143). The following sections will deal with those fucosyltransferases whose genes have recently been cloned.

3.3.1 Human blood group H GDP-Fuc:Galβ-R α2-Fuc-transferase (EC 2.4.1.69)

It is presently believed that there are two GDP-Fuc:Galβ-R α2-Fuc-transferases, encoded respectively by the H and Se (secretory) loci (144, 145). The H enzyme is expressed in haematopoietic tissues and plasma while the Se enzyme is present in the secretory fluids and tissues of ABO/H blood group 'secretors'. The H and Se loci are closely linked on human chromosome 19 (146).

An expression cloning approach (Table 1) was used to isolate the gene for human blood group H GDP-Fuc:Galβ-R α2-Fuc-transferase (147–149). Mouse L cells lack the α2-Fuc-transferase but express on their surfaces substrates which can accept α2-linked fucose. Mouse L cells were stably transfected with total genomic DNA from human A431 cells (147). Transfectants were selected for cell surface H antigen expression by a combination of panning and fluorescence-activated cell sorting. Several clones were isolated showing strong surface expression of H antigen and intracellular α2-Fuc-transferase activity. These clones all showed the same 2.7 and 3.4 kb EcoRI restriction fragments on Southern analysis with a probe for human Alu sequences. Phage libraries were constructed with the 2.7 and 3.4 kb EcoRI fragments and screened with the human Alu probe to obtain pure genomic DNA fragments. The DNA was subcloned into the mammalian expression cosmid vector pWE15 and transiently transfected into COS-1 cells. COS cells transfected with the 3.4 kb fragment expressed significant amounts of α2-Fuc-transferase activity; mock-transfected cells or cells transfected with the 2.7 kb fragment showed no enzyme activity.

The α2-Fuc-transferase activities of stably transfected mouse L cells and transiently transfected COS cells had kinetic properties identical to those of the human A431 enzyme and human serum blood group H α2-Fuc-transferase but differed significantly from the human Se and mouse L cell α2-Fuc-transferases (148).

A 1.2 kb probe from the 3.4 kb EcoRI genomic DNA fragment was used to isolate a 3.37 kb α2-Fuc-transferase cDNA from a human cDNA library in pCDM7 (Table 1) (149). COS-1 cells transiently transfected with this cDNA, designated pCDM7-α(1,2)FT, express H antigen on their surfaces and intracellular α2-Fuc-transferase activity typical of the blood group H enzyme. Northern analysis of A431 cells shows a single 3.6 kb transcript. The protein encoded by this cDNA has the domain structure typical of all glycosyltransferases cloned to date (Figure 5; Table 2) but shows no sequence similarities to any other transferase. The catalytic domain (lacking the cytoplasmic tail and transmembrane domain) was fused to a secreted form of the IgG-binding domain of *Staph. aureus* protein A and placed into a mammalian expression vector. Transient transfection of this construct into COS-1 cells resulted in the secretion of soluble α2-Fuc-transferase activity which was specifically retained on an IgG-Sepharose affinity column. Southern blot analysis of various hybrid cell lines assigned the α2-Fuc-transferase gene to chromosome 19. The definitive assignment of the α2-Fuc-transferase gene to the H locus remains to be done.

3.3.2 Human blood group Lewis GDP-Fuc:Galβ1-4/3GlcNAc (Fuc to GlcNAc) α3/4-Fuc-transferase III (Ec 2.4.1.65)

There are at least four distinct human α3-Fuc-transferases (13–16). These enzymes are named α3-Fuc-transferases III to VI (150, 151); α3-fucosyltransferases I and II are, respectively, the enzymes present in lectin-resistant Chinese hamster ovary somatic cell mutants LEC11 and LEC12 (152–154). These enzymes have been differentiated from one another on the basis of their substrate specificities towards Galβ1-3GlcNAc, Galβ1-4GlcNAc and NeuAcα2-3Galβ1-4GlcNAc-R, inhibition by N-ethylmaleimide, pH optima, kinetic properties, and tissue distribution. The Le-dependent GDP-Fuc:Galβ1-4/3GlcNAc (Fuc to GlcNAc) α3/4-Fuc-transferase III has the broadest substrate specificity among the human α3-Fuc-transferases since it can act on either the type 1 (Galβ1-3GlcNAc) or type 2 (Galβ1-4GlcNAc) structures, whether or not the Gal residue in these structures is substituted by either a Fucα1-2 residue or a sialylα2-3 residue. The enzyme can therefore synthesize several human blood group epitopes (Le^a, Le^b, Le^x, Le^y, sialyl-Le^a, and sialyl-Le^x, Figure 3). The enzyme has been found in gall-bladder, kidney and milk but is absent from plasma (14). The gene encoding the Lewis α3/4-Fuc-transferase has been mapped to the short arm of human chromosome 19 (155).

Expression screening, using a panning technique and a monoclonal antibody specific for the Le^x structure, was used (156) to obtain a 2 kb cDNA for human blood group Lewis α3/4-Fuc-transferase III (Table 1). The recombinant plasmid was named pCDM7-α(1,3/1,4)FT. COS-1 cells transiently transfected with this plasmid express both the Le^a and Le^x antigens on their surfaces (Table 4) and produce an enzyme with both α3- and α4-Fuc-transferase activities capable of synthesizing the Le^a, Le^b, Le^x, and Le^y structures; the data provide strong confirmation that the gene encodes Lewis α3/4-Fuc-transferase III and that this enzyme can make two different glycosidic linkages in violation of the one linkage-one enzyme rule.

Northern analysis of A431 cells showed a single prominent 2.3 kb transcript and a minor 7.5 kb transcript. The protein encoded by this cDNA shows the domain structure typical of all glycosyltransferases cloned to date (Figure 5; Table 2). Although α3-Fuc-transferase III to VI are highly homologous to one another, these enzymes show no sequence similarities to any other glycosyltransferases. The catalytic domain of α3-Fuc-transferase III (lacking the cytoplasmic tail and transmembrane domain) was fused to a secreted form of the IgG-binding domain of *Staph. aureus* protein A and placed into a mammalian expression vector. Transient transfection of this construct into COS-1 cells resulted in the secretion of soluble α3-Fuc-transferase activity which was specifically retained on an IgG-Sepharose affinity column. *In vitro* transcription-translation experiments, carried out in the absence and presence of microsomes, showed the protein to be inserted into microsomal membrane as a type II integral membrane protein. Southern blot analysis of human–mouse hybrid cell lines assigned the α3/4-Fuc-transferase III gene to chromosome 19. The definitive assignment of the α3/4-Fuc-transferase III gene to the Lewis locus will require a demonstration of tight linkage between a molecular defect in this gene and null alleles at the Lewis locus. Caution in assignment is indicated since a Lewis blood group gene-associated fucosyltransferase has been purified from human milk which has a relatively low activity towards Galβ1-4GlcNAc (157, 158).

3.3.3 Human GDP-Fuc:Galβ1-4GlcNAc (Fuc to GlcNAc) α3-Fuc-transferase IV (myeloid)

Human α3-Fuc-transferases IV, V, and VI differ from the Le-dependent α3/4-Fuc-transferase III in that they cannot form the Fucα1-4GlcNAc linkage. An α3-Fuc-transferase found in myeloid tissues (attributed to the Fuc-transferase IV gene, see below) has the narrowest acceptor specificity in this group since it acts poorly, if at all, on NeuAcα2-3Galβ1-4GlcNAc-R and therefore does not appear to be effective in the synthesis of the sialylα2-3Galβ1-4(Fucα1-3)GlcNAc moiety (sialyl-Le^x). The myeloid-type enzyme has been found in myeloid cells and brain (14); less than 10% of the activity in plasma is due to the myeloid enzyme.

Neutrophils express sialyl-Le^x but the myeloid α3-Fuc-transferase found in these cells is not effective in the *in vitro* synthesis of the sialyl-Le^x structure. There are several possible explanations for this paradox (13). The cells may contain a plasma-type α3-Fuc-transferase capable of making sialyl-L^x, or there may be a low level of synthesis of sialyl-Le^x by the myeloid enzyme, or the sialyl-Le^x may have been made in a precursor cell, or the serum enzyme makes the structure on circulating neutrophils, or there is an α2,3-sialyltransferase in neutrophils which can convert Galβ1-4(Fucα1-3)GlcNAc to sialylα2-3Galβ1-4(Fucα1-3)GlcNAc. The origin of the sialyl-Le^x structure on neutrophils has aroused a great deal of recent interest because sialyl-Le^x is the most likely ligand for an endothelial cell receptor called endothelial-leukocyte adhesion molecule-1 (ELAM-1; E-selectin) (13, 29–35, 159).

Human–mouse hybrid cell lines have been used to correlate the presence of chromosome 11q with the antigens Le^x, Le^y, and sialyl-Le^x, and with myeloid-type α3-Fuc-transferase IV activity in cell extracts (15, 16, 160). This indicates that the gene for the myeloid-type enzyme is on chromosome 11q and that the synthesis of Le^x, Le^y, and sialyl-Le^x is regulated by this enzyme.

The existence of several human α3-Fuc-transferases with enzymatic properties similar to the Lewis-dependent α3/4-Fuc-transferase III prompted Lowe and co-workers (31) to use cross-hybridization (homology screening, Table 1) to isolate the gene for the myeloid-type enzyme. A human genomic DNA library was screened at low stringency with a probe from Lewis-dependent α3/4-Fuc-transferase III cDNA. The resulting positive clones were analysed by Southern blots and placed into three groups. The group showing the most intense hybridization signals was attributed to the gene for Lewis α3/4-Fuc-transferase III. A group of clones showing intermediate hybridization signals was characterized in detail and shown to encode the myeloid α3-Fuc-transferase IV. The gene contains a single open reading frame encoding a protein with 405 amino acids and a type II transmembrane protein domain pattern typical for the glycosyltransferases (Table 2, Figure 5).

The mammalian expression vector pCDNA1, carrying a cytomegalovirus immediate early promoter, was used to prove that the gene encodes myeloid-type α3-Fuc-transferase IV. COS-1 and Chinese hamster ovary (CHO) cells were used as hosts since neither line normally expresses detectable α3- or α3/4-Fuc-transferase activities but they do express the required substrates for these enzymes, i.e. Galβ1-4GlcNAc and sialylα2-3Galβ1-4GlcNAc. COS-1 cells, but not CHO cells, also express Galβ1-3GlcNAc and sialylα2-3Galβ1-3GlcNAc. Recombinant plasmids carrying the α3-Fuc-transferase gene in pCDNA1 were used for transient transfection of COS-1 cells and stable transfection of CHO cells. Transfected cell extracts (but not appropriate controls) expressed *in vitro* α3-Fuc-transferase activity capable of converting Galβ1-4GlcNAc to Le^x but were unable to make Le^a, sialyl-Le^a, or sialyl-Le^x (Table 4). The transfected cells were also tested for the presence of these various antigens on their cell surfaces; only Le^x was observed. The data suggest that the gene probably encodes myeloid-type α3-Fuc-transferase.

This paper (31) also made an interesting contribution to a controversy concerning the so-called VIM-2 antigen, sialylα2-3Galβ1-4GlcNAcβ1-3Galβ1-4(Fucα1-3)GlcNAc-R. CHO cells transfected with the myeloid-type α3-Fuc-transferase IV gene expressed the VIM-2 antigen, but not sialyl-Le^x, on their surfaces; yet they were unable to bind to endothelial-leukocyte adhesion molecule-1 (ELAM-1). The data support the hypothesis that the VIM-2 epitope is not a ligand for ELAM-1.

Goelz *et al.* (161) isolated a human HL-60 cell cDNA, which they termed ELFT

Table 4 Substrate specificities of Fuc-transferases (Fuc-T) III, IV, V and VI

Acceptor	Relative enzyme activity (%)			
	Fuc-T III	Fuc-T IV	Fuc-T V	Fuc-T VI
Galβ1-4GlcNAc (Product: Le^x)	100	100	100	100
Galβ1-4Glc	145	3	11	<0.05
Sialylα2-3Galβ1-4GlcNAc (Product: sialyl-Le^x)	56	<1	115	112
Fucα1-2Galβ1-4Glc	254	6	42	<0.05
Galβ1-3GlcNAc (Product: Le^a)	420	<1	10	<0.05
Sialylα2-3Galβ1-4GlcNAc-β1-3Galβ1-4GlcNAc to VIM-2	+++	++	+++	−
Synthesis of difucosyl-sialyl-Le^x	+	−	+	+
Difucosyl-sialyl-Le^x via VIM-2	?	−	?	−
Difucosyl-sialyl-Le^x via sialyl-Le^x	?	−	?	+

A value of 100% was assigned to the activities generated with N-acetyllactosamine as acceptor. All four enzymes were prepared by expression of recombinant DNA (pcDNA1-Fuc-T) in COS cells or CHO cells containing the polyoma large T antigen. No detectable Fuc-T activity was found in extracts from cells transfected with vector alone (pcDNA1). The data on VIM-2 and difucosyl-sialyl-Le^x are based on flow cytometry of intact cells. The VIM-2 antigen is Sialylα2-3Galβ1-4GlcNAcβ1-3Galβ1-4[Fucα1-3]GlcNAc. Difucosyl-sialyl-Le^x is Sialylα2-3Galβ1-4[Fucα1-3]GlcNAcβ1-3Galβ-4[Fucα1-3]GlcNAc. The data are from Weston et al. (150, 151) and Lowe et al. (31).

(ELAM-1 ligand fucosyl transferase), by the technique of expression screening (using the panning technique) of transfected mammalian cells (Table 1). Their procedure was based on an immunological screen for cells which have gained the ability to express surface molecules recognized by ELAM-1. The coding region of ELFT cDNA is colinear with the genomic DNA isolated by Lowe et al. (31) indicating that the two genes encode the same protein and that the entire coding region of the protein is within a single exon.

Goelz et al. (161) detected α3-Fuc-transferase activity in their transfected cells but did not test the cells for the presence of an α3-Fuc-transferase capable of acting on sialylα2-3Galβ1-4GlcNAc to produce sialyl-Le^x. The data of Lowe et al. (31) indicate that ELFT/myeloid α3-Fuc-transferase IV cannot carry out this reaction efficiently. Goelz et al. (161) showed that their ELFT-transfected Chinese hamster ovary (CHO) cells can bind to ELAM-1. Since sialyl-Le^x is the ELAM-1 ligand, it is difficult to understand how ELFT can encode the enzyme which is responsible for ELAM-1 binding. The apparent discrepancy has not been resolved. It has been suggested that the binding of ELFT-transfected CHO cells to ELAM-1 may be due to activation of an endogenous CHO α3-Fuc-transferase and the cloning of ELFT cDNA with an ELAM-1 binding strategy may be due to binding of ELAM-1 to the Le^x epitope at a level high enough to allow cloning (34). It is also possible that α3-Fuc-transferase IV acts before α2,3-sialyltransferase and that CHO cells contain an α2,3-sialyltransferase which can act on fucosylated substrate.

Northern analysis, with an α3-Fuc-transferase IV probe, of HL-60 cells and other cells which express the ELAM-1 ligand detected three transcripts, major bands at 2.3 and 6 kb and a minor band at 3 kb (161); cells which do not express the ELAM-1 ligand did not show a signal on Northern analysis.

The human myeloid-type α3-Fuc-transferase IV gene was also isolated by stable transfection of genomic DNA from HL-60 cells into Chinese hamster ovary (CHO) cells and screening for the appearance of Le^x antigen on the cell surface (34, 162). The human Alu repeat sequence was used to rescue the gene from transfectants. The genomic sequence reported by Kumar et al. (34) is identical to the genomic sequence of Lowe et al. (31) and the ELFT cDNA reported by Goelz et al. (161). Only a single mRNA transcript of 2.3 kb was produced on transfection into CHO cells of 6.3 kb of genomic DNA; transcription in this system is driven by the endogenous promoter. The α3-Fuc-transferase expressed in CHO cells by Kumar et al. (34) showed very high levels of enzyme activity towards Galβ1-4GlcNAc but low activity with sialylα2-3Galβ1-4GlcNAc, as expected for the myeloid-type enzyme. Further, the transfected cells did not express sialyl-Le^x on their surfaces and were unable to bind to ELAM-1. This data supports the conclusion that the myeloid α3-Fuc-transferase IV is not responsible for synthesis of the ELAM-1 ligand.

3.3.4 Human GDP-Fuc:Galβ1-4GlcNAc (Fuc to GlcNAc) α3-Fuc-transferase V

Macher et al. (13) have described two human α3-Fuc-transferases (termed plasma-type and lung carcinoma-type) which can act not only on Galβ1-4GlcNAc-R and

Fucα1-2Galβ1-4GlcNAc-R to form the Le^X and Le^y epitopes respectively, but also on NeuAcα2-3Galβ1-4GlcNAc-R to form the sialylα2-3Galβ1-4(Fucα1-3)GlcNAc moiety (sialyl-Le^X) (Figure 3). The plasma-type α3-Fuc-transferase (EC 2.4.1.152) has been found in plasma, milk and liver (14). The genes for the Lewis α3/4-Fuc-transferase III and the plasma-type α3-Fuc-transferase are closely linked indicating that they are both on human chromosome 19 (14, 16).

A probe prepared from the cDNA encoding human blood group Lewis GDP-Fuc:Galβ1-4/3GlcNAc (Fuc to GlcNAc) α3/4-Fuc-transferase III was used to screen a human genomic DNA library under conditions of low stringency (150). Several weakly hybridizing clones were identified as α3-Fuc-transferase IV (see above). A 1.96 kb fragment from a more strongly hybridizing clone was subcloned into the mammalian expression vector pcDNA1 to yield the construct pcDNA1-FucT V. This genomic DNA contained a single open reading frame encoding a protein with the type II transmembrane domain structure typical of glycosyltransferases (Figure 5; Table 2).

COS-1 cells transiently transfected with pcDNA1-FucT V (and with pcDNA1-FucT VI, see below) express Le^X and sialyl-Le^x, but neither Le^a nor sialyl-Le^a, on their cell surfaces; in contrast, transfection with pcDNA1-FucT III results in the appearance of all four determinants while cells transfected with pcDNA1-FucT IV express only the Le^X determinant (150, 151). Table 4 summarizes the substrate specificity differences between Fuc-transferases III to VI obtained by expression in transiently transfected COS-1 cells, i.e very broad for Fuc-transferase III, very restricted for Fuc-transferase IV and intermediate for Fuc-transferases V and VI. Comparison of these substrate specificities with those of the plasma-type (163) and lung carcinoma-type (13) enzymes indicates that neither corresponds to Fuc-transferase V. The tissue specificity of Fuc-transferase V remains to be determined.

3.3.5 Human GDP-Fuc:Galβ1-4GlcNAc (Fuc to GlcNAc) α3-Fuc-transferase VI

The genes for Fuc-transferases IV and V were isolated from a human genomic DNA phage library screened at low stringency with the cDNA encoding Fuc-transferase III (see above). Eighteen phage from this screening were characterized (151). Nine which exhibited relatively weak hybridization to the Fuc-transferase III probe contained the Fuc-transferase IV gene. The nine remaining phage all exhibited strong hybridization signals; seven contained the Fuc-transferase V gene, one contained the gene for Fuc-transferase III and one encoded a novel protein, i.e. GDP-Fuc:Galβ1-4GlcNAc (Fuc to GlcNAc) α3-Fuc-transferase VI. Fuc-transferase VI is a 358 amino acid protein with the type II transmembrane domain structure typical of glycosyltransferases (Figure 5, Table 2).

Extracts of COS-1 cells transiently transfected with pcDNA1-Fuc-T VI were assayed for enzyme activity *in vitro* and showed the presence of an α3-Fuc-transferase which could make Le^X and sialyl-Le^X but not Le^a (Table 4). Although Fuc-transferases V and VI have similar substrate specificities, Fuc-transferase VI is more restricted (Table 4). Analysis of transfected COS-1 cells with flow cytometry

(151) showed that the *in vivo* substrate specificities of Fuc-transferases III to VI corresponded to their *in vitro* activities (Table 4), i.e. Fuc-transferase III expressed Le^X, sialyl-Le^X, Le^a, and sialyl-Le^a, Fuc-transferase IV expressed only Le^X and Fuc-transferases V and VI expressed Le^X and sialyl-Le^X but neither Le^a nor sialyl-Le^a. Fuc-transferases III, IV, and V, but not VI, can make the VIM-2 antigen (Table 4) whereas Fuc-transferases III, V, and VI, but not IV, can make both sialyl-Le^X and difucosylated-sialyl-Le^X structures.

The substrate specificity of Fuc-transferase VI most closely corresponds to the lung carcinoma-type enzyme (163) in that both enzymes can convert sialylα2-3Galβ1-4GlcNAcβ1-3Galβ1-4GlcNAc to sialylα2-3Galβ1-4[Fucα1-3]GlcNAcβ1-3Galβ1-4GlcNAc to sialylα2-3Galβ1-4[Fucα1-3]GlcNAcβ1-3Galβ1-4[Fucα1-3]GlcNAc (difucosyl-sialyl-Le^X, Table 4) but cannot make the VIM-2 antigen. No gene has as yet been allocated to the plasma-type enzyme.

Fuc-transferase VI cDNA has been isolated by a PCR approach from human A431 and HL60 cells (164). The presence of message in these cells suggests that Fuc-transferase VI may be responsible for the synthesis of sialyl-Le^X on leukocytes; this structure serves as the ligand for E-selectin on endothelial cell walls and is involved in inflammation-activated homing of leukocytes to endothelial cells.

3.3.6 Comparative analysis of human Fuc-transferases III, IV, V, and VI

The substrate specificities of Fuc-transferases III to VI were discussed above and are summarized in Table 4. The entire coding regions of Fuc-transferases III, IV, V, and VI are all located in single exons (Figure 12) (150, 151). The presence of a consensus sequence for a splice acceptor site upstream of the Met initiation codon of Fuc-transferase VI (151) suggests that there may be one or more extra exons in the 5'-untranslated region, as is the case with some glycosyltransferases (Figure 12). Fuc-transferase IV is located on chromosome 11 whereas the other three enzymes are all on chromosome 19. The gene for Fuc-transferase VI is 13 kb downstream of the gene for Fuc-transferase III (165). The data suggest that a relatively distant gene duplication event generated two distinct genes on two different chromosomes and that the sequences have diverged in the interim.

Table 2 compares the domain structures of these four enzymes. Fuc-transferase IV is appreciably longer (31 to 47 amino acid residues) than Fuc-transferases III, V, and VI. The catalytic domain of Fuc-transferase IV shows a high degree of amino acid sequence similarity (58% identity) to the analogous region of Fuc-transferase III. The similarity is appreciably lower in the cytoplasmic, transmembrane and 'stem' regions. Interspecies comparisons for UDP-Gal:GlcNAc-R β4-Gal-transferase (Figure 8) and UDP-GlcNAc:Manα1-3R (GlcNAc to Manα1-3) β2-GlcNAc-transferase I (Figure 13) show a similar pattern suggesting that catalytic domains tend to be more highly conserved.

Fuc-transferases III, V, and VI share 85–90% amino acid identity over their entire lengths. Fuc-transferase V is 13 and 16 amino acids longer than Fuc-transferases III and VI respectively (Table 2). Most of the differences between the three enzymes are in the stem region and at the amino-terminal end of the catalytic domain. The

```
Human   MLKKQSAGLVLWGAILFVAWNALLLLFFWTRPAPGRPPSVSALDGDPASLTREVIRLAQD
Mouse   -----T---------I--G----------------L--D---GD----------H--E-
Rabbit  -------------------------------V-S-L--DN---D---------------

                                                                 *
Human   AEVELERQRGLLQQIGD..ALSSQRGRVPTAAPPAQPRVPVTPAPAVIPILVIACDRSTV
Mouse   --A------------KEhy--WR--W----V----W--------S-VQ-----------
Rabbit  ---------------REhh--W---WK---------H-----P----------------

             *                *
Human   RRCLDKLLHYRPSAELFPIIVSQDCGHEETAQAIASYGSAVTHIRQPDLSSIAVPPDHRK
Mouse   --------------R-----------------V-----T-----------N---Q-----
Rabbit  --------------------------------V-----------------N---Q-----

                                                          *
Human   FQGYYKIARHYRWALGQVFRQFRFPAAVVVEDDLEVAPDFFEYFRATYPLLKADPSLWCV
Mouse   ----------------I-NK-K--------------------Q-----RT---------
Rabbit  ----------------I-HN-NY-------------------Q----------------

Human   SAWNDNGKEQMVDASRPELLYRTDFFPGLGWLLLAELWAELEPKWPKAFWDDWMRRPEQR
Mouse   -------------S-K---------------------D---------------------
Rabbit  -------------S-K-------------------------------------------

             *
Human   QGRACIRPEISRTMTFGRKGVSHGQFFDQHLKFIKLNQQFVHFTQLDLSYLQREAYDRDF
Mouse   K---------------------------------------P----------Q-------
Rabbit  K----V----------------------------------P----------Q-------

Human   LARVYGAPQLQVEKVRTNDRKELGEVRVQYTGRDSFKAFAKALGVMDDLKSGVPRAGYRG
Mouse   --Q--------------Q----------S-------------------------------
Rabbit  ------------------------------------------------------------

Human   IVTFQFRGRRVHLAPPPTWEGYDPSWN
Mouse   ----------------Q--T-------
Rabbit  ----L-----------Q--D------T
```

Fig. 13 Comparison of the amino acid sequences of human, mouse and rabbit UDP-GlcNAc:Manα1-3R (GlcNAc to Manα1-3) β2-GlcNAc-transferase I, using the l-letter code. The transmembrane segment is under- and overlined. The 5 Cys residues are indicated by an asterisk. See Table 1 for references

substrate specificity differences between these enzymes (Table 4) must therefore depend primarily on the amino-terminal portion of the catalytic domain.

Fuc-transferase V contains a 33 bp segment which is absent in Fuc-transferases III and VI. Oligonucleotides flanking this region were used as primers in the polymerase chain reaction (PCR) to analyse genomic DNA from 50 individuals, including 5 Lewis-negative subjects (150). All Lewis-positive samples showed amplified DNA bands characteristic of both Fuc-transferases III and V. The data indicate that the gene for Fuc-transferase V is a distinct gene and not an allele on the Lewis locus.

PCR analysis was carried out on a series of somatic cell hybrid DNAs using primers which give diagnostic DNA fragments. Southern analysis showed that bands characteristic of Fuc-transferases III, V, and VI all localized to human chromosome 19. PCR analysis of these hybrid DNAs with primers specific for Fuc-transferase IV showed this gene to be localized to human chromosome 11.

Human DNA was subjected to Southern analysis with a probe from the catalytic domain of Fuc-transferase III that was 95–6% identical to the homologous region of Fuc-transferases V and VI and 62% identical to the homologous region of Fuc-transferase IV (151). Even at low stringency, the only bands detected could be assigned to one of the four known enzymes as determined by the use of gene-specific probes. When the experiment was repeated with a probe from the catalytic domain of Fuc-transferase IV, most bands could again be attributed to one of the four known enzymes but there were in addition some unidentified bands. The data suggest that there may be as yet uncloned fucosyltransferases more similar to Fuc-transferase IV than to Fuc-transferases III, V, and VI.

3.4 N-Acetylgalactosaminyltransferases

3.4.1 Human blood group A UDP-GalNAc:Fucα1-2Gal-R (GalNAc to Gal) α3-GalNAc-transferase (EC 2.4.1.40)

Serum from blood group B individuals contain two types of antibody, anti-A, and anti-A_1. Erythrocytes which react with both antibodies are classified as A_1 whereas those which react only with anti-A are classified as A_2. The incidence of the A_2 phenotype varies among different races (for example, 20% of the A population in Caucasians but less than 1% in Japanese) (166). The A^1 and A^2 α3-GalNAc-transferases differ from one another in cation requirements, pH optima and kinetic constants (167, 168) and are therefore different enzymes. The A^2 α3-GalNAc-transferase has an appreciably weaker activity and a more restricted acceptor substrate requirement.

Blood group A_1 UDP-GalNAc:Fucα1-2Gal-R (GalNAc to Gal) α3-GalNAc-transferase was purified to homogeneity from human lung tissue to a final specific activity of 5.7 μmol/min/mg and partial amino acid sequences were obtained (169). Polymerase chain reaction amplification of cDNA from a human stomach endothelial cell line (MKN45), using degenerate oligonucleotides based on the amino acid sequences as primers, provided a 98 bp fragment (170). This fragment was used to screen a human cDNA library (Table 1) and positive clones were used to re-screen the library. Although many cloning problems were encountered, a 1.6 kb clone was eventually isolated. An open reading frame was found with the type II integral membrane protein domain structure typical of glycosyltransferases (Table 2; Figure 5). A repetitive sequence was found downstream of the coding region which may have caused some of the subcloning problems. The stem region has 9 Pro residues in a 60 amino acid segment. Southern analysis of human genomic DNA indicated only a single gene copy; further, bands of the same size were found in A, B, and O individuals. Northern analysis revealed similar multiple bands of mRNA from A, B, and O phenotypes. The sequence of the A^1 α3-GalNAc-transferase shows sequence similarity to the UDP-Gal:Galβ1-4GlcNAc-R (Gal to Gal) α3-Gal-transferase and differs at only four amino acid sites from the allelic blood group B UDP-Gal:Fucα1-2Gal-R (Gal to Gal) α3-Gal-transferase (see above).

The molecular basis of the ABO genotypes was determined by Yamamoto et al. (104); this work was described above in the section dealing with blood group B α3-Gal-transferase.

The polymerase chain reaction was carried out using genomic DNA from several A^2O individuals as template and primers based on the A^1 transferase sequence (166). A partial sequence for the C-terminal end of the A^2 transferase was obtained. All samples tested showed a single base substitution at amino acid position 156 (Pro to Leu conversion) and a single base deletion three bases upstream of the stop codon which resulted in an additional 21 amino acids at the C-terminal end of the protein. The single base deletion was introduced into the complete A^1 transferase cDNA and transient expression of this cDNA in HeLa cells resulted in a 30–50-fold reduction in A transferase activity. These data suggest that the additional 21 amino acids are responsible for the qualitative differences between the A^1 and A^2 α3-GalNAc-transferases (see above).

3.4.2 Human UDP-GalNAc:G_{M3}/G_{D3}β4-GalNAc-transferase (EC 2.4.1.92)

Transient expression cloning (171) has been used to clone the gene for human UDP-GalNAc:G_{M3}/G_{D3} β4-GalNAc-transferase (EC 2.4.1.92) (Table 1). The host cell was a B16 melanoma cell line which expressed G_{M3} but not G_{M2}. This cell line was stably transfected with the gene for polyoma T antigen to permit efficient replication of the mammalian expression vector pCDM8. A human cDNA library in pCDM8 was transiently transfected into this host and monoclonal antibody to G_{M2} was used in repetitive screening procedures designed to select cells with cell surface G_{M2}. A 2.5 kb cDNA was isolated encoding a type II integral membrane protein with domain structure typical of glycosyltransferases (Table 2; Figure 5). Cells stably transfected with this cDNA acquired the ability to synthesize both G_{M2} and G_{D2}.

3.5 N-Acetylglucosaminyltransferases

3.5.1 UDP-GlcNAc:Manα1-3R(GlcNAc to Manα1-3) β2-GlcNAc-transferase I (EC 2.4.1.101)

As discussed above, β2-GlcNAc-transferase I plays a key role in controlling the conversion of high mannose N-glycans to hybrid and complex structures. Chinese hamster ovary mutant cell lines lacking β2-GlcNAc-transferase I are unable to make hybrid and complex N-glycans but are nevertheless viable and have normal growth characteristics (172). The availability of the gene for this enzyme should permit perturbation of this key control step in intact organisms such as the mouse thereby providing insight into the biological roles of the hybrid and complex N-glycans in processes such as development and differentiation.

Rabbit β2-GlcNAc-transferase I

Rabbit liver UDP-GlcNAc:Manα1-3R (GlcNAc to Manα1-3) β2-GlcNAc-transferase I was purified to homogeneity (specific activity of about 20 μmoles/min/mg) (47) and partial amino acid sequence data was obtained. Polymerase chain reaction amplification of rabbit liver cDNA using degenerate oligonucleotide primers based on the amino acid sequence yielded two fragments of 0.45 and 0.50 kb (173–175). These fragments were further amplified in a plasmid vector and used to probe a rabbit liver cDNA library (Table 1). A 2.5 kb cDNA was eventually obtained with an open reading frame encoding a protein with the type II integral membrane protein domain structure typical of all the glycosyltransferases cloned to date (Figures 5 and 13; Table 2). The stem region is enriched in Pro residues. The GlcNAc-transferase I purified from rabbit liver is an enzymatically active truncated form of the enzyme which was cleaved in the stem region. The protein is not N-glycosylated because there are no Asn-X-Ser(Thr) sequons. There is no sequence homology to any other known glycosyltransferase. Northern analysis of rabbit liver showed a major band at about 3 kb.

Human β2-GlcNAc-transferase I

A human genomic DNA library from chronic myeloid leukaemia cells (Table 1) was screened with a rabbit liver β2-GlcNAc-transferase I probe and a 4.5 kb hybridizing DNA fragment was isolated and sequenced (175–177). There is 92% identity between the amino acid sequences of the human and rabbit enzymes (Table 2; Figure 13). Northern analysis of human liver (177) and HL-60 cells (178) showed a 3 kb message. Southern analysis indicated that there is only a single copy of the gene in the haploid human genome (176).

The gene encoding human β2-GlcNAc-transferase I was also isolated by an expression screening strategy (Table 1) involving sequential stable transfection of human genomic DNA into Lec1 Chinese hamster ovary (CHO) mutant cells which lack β2-GlcNAc-transferase I activity (178). A genomic DNA library was constructed from a tertiary CHO transfectant and screened with a probe for human repetitive Alu sequences that are not present in CHO DNA. DNA from positive clones was tested for the presence of the β2-GlcNAc-transferase I gene by stable transfection into Lec1 CHO cells followed by assay for cell surface lectin binding characteristic of GlcNAc-transferase I activity. This approach permitted rescue of the β2-GlcNAc-transferase I gene. A human A431 cDNA library in pCDM7 was then screened with this genomic DNA and a 2.6 kb cDNA was isolated and sequenced.

Northern analysis of RNA from both wild type and Lec1 Chinese hamster ovary cells showed a message at about 3 kb suggesting that the mutation in the GlcNAc-transferase I gene of Lec1 cells is a point mutation (178).

Mouse β2-GlcNAc-transferase I

A murine genomic DNA library was screened (Table 1) with a 1.7 kb probe from human genomic DNA encoding β2-GlcNAc-transferase I (179). A 4.1 kb fragment containing the entire coding region of β2-GlcNAc-transferase I was isolated. The coding region of the mouse gene (designated *Mgat-1*) is on a single exon and is 92% identical with the rabbit sequence at the amino acid level (Table 2; Figure 13). Northern analysis of RNA from various mouse tissues and embryonic stem cells (179, 180) revealed a major transcript at about 3 kb in most tissues; there was, however, significant variation in expression levels with the highest being in kidney. Some tissues showed two transcripts of about 3.0 and 3.4 kb whereas brain showed a single transcript at 3.4 kb. The functional significance of this tissue-specific variation in transcription remains to be determined.

A mouse F9 teratocarcinoma cell cDNA library was screened with a human β2-GlcNAc-transferase I DNA probe and a 2.7 kb cDNA was isolated (180) containing a 363 bp 5'-untranslated sequence and the complete coding region and 3'-untranslated sequences. Southern analysis of mouse genomic DNA indicated that β2-GlcNAc-transferase I is present as a single copy in the haploid mouse genome, as was shown (above) for the human gene.

Comparison of the rabbit, human and mouse sequences (Figure 13) indicate that most of the differences between the species occur in the stem region. The transmembrane and catalytic domains are highly conserved. A similar observation was made for β4-Gal-transferase (Figure 8). It is interesting that the catalytic domains of both β2-GlcNAc-transferase I and β4-Gal-transferase have 5 Cys residues which are conserved between species.

Genomic structure

Comparison (Figure 14) of the human genomic DNA sequence (175–177) with the human cDNA sequence (178), and of the mouse genomic DNA sequence (179) with the mouse cDNA sequence (180) shows that 126 bp and 121 bp, respectively, of the 5'-untranslated region and all of the coding and 3'-untranslated regions are on a single 2.46 kb exon. The remaining 5'-untranslated sequence of the human gene is on a different exon (or exons) at least 5 kb upstream (177); the presence of at least 2 upstream exons has recently been shown for the mouse gene (180). Analysis of human and mouse sequences 5'-upstream of the ATG start codon (Figure 14) shows that there is 76% identity in the 5'-untranslated regions present in the downstream exon but only 29% identity in the 34 bp available for comparison from the upstream exon. Also of interest is the 50% identity in the 84 bp of intron shown in Figure 14.

The human gene appears to have at least two promoters (see below). The genes for α2,6-sialyltransferase, β4-Gal-transferase and α3-Gal-transferase (Figures 7, 10, 11, and 12) also have multiple promoters and exons several kb upstream of the coding region. Unlike the β2-GlcNAc-transferase I gene, however, the coding regions cover several exons and the genes span 35 kb or more of genomic DNA.

The β2-GlcNAc-transferase I gene has been localized to human chromosome 5 by Southern analysis of human–hamster somatic cell hybrids (177). Fluorescent *in situ* hybridization (180, 181) has localized the gene to human chromosome 5q35. Southern analysis (179) of the DNA from the progeny of various mouse genetic

```
Human gen.    -210  tttcctgccc ccatttcctc tacctgtggg gcatggagcc acgagcttt
Human cDNA                                                              g
Mouse gen.          -ca----tg- ---gcct--- cc-actgct- tgtg-a-tta tac-aa---g
Mouse cDNA          c-aatccaga -atcaaggg- ggtggagaa- a--aaagt-g gttgt-tg--

Human gen.    -160  gtgtgacggt ttgctttctc tctcctgtct ttaggtgcat ggctgcctcc
Human cDNA          -ccaagttcg gg--cagga- gtcgggagga cct------- ----------
Mouse gen.          ag--a--aa- -g-t------ ***t--c--- -----a--- ----t-t---
Mouse cDNA          taca--a-ag ggaac-gagg cagtggt--- -c----a--- ----t-t---

Human gen.    -110  taatcccata gtccagagga ggcatccota ggactgcggg caagggagcc
Human cDNA          ---------- ---------- ---------- ---------- ----------
Mouse gen.          --c--t--gc c-**---a-- ata-a----t a-**--t--- g-cct-----
Mouse cDNA          --c--t--gc c-**---a-- ata-a----t a-**--t--- g-cct-----

Human gen.    -60   gggcaagccc agggcagcct tgaaccgtcc cctggcctgc cctccccggt
Human cDNA          ---------- ---------- ---------- ---------- ----------
Mouse gen.          a--------a -a-------- ---g--c-*- ---cc----- -------t-c
Mouse cDNA          a--------a -a-------- ---g--c-*- ---cc----- -------t-c

Human gen.    -10   ggggggccagg ATGCTGAAGA AGCAGTCTG
Human cDNA          ---------- ---------- ---------
Mouse gen.          ---------- ---------- -----A---
Mouse cDNA          ---------- ---------- -----A---
```

Fig. 14 Comparison of the genomic and cDNA sequences of human and mouse UDP-GlcNAc:Manα1-3R (GlcNAc to Manα1-3) β2-GlcNAc-transferase I. The upper case bases represent the start of the coding region. The asterisks represent gaps introduced to optimize sequence alignment. The dashes represent base positions which are identical to the human genomic DNA sequence. A single 2.46 kb exon contains 121–126 bp of the 5'-untranslated region and all of the coding and 3'-untranslated regions. Sequences typical of an intron-exon junction at the 5'-end of this exon are underlined. The remaining 5'-untranslated cDNA sequences are on one or more upstream exons. Data from references 175–180

crosses indicated linkage of the mouse β2-GlcNAc-transferase I gene (*Mgat-1*) with markers on mouse chromosome 11. No recombinants were detected between *Mgat-1* and the gene *Il-3* encoding interleukin-3 indicating that these two genes are closely linked.

Expression of β2-GlcNAc-transferase I
In vitro transcription-translation of the rabbit liver cDNA resulted in the synthesis of a 52 kDa protein with β2-GlcNAc-transferase I activity (174). Transient transfection of Lec1 Chinese hamster ovary cells, which lack β2-GlcNAc-transferase I, with a human genomic DNA fragment encoding β2-GlcNAc-transferase I gave expression of enzyme activity (177). Expression was driven by an endogenous promoter in the intron between the upstream exon and the exon containing the coding region indicating the presence of at least one other upstream promoter.

The bacterial expression vector pGEX was used to generate an in-frame construct encoding a fusion protein comprising the lumenal domain of rabbit β2-GlcNAc-transferase I fused to the carboxyl terminus of glutathione-S-transferase (182, 183). Triton X-100 extraction of *E. coli* transformed with this construct solubilized only 5% of the fusion protein. Enzymatically active soluble β2-GlcNAc-transferase I was obtained by purification on a glutathione-agarose affinity column. The bacterial expression vector pET.3b was used to express a fusion protein containing the lumenal domain of mouse β2-GlcNAc-transferase I (180); NP-40 extraction of *E. coli* transformed with this construct yielded soluble active enzyme although the bulk of the fusion protein appeared to be present as an inactive intracellular aggregate.

High-level expression of enzymatically active β2-GlcNAc-transferase I has been achieved using an insect cell expression system (184). A baculovirus shuttle vector was constructed in which the short N-terminal cytoplasmic tail and transmembrane segment of the rabbit β2-GlcNAc-transferase I cDNA was replaced with an in-frame cleavable signal sequence. The modified β2-GlcNAc-transferase I gene was inserted into the genome of *Autographa californica* nuclear polyhedrosis baculovirus (AcNPV) and *Spodoptera frugiperda* (Sf9) insect cells were infected with the recombinant baculovirus. High level expression of enzymatically active and soluble enzyme was achieved under the control of the polyhedrin promoter. About 90% of the enzyme was secreted into the growth medium at a concentration of about 1–5 mg/l, indicating that the cytoplasmic tail and transmembrane domain are probably required for Golgi retention (see below). Recombinant enzyme has been purified to near homogeneity by affinity chromatography. The recombinant enzyme was shown to incorporate GlcNAc in β1-2 linkage to the Manα1-3Manβ1-4GlcNAc arm of the *N*-glycan core (Figure 1) (185).

Lec1 Chinese hamster ovary cells which lack a functional GlcNAc-transferase I are viable and grow with a normal doubling time indicating that complex *N*-glycans are not essential for the survival and proliferation of mammalian cells in culture. To determine whether complex *N*-glycans play a role in development and differentiation in the living animal, the mouse GlcNAc-transferase I gene locus has been mutated via homologous recombination in embryonic stem (ES) cells and the mutated allele has been introduced into the mouse germ line (M. Metzler, A. Gertz, M. Sarkar, H. Schachter, J. W. Schrader, and J. D. Marth, *EMBO J.*, in press, 1994). Live-born heterozygotes were obtained and mated but no living homozygous mutant mice were obtained. Homozygous mutant embryos die prenatally at about nine days of gestation. These experiments suggest that complex *N*-glycans do indeed play a crucial role in embryogenesis.

Targeting to the Golgi apparatus
A polyclonal antibody was generated by injecting mice with purified glutathione-S-transferase/rabbit β2-GlcNAc-transferase I fusion protein (182). Murine L cells stably transfected with rabbit cDNA for β2-GlcNAc-transferase I showed localization of the transferase in Golgi cisternae by immunofluorescence; no cell surface expression was detected. Immunoelectron microscopy localized the transferase in medial-Golgi cisternae. A hybrid construct encoding a protein with the 31 amino-terminal amino acids of β2-GlcNAc-transferase I fused to ovalbumin was expressed

transiently in COS-7 cells and stably in murine L cells. Ovalbumin is a secretory protein; however, the fusion protein was arrested in the medial-Golgi cisternae. These experiments show that the cytoplasmic tail and transmembrane segment of β2-GlcNAc-transferase I cause retention in medial-Golgi cisternae. It will be of great interest to determine the precise retention signal and how it differs from the transmembrane signals that target α2,6-sialyltransferase and β4-Gal-transferase to *trans*-Golgi cisternae.

Expression of a hybrid protein in which the transmembrane domain of a type II membrane surface protein was replaced with the transmembrane domain of human β2-GlcNAc-transferase I showed that the transferase transmembrane domain was by itself sufficient to retain a polypeptide in the Golgi (186). Electron microscopy was not performed in this study and the exact cisternal location of the fusion protein could not be determined.

HeLa cells were stably transfected with a DNA expression vector encoding the complete human GlcNAc-transferase I protein tagged at its COOH-terminal end with an 11 amino acid *myc* epitope (187). This cell line expressed GlcNAc-transferase I that could be detected with monoclonal antibody to the *myc* epitope. Double labelling of this cell line was achieved by using the anti-*myc* antibody and an antibody to UDP-Gal:GlcNAc β4-Gal-transferase. Immunogold electron microscopy and confocal microscopy showed that GlcNAc-transferase I and Gal-transferase were located predominantly in the medial- and *trans*-Golgi cisternae respectively but that there was appreciable labelling of GlcNAc-transferase I in the *trans*-Golgi cisterna.

3.5.2 UDP-GlcNAc:Manα1-6R (GlcNAc to Manα1-6) β2-GlcNAc-transferase II (EC 2.4.1.143)

Rat liver β2-GlcNAc-transferase II

UDP-GlcNAc:Manα1-6R (GlcNAc to Manα1-6) β2-GlcNAc-transferase II (EC 2.4.1.143) was purified from rat liver (48) and partial amino acid sequences were used to design synthetic oligonucleotide probes. Screening of rat liver cDNA libraries yielded a cDNA clone encoding a 430 amino acid protein with the type II transmembrane protein domain structure typical of glycosyltransferases (Figure 5) (188). The C-terminal 389 amino acids were linked in frame to the cleavable signal sequence of the Il-2 receptor and expressed under control of the Rous sarcoma virus promoter in COS-7 cells. A similar construct carrying GlcNAc-transferase II cDNA out of frame was used as a negative control. A 77-fold enhancement of enzyme activity in the culture medium over the negative control was detected at 72 h after transfection in transient expression experiments. These data verify the identity of the cloned cDNA sequences for rat GlcNAc-transferase II and demonstrate that the COOH-terminal region of the polypeptide chain includes the catalytic site.

A rat GlcNAc-transferase II cDNA probe was used to screen a rat genomic DNA library and several clones were isolated (189). The nucleotide sequence of a genomic DNA region spanning 3.3 kb was determined revealing a 1290-nucleotide open reading frame that codes for the entire GlcNAc-transferase II polypeptide chain and is not interrupted by introns. The single-exon GlcNAc-transferase II gene is flanked by a GC-rich 5'-untranslated region. The 3'-untranslated region contains multiple polyadenylation signals and ATTTA motifs.

Human β2-GlcNAc-transferase II

A 1.2 kb probe from rat liver cDNA encoding GlcNAc-transferase II (above) was used to screen a human genomic DNA library (190). Two overlapping subclones containing 5.5 kb of genomic DNA were isolated and analysed. One of the clones contains a 1341 bp open reading frame encoding a 447 amino acid protein with the type II transmembrane protein domain structure typical of all previously cloned glycosyltransferases (Figure 5, Table 2). Northern analyses of several human lymphocyte lines showed a message size at 2.6 kb. There is no sequence homology to any previously cloned glycosyltransferase including human β2-GlcNAc-transferase I. There is a 90% identity between the amino acid sequences of the catalytic domains of human and rat GlcNAc-transferase II. The entire coding regions of the human and rat enzymes are on single exons. The human GlcNAc-transferase II gene is on chromosome 14q21 (181).

3.5.3 UDP-GlcNAc:R_1-Manα1-6[GlcNAcβ1-2Manα1-3]Manβ1-4R_2 (GlcNAc to Manβ1-4) β4-GlcNAc-transferase III (EC 2.4.1.144)

GlcNAc-transferase III was purified from rat kidney (191) by a 9-step procedure giving a 1.5% yield. The most effective purification steps were affinity chromatography on agarose columns with immobilized UDP or oligosaccharide substrate. The final purification was 153 000-fold with a specific activity of 5 μmol/min/mg. Tryptic peptides were purified by HPLC and sequenced. Polymerase chain reaction amplification of rat kidney cDNA using degenerate oligonucleotide primers based on the amino acid sequences yielded three fragments of 0.16, 0.49, and 0.63 kb (191). These fragments were further amplified in a plasmid vector and one was used to probe a rat kidney cDNA library (Table 1). A 2.7 kb cDNA sequence was obtained (from three overlapping clones) with a 1608 bp open reading frame encoding a 536 amino acid protein with the type II integral membrane protein domain structure typical of all the glycosyltransferases cloned to date (Figure 5; Table 2). Transient expression of this cDNA in both COS-1 and HeLa cells yielded GlcNAc-transferase III activity. There is no apparent sequence similarity to any known glycosyltransferase. The stem region is enriched in Pro residues as is the case for several other glycosyltransferases. The enzyme has three putative N-glycosylation sites and adheres to Con A-Sepharose suggesting that it is a glycoprotein. Northern analysis showed a strong 4.7 kb message in rat kidney and a weak message in rat liver.

3.5.4 Core 2 UDP-GlcNAc:Galβ1-3GalNAc-R (GlcNAc to GalNAc) β6-GlcNAc-transferase (EC 2.4.1.102)

Increased activity of the core 2 UDP-GlcNAc:Galβ1-3GalNAc-R (GlcNAc to GalNAc) β6-GlcNAc-transferase (EC 2.4.1.102) has been implicated in T-cell activation, leukaemias, immunodeficiency, metastasis, and other important biological processes (192–195). It has also been shown that the core 2 GlcNAc-transferase regulates the level of poly-N-acetyllactosamine synthesis on O-glycans (195). This important enzyme has recently been cloned (196) using a transient expression cloning approach (Table 1). A Chinese hamster ovary cell line was first established which stably expressed both the polyoma virus large T antigen and human leukosialin (CHO-Py.leu). This cell line does not express core 2 GlcNAc-transferase. The CHO-Py.leu cell line was transiently transfected with a human HL-60 cDNA library in the plasmid expression vector pcDNAI. HL-60 cells are known to contain relatively high levels of core 2 GlcNAc-transferase. The presence of the large T antigen in the host cells permits efficient replication of pcDNAI. The transfected cells were screened by a panning procedure using monoclonal antibody T305 which recognizes a core 2-containing hexasaccharide O-glycan on human leukosialin (sialylα2-3Galβ1-4GlcNAcβ1-6[sialylα2-3Galβ1-3]GalNAc-R). Cells adherent to goat antimouse IgG were recovered. Plasmid DNA was prepared, amplified in bacteria, and used for further cycles of transfection of CHO-Py.leu cells and screening for the presence of cell surface antigen reactive with T305.

The procedure resulted in the isolation of a 2.1 kb cDNA encoding a 428 amino acid type II integral membrane protein (N_{in}/C_{out} orientation) typical of the glycosyltransferases (Table 2). The enzyme showed homology to the blood group I β6-GlcNAc-transferase (see below). Northern blots of HL-60 cells showed a major transcript at 2.1 kb and minor bands at 3.3 and 5.4 kb. The putative catalytic domain (lacking the cytoplasmic and transmembrane domains) was fused in frame to cDNA encoding a signal peptide sequence and the IgG binding domain of protein A in an expression plasmid. COS-1 cells transfected with this plasmid secreted a fusion protein with catalytic activity. The recombinant enzyme was specific for the substrate Galβ1-3GalNAc and was unable to synthesize O-glycan core 4 [GlcNAcβ1-6(GlcNAcβ1-3)GalNAc-] or the blood group I antigen [GlcNAcβ1-6(GlcNAcβ1-3)Gal-] nor did it possess β6-GlcNAc-transferase V activity. It has been suggested that there is a single β6-GlcNAc-transferase with three enzymatic activities (the synthesis of cores 2 and 4 and the I antigen) (197, 198). However, the substrate specificity of the recombinant myeloid cell enzyme is consistent with the finding that myeloid cell extracts can make core 2 but not core 4 nor the I antigen (194). This suggests there are other β6-GlcNAc-transferases responsible for the synthesis of core 4 and the I antigen (see below).

3.5.5 Blood group I UDP-GlcNAc:GlcNAcβ1-3Galβ1-4GlcNAc-R (GlcNAc to Gal) β6-GlcNAc-transferase

The blood group i/I antigens (linear and branched poly-N-acetyllactosamines) undergo dramatic changes during development. The initiation of poly-N-acetyllactosamine synthesis is controlled by two β6-GlcNAc-transferases (195), core 2 β6-GlcNAc-transferase in O-glycans and β6-GlcNAc-transferase V in N-glycans. The conversion of i to I is controlled by a third β6-GlcNAc-transferase, blood group I UDP-GlcNAc:GlcNAcβ1-3Galβ1-4GlcNAc-R (GlcNAc to Gal) β6-

GlcNAc-transferase. An enzyme that carries out this reaction *in vivo* has been cloned by the transient expression approach (199). CHO-Py.leu cells (which do not express I antigen) were transfected with an expression library containing PA-1 human teratocarcinoma cDNA in the vector pcDNA1 (see above). PA-1 cells express a large amount of I antigen. Transfected cells expressing the I antigen on their cell surfaces were detected by panning or by immunofluorescence using an anti-I antibody. Repeated cycles of transfection and screening yielded a single plasmid capable of directing expression of the I antigen, pcDNA1-IGnT. The cDNA insert contains an open reading frame encoding a 400 amino acid protein with the domain structure typical of glycosyltransferases (Figure 5, Table 2). Northern analysis revealed a 4.4 kb transcript in PA-1 cells but not in CHO-Py.leu or HL-60 cells, consistent with chemical data on the presence or absence of the I antigen.

Cells transfected with pcDNA1-IGnT express I antigen on their cell surfaces as determined by immunofluorescence. Metabolic labelling of these transfected cells with [^{3}H]galactose enabled the demonstration by chemical analysis that the I-antigen had been made *in vivo*. However, the synthesis of I antigen *in vitro* using extracts from the transfected cells has not as yet been demonstrated.

The I GlcNAc-transferase shows three regions of sequence homology to the core 2 β6-GlcNAc-transferase but neither enzyme shows homology to any other known glycosyltransferase. Both enzymes are located on human chromosome 9q21. The coding region of the core 2 β6-GlcNAc-transferase is on a single exon but there are 3 exons in the coding region of I β6-GlcNAc-transferase (Bierhuizen and M. Fukuda, personal communication). Southern analysis with probes from these enzymes under low stringency hybridization conditions indicate the presence of other homologous genes. One of these genes may encode the β6-GlcNAc-transferase that synthesizes core 2, core 4, and I antigen (197, 198).

4. Conclusions

The data summarized in Tables 1 and 2 indicate the dramatic advances that have been made in recent years in the cloning of glycosyltransferase genes. However, this data covers only 18 of the more than a hundred glycosyltransferases required for the synthesis of the known mammalian oligosaccharide structures and attempts at generalization (12, 24, 200) must of necessity be limited until more information becomes available.

4.1 Genomic organization

Joziasse (200) has summarized the genomic organization of the mammalian glycosyltransferase genes that have been studied to date. The genes fall into two classes based on the organization of the coding region (Figure 12): (i) single-exonic and (ii) multi-exonic. The entire coding regions for human α3-Fuc-transferases III to VI, human and mouse β2-GlcNAc-transferase I, human and rat β2-GlcNAc-transferase

II and human core 2 β6-GlcNAc-transferase are within a single exon whereas the coding regions for human (Figure 11) and rat (Figure 10) α2,6-sialyltransferase, human and murine β4-Gal-transferase (Figure 7) and murine α3-Gal-transferase are distributed over five or more exons. Many of these genes have a single long (2–3 kb) 3'-terminal exon which carries the entire relatively long 3'-untranslated region, the translational stop site and all or part of the 3'-terminal coding region. Some of these genes have at least one 5'-terminal exon several kilobases upstream of the translational start site.

4.2 Glycosyltransferase gene families

There are at least four glycosyltransferase gene families (200): (i) the α3-Fuc-transferases, (ii) the α3-Gal-transferases, (iii) the β6-GlcNAc-transferases, and (iv) the sialyltransferases. The α3-Fuc-transferase family presently includes four members, human α3-Fuc-transferases III, IV, V, and VI (Tables 1 and 2), which are similar in primary sequence over most of their length. Three of the genes (α3-Fuc-transferases III, V, and VI) are syntenic on human chromosome 19 suggesting that the family arose by gene duplication and subsequent divergence.

The α3-Gal-transferase family includes the enzymes which transfer Gal in α1-3 linkage to the Gal residue of Galβ1-4GlcNAc-R (murine and bovine) and Fucα1-2Gal-R (human blood group B) and the human blood group A α3-GalNAc-transferase (Tables 1 and 2). The family also includes three human nonfunctional pseudogenes (HGT-2, HGT-10, and blood group O, discussed above). The genes for the human blood group A and B transferases and for the human HGT-10 pseudogene are all located on human chromosome 9q33-34 suggesting evolution via gene duplication and subsequent divergence. However, Joziasse (200) points out that the exon structure of UDP-Gal:Galβ1-4GlcNAc-R α3-Gal-transferase is consistent with the involvement of exon shuffling in the evolution of these genes.

The β6-GlcNAc-transferase family has at least two, and possibly three, members (196, 199). Two of these enzymes have been localized to human chromosome 9q21 suggesting gene duplication and divergence. The sialyltransferase gene family was discussed in section 3.2.3.

4.3 The domain structure of glycosyltransferases

With the exception of the similarities in primary sequence observed in the four gene families discussed above, there is little or no sequence similarity between the glycosyltransferases cloned to date. However, all these enzymes (Table 2) are of approximately the same length and share the same type II integral membrane glycoprotein domain structure (Figure 5) (12, 24). This structure is eminently suited for the functions of these enzymes. The hydrophobic sequence near the amino-terminus serves to target the enzyme to the Golgi membrane and to anchor the enzyme into the membrane. The short amino-terminal cytoplasmic domain carries positive charges which are believed to be required for anchoring the enzyme in a correct orientation. The large globular catalytic domain thus becomes available within the Golgi lumen to act on glycoproteins traversing the endomembrane assembly line. The stem region is often rich in Pro residues and may serve as a flexible tether for the catalytic domain. The lack of primary sequence similarity suggests that the transferase domain structure is the product of convergent evolution.

Comparison of protein domain and exon structures (Figure 12) (200) indicates that there is no common pattern of intron–exon structure between the glycosyltransferase genes and no correlation between exons and protein domains. This suggests that exon shuffling did not play a role in the development of these genes. However, little is as yet known on the functional domains of the transferases. An accurate assessment of the correlation between exons and functional domains must await elucidation of the three-dimensional structures of the transferases and an understanding of how amino acid residues that are widely separated from one another come together to create the substrate-binding sites and catalytic sites of the enzymes.

4.4 Control of glycosyltransferase activities in time and space

There is a large body of evidence which documents the variations that occur in complex carbohydrate structures during development and differentiation, in disease processes and between different normal tissues. Substrate-level factors such as the organization of the transferases on the Golgi membrane and the precise substrate specificities of the transferases are known to play major roles in regulating the synthesis of complex carbohydrates (1, 4, 5). Information on the regulation of glycosyltransferase activity at the transcriptional and translational levels has recently become available for rat and human α2,6-sialyltransferase and murine β4-Gal-transferase (Figures 6, 7, 10, and 11; Tables 1 and 2). Transcription of both genes appears to be under the control of multiple promoters (multiple transcriptional start sites). One of the promoters for the rat α2,6 sialyltransferase is liver-specific and the other is kidney-specific; the β4-Gal-transferase gene has a germ cell-specific promoter. There is also evidence for multiple promoters in the genes for UDP-Gal:Galβ1-4GlcNAc-R α3-Gal-transferase and β2-GlcNAc-transferase I but the tissue-specificity and functional aspects of these promoters are not known. Both multiple transcriptional starts and alternative transcript splicing have been implicated in the production of different transcripts from the same gene. Different transcripts may produce either the same protein or several isoforms (200).

Detailed data on the transcriptional and translational controls of the transferases is essential if we are to understand the synthesis of complex carbohydrates in normal and diseased tissue. It is already clear that perturbation of the glycosylation machinery (for example the absence of the β2-GlcNAc-transferase I gene) has little effect on the ability of cells to survive in culture but stops normal mouse embryogenesis. Perturbation experiments must therefore be carried out on the intact organism, for example by using the transgenic mouse model. Such studies will not only elucidate the mechanisms whereby complex carbohydrate structures are synthesized but also the more difficult and elusive problem of complex carbohydrate function.

Please refer to Appendix to Chapter 3 at the end of the book for material added in proof.

Acknowledgements

The author wishes to thank Drs M. N. Fukuda, J. T. Y. Lau, J. B. Lowe, J. C. Paulson, P. D. Qasba, J. H. Shaper, and B. D. Shur for access to manuscripts prior to publication and for advice during the writing of this chapter.

References

1. Kornfeld, R. and Kornfeld, S. (1985) Assembly of asparagine-linked oligosaccharides. *Annu. Rev. Biochem.*, **54**, 631.
2. Basu, S., Basu, M., Das, K. K., Daussin, F., Schaeper, R. J., Banerjee, P. *et al.* (1988) Solubilized glycosyltransferases and biosynthesis 'in vitro' of glycolipids. *Biochimie*, **70**, 1551.
3. Basu, S. C. (1991) The serendipity of ganglioside biosynthesis: pathway to CARS and HY-CARS glycosyltransferases. *Glycobiology*, **1**, 469.
4. Schachter, H. (1986) Biosynthetic controls that determine the branching and microheterogeneity of protein-bound oligosaccharides. *Biochem. Cell Biol.*, **64**, 163.
5. Schachter, H. (1991) The 'yellow brick road' to branched complex N-glycans. *Glycobiology*, **1**, 453.
6. Snider, M. D. (1984) Biosynthesis of glycoproteins: formation of N-linked oligosaccharides. In *Biology of carbohydrates* (ed. V. Ginsburg and P. W. Robbins), Vol. 2, p. 163. Wiley, New York, NY.
7. Sadler, J. E. (1984) Biosynthesis of glycoproteins: formation of O-linked oligosaccharides. In *Biology of carbohydrates* (ed. V. Ginsburg, and P. W. Robbins), Vol. 2, p. 199. Wiley, New York, NY.
8. Schachter, H., Narasimhan, S., Gleeson, P., Vella, G., and Brockhausen, I. (1985) Glycosyltransferases involved in the biosynthesis of protein-bound oligosaccharides of the asparagine-N-acetyl-D-glucosamine and serine(threonine)-N-acetyl-D-galactosamine types. In *The enzymes of biological membranes* (ed. A. N. Martonosi) *Biosynthesis and metabolism*, Vol. 2, p. 227. Plenum Press, New York, NY.
9. Schachter, H., Brockhausen, I., and Hull, E. (1989) High-performance liquid chromatography assays for N-acetylglucosaminyltransferases involved in N- and O-glycan synthesis. *Methods Enzymol.*, **179**, 351.
10. Schachter, H. and Brockhausen, I. (1989) The biosynthesis of branched O-glycans. In *Mucus and related topics* (ed. E. Chantler and N. A. Ratcliffe), Vol. 43, p. 1. Society for Experimental Biology, Cambridge.
11. Schachter, H., Narasimhan, S., Gleeson, P., and Vella, G. (1983) Control of branching during the biosynthesis of asparagine-linked oligosaccharides. *Can. J. Biochem. Cell Biol.*, **61**, 1049.
12. Paulson, J. C. and Colley, K. J. (1989) Glycosyltransferases. Structure, localization, and control of cell type-specific glycosylation. *J. Biol. Chem.*, **264**, 17 615.

13. Macher, B. A., Holmes, E. H., Swiedler, S. J., Stults, C. L. M., and Srnka, C. A. (1991) Human α1-3-fucosyltransferases. *Glycobiology*, **1**, 577.
14. Mollicone, R., Gibaud, A., François, A., Ratcliffe, M., and Oriol, R. (1990) Acceptor specificity and tissue distribution of three human alpha-3-fucosyltransferases. *Eur. J. Biochem.*, **191**, 169.
15. Tetteroo, P. A. T., de Heij, H. T., van den Eijnden, D. H., Visser, F. J., Schoenmaker, E., and van Kessel, A. H. M. G. (1987) A GDP-fucose: [Galβ1-4]GlcNAc α1-3-fucosyltransferase activity is correlated with the presence of human chromosome 11 and the expression of the Le^x, Le^y, and sialyl-Le^x antigens in human–mouse cell hybrids. *J. Biol. Chem.*, **262**, 15 984.
16. Couillin, P., Mollicone, R., Grisard, M. C., Gibaud, A., Ravisé, N., Feingold, J. *et al.* (1991) Chromosome 11q localization of one of the three expected genes for the human alpha-3-fucosyltransferases, by somatic hybridization. *Cytogenet. Cell Genet.*, **56**, 108.
17. Greenwell, P., Yates, A. D., and Watkins, W. M. (1986) UDP-N-acetyl-D-galactosamine as a donor substrate for the glycosyltransferase encoded by the B gene at the human blood group ABO locus. *Carbohyd. Res.*, **149**, 149.
18. Gross, H. J., Rose, U., Krause, J. M., Paulson, J. C., Schmid, K., Feeney, R. E. *et al.* (1989) Transfer of synthetic sialic acid analogues to *N*- and *O*-linked glycoprotein glycans using four different mammalian sialyltransferases. *Biochemistry*, **28**, 7386.
19. Van Pelt, J., Dorland, L., Duran, M., Hokke, C. H., Kamerling, J. P., and Vliegenthart, J. F. G. (1989) Transfer of sialic acid in alpha2-6 linkage to mannose in Manbetal-4GlcNAc and Manbetal-4GlcNAcbetal-4GlcNAc by the action of Galbetal-4GlcNAc alpha2-6-sialyltransferase. *FEBS Lett.*, **256**, 179.
20. Srivastava, G., Alton, G., and Hindsgaul, O. (1990) Combined chemical-enzymic synthesis of deoxygenated oligosaccharide analogs: Transfer of deoxygenated D-GlcpNAc residues from their UDP-GlcpNAc derivatives using *N*-acetylglucosaminyltransferase I. *Carbohyd. Res.*, **207**, 259.
21. Gokhale, U. B., Hindsgaul, O., and Palcic, M. M. (1990) Chemical synthesis of GDP-fucose analogs and their utilization by the Lewis alpha(1→4) fucosyltransferase. *Can. J. Chem.*, **68**, 1063.
22. Gross, H. J., Bunsch, A., Paulson, J. C., and Brossmer, R. (1987) Activation and transfer of novel synthetic 9-substituted sialic acids. *Eur. J. Biochem.*, **168**, 595.
23. Gross, H. J. and Brossmer, R. (1988) Enzymatic introduction of a fluorescent sialic acid into oligosaccharide chains of glycoproteins. *Eur. J. Biochem.*, **177**, 583.
24. Schachter, H. (1991) Enzymes associated with glycosylation. *Curr. Opin. Struct. Biol.*, **1**, 755.
25. Struck, D. K., and Lennarz, W. J. (1980) The function of saccharide-lipids in synthesis of glycoproteins. In *The biochemistry of glycoproteins and proteoglycans* (ed. W. J. Lennarz), p. 35. Plenum Press, New York, NY.
26. Abeijon, C. and Hirschberg, C. B. (1992) Topography of glycosylation reactions in the endoplasmic reticulum. *Trends Biochem. Sci.*, **17**, 32.
27. Potvin, B. and Stanley, P. (1991) Activation of two new alpha(1,3)fucosyltransferase activities in Chinese hamster ovary cells by 5-azacytidine. *Cell Regul.*, **2**, 989.
28. Gooi, H. C., Feizi, T., Kapadia, A., Knowles, B. B., Solter, D., and Evans, M. J. (1981) Stage-specific embryonic antigen involves α1-3 fucosylated type 2 blood group chains. *Nature*, **292**, 156.
29. Walz, G., Aruffo, A., Kolanus, W., Bevilacqua, M., and Seed, B. (1990) Recognition

by ELAM-1 of the sialyl-Le^X determinant on myeloid and tumor cells. *Science*, **250**, 1132.
30. Lowe, J. B., Stoolman, L. M., Nair, R. P., Larsen, R. D., Berhend, T. L., and Marks, R. M. (1990) ELAM-1-dependent cell adhesion to vascular endothelium determined by a transfected human fucosyltransferase cDNA. *Cell*, **63**, 475.
31. Lowe, J. B., Kukowska-Latallo, J. F., Nair, R. P., Larsen, R. D., Marks, R. M., Macher, B. A. *et al.* (1991) Molecular cloning of a human fucosyltransferase gene that determines expression of the Lewis X and VIM-2 epitopes but not ELAM-1-dependent cell adhesion. *J. Biol. Chem.*, **266**, 17 467.
32. Phillips, M. L., Nudelman, E., Gaeta, F., Perez, M., Singhal, A. K., Hakomori, S., and Paulson, J. C. (1990) ELAM-1 mediates cell adhesion by recognition of a carbohydrate ligand, sialyl-Le^X. *Science*, **250**, 1130.
33. Polley, M. J., Phillips, M. L., Wayner, E., Nudelman, E., Singhal, A. K., Hakomori, S. *et al.* (1991) CD62 and endothelial cell-leukocyte adhesion molecule 1 (ELAM-1) recognize the same carbohydrate ligand, sialyl-Lewis X. *Proc. Natl Acad. Sci. USA*, **88**, 6224.
34. Kumar, R., Potvin, B., Muller, W. A., and Stanley, P. (1991) Cloning of a human alpha(1,3)-fucosyltransferase gene that encodes ELFT but does not confer ELAM-1 recognition on Chinese hamster ovary cell transfectants. *J. Biol. Chem.*, **266**, 21 777.
35. Brandley, B. K., Swiedler, S. J., and Robbins, P. W. (1990) Carbohydrate ligands of the LEC cell adhesion molecules. *Cell*, **63**, 861.
36. Etzioni, A., Frydman, M., Pollack, S., Avidor, I., Phillips, M. L., Paulson, J. C. *et al.* (1992) Brief report: recurrent severe infections caused by a novel leukocyte adhesion deficiency. *N. Engl. J. Med.*, **327**, 1789.
37. Narimatsu, H., Sinha, S., Brew, K., Okayama, H., and Qasba, P. K. (1986) Cloning and sequencing of cDNA of bovine N-acetylglucosamine (β1-4)galactosyltransferase. *Proc. Natl Acad. Sci. USA*, **83**, 4720.
38. D'Agostaro, G., Bendiak, B., and Tropak, M. (1989) Cloning of cDNA encoding the membrane-bound form of bovine beta 1,4-galactosyltransferase. *Eur. J. Biochem.*, **183**, 211.
39. Shaper, N. L., Shaper, J. H., Meuth, J. L., Fox, J. L., Chang, H., Kirsch, I. R. *et al.* (1986) Bovine galactosyltransferase: a clone identified by direct immunological screening of a cDNA expression library. *Proc. Natl Acad. Sci. USA*, **83**, 1573.
40. Masibay, A. S. and Qasba, P. K. (1989) Expression of bovine beta-1,4-galactosyltransferase cDNA in COS-7 cells. *Proc. Natl Acad. Sci. USA*, **86**, 5733.
41. Russo, R. N., Shaper, N. L., and Shaper, J. H. (1990) Bovine beta1→4-galactosyltransferase: Two sets of mRNA transcripts encode two forms of the protein with different amino-terminal domains. *In vitro* translation experiments demonstrate that both the short and the long forms of the enzyme are type II membrane-bound glycoproteins. *J. Biol. Chem.*, **265**, 3324.
42. Schachter, H. and Roseman, S. (1980) Mammalian glycosyltransferases: their role in the synthesis and function of complex carbohydrates and glycolipids. In *Biochemistry of glycoproteins and proteoglycans* (ed. W. J. Lennarz), p. 85. Plenum Press, New York, NY.
43. Beyer, T. A., Sadler, J. E., Rearick, J. I., Paulson, J. C., and Hill, R. L. (1981) Glycosyltransferases and their use in assessing oligosaccharide structure and structure-function relationships. In *Advances in enzymology* (ed. A. Meister), Vol. 52, p. 23. Wiley, New York.
44. Elhammer, A. and Kornfeld, S. (1986) Purification and characterization of UDP-N-acetylgalactosamine:polypeptide N-acetylgalactosaminyltransferase from bovine colostrum and murine lymphoma BW5147 cells. *J. Biol. Chem.*, **261**, 5249.
45. Sugiura, M., Kawasaki, T., and Yamashina, I. (1982) Purification and characterization

of UDP-GalNAc:polypeptide N-acetylgalactosamine transferase from an ascites hepatoma, AH 66. *J. Biol. Chem.*, **257**, 9501.

46. Weinstein, J., DeSouza-e-Silva, U., and Paulson, J. C. (1982) Sialylation of glycoprotein oligosaccharides N-linked to asparagine. Enzymatic characterization of a Galβ1-3(4)GlcNAc α2-3-sialyltransferase and a Galβ1-4GlcNAc α2-6-sialyltransferase from rat liver. *J. Biol. Chem.*, **257**, 13 845.

47. Nishikawa, Y., Pegg, W., Paulsen, H., and Schachter, H. (1988) Control of glycoprotein synthesis. XV. Purification and characterization of rabbit liver UDP-N-acetylglucosamine:α-3-D-mannoside β-1,2,N-acetylglucosaminyltransferase I. *J. Biol. Chem.*, **263**, 8270.

48. Bendiak, B. and Schachter, H. (1987) Control of glycoprotein synthesis. XII. Purification of UDP-GlcNAc:α-D-mannoside β1-2-N-acetylglucosaminyltransferase II from rat liver. *J. Biol. Chem.*, **262**, 5775.

49. Appert, H. E., Rutherford, T. J., Tarr, G. E., Wiest, J. S., Thomford, N. R., and McCorquodale, D. J. (1986) Isolation of a cDNA coding for human galactosyltransferase. *Biochem. Biophys. Res. Commun.*, **139**, 163.

50. Appert, H. E., Rutherford, T. J., Tarr, G. E., Thomford, N. R., and McCorquodale, D. J. (1986) Isolation of galactosyltransferase from human milk and the determination of its N-terminal amino acid sequence. *Biochem. Biophys. Res. Commun.*, **138**, 224.

51. Masri, K. A., Appert, H. E., and Fukuda, M. N. (1988) Identification of the full-length coding sequence for human galactosyltransferase (β-N-acetylglucosaminide: β1,4-galactosyltransferase). *Biochem. Biophys. Res. Commun.*, **157**, 657.

52. Shaper, N. L., Hollis, G. F., Douglas, J. G., Kirsch, I. R., and Shaper, J. H. (1988) Characterization of the full length cDNA for murine β-1,4-galactosyltransferase. Novel features at the 5'-end predict two translational start sites at two in-frame AUGs. *J. Biol. Chem.*, **263**, 10 420.

53. Nakazawa, K., Ando, T., Kimura, T., and Narimatsu, H. (1988) Cloning and sequencing of a full-length cDNA of mouse N-acetylglucosamine (β1-4)galactosyltransferase. *J. Biochem. (Tokyo)*, **104**, 165.

54. Hollis, G. F., Douglas, J. G., Shaper, N. L., Shaper, J. H., Stafford-Hollis, J. M., Evans, R. J. *et al.* (1989) Genomic structure of murine beta-1,4-galactosyltransferase. *Biochem. Biophys. Res. Commun.*, **162**, 1069.

55. Shaper, J. H., Hollis, G. F., and Shaper, N. L. (1988) Evidence for two forms of murine β-1,4-galactosyltransferase based on cloning studies. *Biochimie*, **70**, 1683.

56. Mengle-Gaw, L., McCoy-Haman, M. F., and Tiemeier, D. C. (1991) Genomic structure and expression of human beta-1,4-galactosyltransferase. *Biochem. Biophys. Res. Commun.*, **176**, 1269.

57. Shaper, N. L., Shaper, J. H., Hollis, G. F., Chang, H., Kirsch, I. R., and Kozak, C. A. (1987) The gene for galactosyltransferase maps to mouse chromosome 4. *Cytogenet. Cell. Genet.*, **44**, 18.

58. Shaper, N. L., Shaper, J. H., Peyser, M., and Kozak, C. A. (1990) Localization of the gene for beta1,4-galactosyltransferase to a position in the centromeric region of mouse Chromosome 4. *Cytogenet. Cell. Genet.*, **54**, 172.

59. Shaper, N. L., Shaper, J. H., Bertness, V., Chang, H., Kirsch, I. R., and Hollis, G. F. (1986) The human galactosyltransferase gene is on chromosome 9 at band p13. *Somat. Cell. Molec. Genet.*, **12**, 633.

60. Duncan, A. M. V., McCorquodale, M. M., Morgan, C., Rutherford, T. J., Appert,

H. E., and McCorquodale, D. J. (1986) Chromosomal localization of the gene for a human galactosyltransferase (GT-1). *Biochem. Biophys. Res. Commun.*, **141**, 1185.
61. Nakazawa, K., Furukawa, K., Kobata, A., and Narimatsu, H. (1991) Characterization of a murine beta1-4 galactosyltransferase expressed in COS-1 cells. *Eur. J. Biochem.*, **196**, 363.
62. Strous, G. J. (1986) Golgi and secreted galactosyltransferase. *Crit. Rev. Biochem.*, **21**, 119.
63. Aoki, D., Appert, H. E., Johnson, D., Wong, S. S., and Fukuda, M. N. (1990) Analysis of the substrate binding sites of human galactosyltransferase by protein engineering. *EMBO J.*, **9**, 3171.
64. Lopez, L. C. and Shur, B. D. (1988) Comparison of two independent cDNA clones reported to encode β1,4 galactosyltransferase. *Biochem. Biophys. Res. Commun.*, **156**, 1223.
65. Lopez, L. C., Maillet, C. M., Oleszkowicz, K., and Shur, B. D. (1989) Cell surface and Golgi pools of beta-1,4-galactosyltransferase are differentially regulated during embryonal carcinoma cell differentiation. *Mol. Cell. Biol.*, **9**, 2370.
66. Lopez, L. C., Youakim, A., Evans, S. C., and Shur, B. D. (1991) Evidence for a molecular distinction between Golgi and cell surface forms of beta1,4-galactosyltransferase. *J. Biol. Chem.*, **266**, 15 984.
67. Shur, B. D. (1991) Cell surface β1,4-galactosyltransferase: twenty years later. *Glycobiology*, **1**, 563.
68. Evans, S. C., Lopez, L. C., and Shur, B. D. (1993) Dominant negative mutation in cell surface β1,4-galactosyltransferase inhibits cell–cell and cell–matrix interactions. *J. Cell. Biol.*, **120**, 1045.
69. Nilsson, T., Lucocq, J. M., Mackay, D., and Warren, G. (1991) The membrane spanning domain of beta-1,4-galactosyltransferase specifies *trans* Golgi localization. *EMBO J.*, **10**, 3567.
70. Teasdale, R. D., D'Agostaro, G., and Gleeson, P. A. (1992) The signal for Golgi retention of bovine β1,4-galactosyltransferase is in the transmembrane domain. *J. Biol. Chem.*, **267**, 4084.
71. Aoki, D., Lee, N., Yamaguchi, N., Dubois, C., and Fukuda, M. N. (1992) Golgi retention of a *trans*-Golgi membrane protein, galactosyltransferase, requires cysteine and histidine residues within the membrane-anchoring domain. *Proc. Natl Acad. Sci. USA*, **89**, 4319.
72. Masibay, A. S., Balaji, P. V., Boeggeman, E. E., and Qasba, P. K. (1993) Mutational analysis of the Golgi retention signal of bovine β-1,4-galactosyltransferase. *J. Biol. Chem.*, **268**, 9908.
73. Russo, R. N., Shaper, N. L., Taatjes, D. J., and Shaper, J. H. (1992) β1,4-Galactosyltransferase: a short NH_2-terminal fragment that includes the cytoplasmic and transmembrane domain is sufficient for Golgi retention. *J. Biol. Chem.*, **267**, 9241.
74. Harduin-Lepers, A., Shaper, J. H., and Shaper, N. L. (1993) Characterization of two *cis*-regulatory regions in the murine β1,4-galactosyltransferase gene: evidence for a negative regulatory element that controls initiation at the proximal site. *J. Biol. Chem.*, **268**, 14 348.
75. Shaper, J. H. and Shaper, N. L. (1992) Enzymes associated with glycosylation. *Curr. Opin. Struct. Biol.*, **2**, 701.
76. Miller, D. J., Macek, M. B., and Shur, B. D. (1992) Complementarity between sperm surface beta-1,4-galactosyl-transferase and egg-coat ZP3 mediates sperm-egg binding. *Nature*, **357**, 589.

77. Hathaway, H. J., and Shur, B. D. (1992) Cell surface beta1,4-galactosyltransferase functions during neural crest cell migration and neurulation *in vivo*. *J. Cell. Biol.*, **117**, 369.

78. Moremen, K. W. and Robbins, P. W. (1991) Isolation, characterization, and expression of cDNAs encoding murine alpha-mannosidase II, a Golgi enzyme that controls conversion of high mannose to complex *N*-glycans. *J. Cell. Biol.*, **115**, 1521.

79. Masibay, A. S., Boeggeman, E., and Qasba, P. K. (1992) Deletion analysis of the NH2-terminal region of beta-1,4-galactosyltransferase. *Mol. Biol. Rep.*, **16**, 99.

80. Munro, S. (1991) Sequences within and adjacent to the transmembrane segment of alpha-2,6-sialyltransferase specify Golgi retention. *EMBO J.*, **10**, 3577.

81. Humphreys-Beher, M. G., Bunnell, B., van Tuinen, P., Ledbetter, D. H., and Kidd, V. J. (1986) Molecular cloning and chromosomal localization of human 4-β-galactosyltransferase. *Proc. Natl Acad. Sci. USA*, **83**, 8918.

82. Bunnell, B. A., Adams, D. E., and Kidd, V. J. (1990) Transient expression of a p58 protein kinase cDNA enhances mammalian glycosyltransferase activity. *Biochem. Biophys. Res. Commun.*, **171**, 196.

83. Delves, P. J., Lund, T., Axford, J. S., Alavi-Sadrieh, A., Lydyard, P. M., MacKenzie, L. *et al.* (1990) Polymorphism and expression of the galactosyltransferase-associated protein kinase gene in normal individuals and galactosylation-defective rheumatoid arthritis patients. *Arthritis Rheum.*, **33**, 1655.

84. Shaper, N. L., Wright, W. W., and Shaper, J. H. (1990) Murine beta1,4-galactosyltransferase: both the amounts and structure of the mRNA are regulated during spermatogenesis. *Proc. Natl Acad. Sci. USA*, **87**, 791.

85. Harduin-Lepers, A., Shaper, N. L., Mahoney, J. A., and Shaper, J. H. (1992) Murine β1,4-galactosyltransferase: round spermatid transcripts are characterized by an extended 5'-untranslated region. *Glycobiology*, **2**, 361.

86. Galili, U., Shohet, S. B., Kobrin, E., Stults, C. L. M., and Macher, B. A. (1988) Man, apes, and old world monkeys differ from other mammals in the expression of α-galactosyl epitopes on nucleated cells. *J. Biol. Chem.*, **263**, 17755.

87. Galili, U. (1989) Abnormal expression of alpha galactosyl epitopes in man. A trigger for autoimmune processes. *Lancet*, **2**, 358.

88. Galili, U. and Swanson, K. (1991) Gene sequences suggest inactivation of α-1,3-galactosyltransferase in catarrhines after the divergence of apes from monkeys. *Proc. Natl Acad. Sci. USA*, **88**, 7401.

89. Galili, U. (1992) The natural anti-Gal antibody: evolution and autoimmunity in man. In *Molecular immunobiology of self-reactivity* (ed. C. A. Bona and A. K. Kaushik), p. 355. Marcel Dekker, New York.

90. Joziasse, D. H., Shaper, J. H., van den Eijnden, D. H., van Tunen, A. J., and Shaper, N. L. (1989) Bovine α1→3-galactosyltransferase: isolation and characterization of a cDNA clone. Identification of homologous sequences in human genomic DNA. *J. Biol. Chem.*, **264**, 14290.

91. Larsen, R. D., Rajan, V. P., Ruff, M. M., Kukowska-Latallo, J., Cummings, R. D., and Lowe, J. B. (1989) Isolation of a cDNA encoding a murine UDP-galactose:beta-D-galactosyl-1,4-N-acetyl-D-glucosaminide α-1,3-galactosyltransferase: expression cloning by gene transfer. *Proc. Natl Acad. Sci. USA*, **86**, 8227.

92. Aruffo, A. and Seed, B. (1987) Molecular cloning of a CD28 cDNA by a high-efficiency COS cell expression system. *Proc. Natl Acad. Sci. USA*, **84**, 8573.

93. Seed, B. and Aruffo, A. (1987) Molecular cloning of the CD2 antigen, the T-cell

erythrocyte receptor, by a rapid immunoselection procedure. *Proc. Natl Acad. Sci. USA*, **84**, 3365.
94. Hirt, B. J. (1967) Selective extraction of polyoma DNA from infected mouse cell cultures. *J. Mol. Biol.*, **26**, 365.
95. Joziasse, D. H., Shaper, N. L., Kim, D., van den Eijnden, D. H., and Shaper, J. H. (1992) Murine α-1,3-galactosyltransferase. A single gene locus specifies four isoforms of the enzyme by alternative splicing. *J. Biol. Chem.*, **267**, 5534.
96. Yamamoto, F. and Hakomori, S. (1990) Sugar-nucleotide donor specificity of histo-blood group A and B transferases is based on amino acid substitutions. *J. Biol. Chem.*, **265**, 19 257.
97. Larsen, R. D., Rivera-Marrero, C. A., Ernst, L. K., Cummings, R. D., and Lowe, J. B. (1990) Frame-shift and nonsense mutations in a human genomic sequence homologous to a murine UDP-Gal:β-D-Gal(1,4)-D-GlcNAc α(1,3)-galactosyltransferase cDNA. *J. Biol. Chem.*, **265**, 7055.
98. Joziasse, D. H., Shaper, J. H., Jabs, E. W., and Shaper, N. L. (1991) Characterization of an α1→3-galactosyltransferase homologue on human chromosome 12 that is organized as a processed pseudogene. *J. Biol. Chem.*, **266**, 6991.
99. Joziasse, D. H., Shaper, N. L., Shaper, J. H., and Kozak, C. A. (1991) Gene for murine α1→3-galactosyltransferase is located in the centromeric region of chromosome 2. *Somatic Cell Mol. Genet.*, **17**, 201.
100. Shaper, N. L., Lin, S., Joziasse, D. H., Kim, D., and Yang-Feng, T. L. (1992) Assignment of two human α-1,3-galactosyltransferase gene sequences (GGTA1 and GGTA1P) to chromosomes 9q33-q34 and 12q14-q15. *Genomics*, **12**, 613.
101. Joziasse, D. H., Shaper, N. L., Salyer, L. D., van den Eijnden, D. H., van der Spoel, A. C., and Shaper, J. H. (1990) α1→3-galactosyltransferase: The use of recombinant enzyme for the synthesis of α-galactosylated glycoconjugates. *Eur. J. Biochem.*, **191**, 75.
102. Cummings, R. D. and Mattox, S. A. (1988) Retinoic acid-induced differentiation of the mouse teratocarcinoma cell line F9 is accompanied by an increase in the activity of UDP-galactose: β-D-galactosyl-α1,3-galactosyltransferase. *J. Biol. Chem.*, **263**, 511.
103. Smith, D. F., Larsen, R. D., Mattox, S., Lowe, J. B., and Cummings, R. D. (1990) Transfer and expression of a murine UDP-Gal:β-D-Gal α-1,3-galactosyltransferase gene in transfected Chinese hamster ovary cells. Competition reactions between the α-1,3-galactosyltransferase and the endogenous α2,3-sialyltransferase. *J. Biol. Chem.*, **265**, 6225.
104. Yamamoto, F., Clausen, H., White, T., Marken, J., and Hakomori, S. (1990) Molecular genetic basis of the histo-blood group ABO system. *Nature*, **345**, 229.
105. Ugozolli, L. and Wallace, R. B. (1992) Application of an allele-specific polymerase chain reaction to the direct determination of ABO blood group genotypes. *Genomics*, **12**, 670.
106. Yamamoto, F., McNeill, P. D., and Hakomori, S. (1991) Identification in human genomic DNA of the sequence homologous but not identical to either the histo-blood group ABH genes or α1→3 galactosyltransferase pseudogene. *Biochem. Biophys. Res. Commun.*, **175**, 986.
107. Weinstein, J., Lee, E. U., McEntee, K., Lai, P.-H., and Paulson, J. C. (1987) Primary structure of β-galactoside α-2,6-sialyltransferase. Conversion of membrane-bound enzyme to soluble forms by cleavage of the NH_2-terminal signal anchor. *J. Biol. Chem.*, **262**, 17 735.

108. Paulson, J. C., Weinstein, J., and Schauer, A. (1989) Tissue-specific expression of sialyltransferases. *J. Biol. Chem.*, **264**, 10 931.
109. Grundmann, U., Nerlich, C., Rein, T., and Zettlmeissl, G. (1990) Complete cDNA sequence encoding human β-galactoside α-2,6-sialyltransferase. *Nucleic Acids Res.*, **18**, 667.
110. Lance, P., Lau, K. M., and Lau, J. (1989) Isolation and characterization of a partial cDNA for a human sialyltransferase. *Biochem. Biophys. Res. Commun.*, **164**, 225.
111. Stamenkovic, I., Asheim, H. C., Deggerdal, A., Blomhoff, H. K., Smeland, E. B., and Funderud, S. (1990) The B cell antigen CD75 is a cell surface sialyltransferase. *J. Exp. Med.*, **172**, 641.
112. Bast, B. J. E. G., Zhou, L. J., Freeman, G. J., Colley, K. J., Ernst, T. J., Munro, J. M. *et al.* (1992) The HB-6, CDw75, and CD76 differentiation antigens are unique cell-surface carbohydrate determinants generated by the β-galactoside α-2,6-sialyltransferase. *J. Cell. Biol.*, **116**, 423.
113. Svensson, E. C., Soreghan, B., and Paulson, J. C. (1990) Organization of the β-galactoside α-2,6-sialyltransferase gene. Evidence for the transcriptional regulation of terminal glycosylation. *J. Biol. Chem.*, **265**, 20 863.
114. Wen, D. X., Svensson, E. C., and Paulson, J. C. (1992) Tissue-specific alternative splicing of the β-galactoside α-2,6-sialyltransferase gene. *J. Biol. Chem.*, **267**, 2512.
115. Wang, X.-C., O'Hanlon, T. P., Young, R. F., and Lau, J. T. Y. (1990) Rat β-galactoside α2,6-sialyltransferase genomic organization: alternate promoters direct the synthesis of liver and kidney transcripts. *Glycobiology*, **1**, 25.
116. O'Hanlon, T. P., Lau, K. M., Wang, X. C., and Lau, J. T. Y. (1989) Tissue-specific expression of β-galactoside α-2,6-sialyltransferase. *J. Biol. Chem.*, **264**, 17 389.
117. O'Hanlon, T. P. and Lau, J. T. Y. (1992) Analysis of kidney mRNAs expressed from the rat β-galactoside α2,6-sialyltransferase gene. *Glycobiology*, **2**, 257.
118. Wang, X.-C., Vertino, A., Eddy, R. L., Byers, M. G., Jani-Sait, S. N., Shows, T. B. *et al.* (1993) Chromosome mapping and organization of the human β-galactoside α2,6-sialyltransferase gene. Differential and cell-type specific usage of upstream exon sequences in B-lymphoblastoid cells. *J. Biol. Chem.*, **268**, 4355.
119. Munro, S., Bast, B. J. E. G., Colley, K. J., and Tedder, T. F. (1992) The B lymphocyte surface antigen CD75 is not an α-2,6-sialyltransferase but is a carbohydrate antigen, the production of which requires the enzyme. *Cell*, **68**, 1003.
120. Stamenkovic, I., Sgroi, D., and Aruffo, A. (1992) CD22 binds to α-2,6-sialyltransferase-dependent epitopes on COS cells. *Cell*, **68**, 1003.
121. Svensson, E. C., Conley, P. B., and Paulson, J. C. (1992) Regulated expression of α-2,6-sialyltransferase by the liver-enriched transcription factors HNF-1, DBP, and LAP. *J. Biol. Chem.*, **267**, 3466.
122. Lammers, G. and Jamieson, J. C. (1988) The role of a cathepsin D-like activity in the release of Galβ1-4GlcNAc α2-6-sialyltransferase from rat liver Golgi membranes during the acute-phase response. *Biochem. J.*, **256**, 623.
123. Lammers, G. and Jamieson, J. C. (1989) Studies on the effect of lysosomotropic agents on the release of Galbeta1-4GlcNAc α-2,6-sialyltransferase from rat liver slices during the acute-phase response. *Biochem. J.*, **261**, 389.
124. Lammers, G. and Jamieson, J. C. (1990) Cathepsin D-like activity in the release of Galβ1-4GlcNAc α-2-6sialyltransferase from mouse and guinea pig liver Golgi membranes during the acute phase response. *Comp. Biochem. Physiol.*, **95B**, 327.

125. Kaplan, H. A., Woloski, B. M. R. N. J., Hellman, M., and Jamieson, J. C. (1983) Studies on the effect of inflammation on rat liver and serum sialyltransferase. Evidence that inflammation causes release of Galβ1-4GlcNAc α-2-6 sialyltransferase from liver. *J. Biol. Chem.*, **258**, 11 505.
126. Lombart, C., Sturgess, J., and Schachter, H. (1980) The effect of turpentine-induced inflammation on rat liver glycosyltransferases and Golgi complex ultrastructure. *Biochim. Biophys. Acta*, **629**, 1.
127. Wang, X. C., O'Hanlon, T. P., and Lau, J. T. Y. (1989) Regulation of β-galactoside α2,6-sialyltransferase gene expression by dexamethasone. *J. Biol. Chem.*, **264**, 1854.
128. Wang, X. C., Smith, T. J., and Lau, J. (1990) Transcriptional regulation of the liver β-galactoside α-2,6-sialyltransferase by glucocorticoids. *J. Biol. Chem.*, **265**, 17 849.
129. Lee, E. U., Roth, J., and Paulson, J. C. (1989) Alteration of terminal glycosylation sequences on N-linked oligosaccharides of Chinese hamster ovary cells by expression of β-galactoside α-2,6-sialyltransferase. *J. Biol. Chem.*, **264**, 13 848.
130. Colley, K. J., Lee, E. U., Adler, B., Browne, J. K., and Paulson, J. C. (1989) Conversion of a Golgi apparatus sialyltransferase to a secretory protein by replacement of the NH_2-terminal signal anchor with a signal peptide. *J. Biol. Chem.*, **264**, 17 619.
131. Paulson, J. C., Weinstein, J., Ujita, E. L., Riggs, K. J., and Lai, P.-H. (1987) The membrane-binding domain of a rat liver Golgi sialyltransferase. *Biochem. Soc. Trans.*, **15**, 618.
132. Colley, K. J., Lee, E. U., and Paulson, J. C. (1992) The signal anchor and stem regions of the β-galactoside α-2,6-sialyltransferase may each act to localize the enzyme to the Golgi apparatus. *J. Biol. Chem.*, **267**, 7784.
133. Wong, S. H., Low, S. H., and Hong, W. (1992) The 17-residue transmembrane domain of β-galactoside α2,6-sialyltransferase is sufficient for Golgi retention. *J. Cell. Biol.*, **117**, 245.
134. Roth, J. (1987) Subcellular organization of glycosylation in mammalian cells. *Biochim. Biophys. Acta*, **906**, 405.
135. Gillespie, W., Kelm, S., and Paulson, J. C. (1992) Cloning and expression of the Galβ1,3GalNAc α-2,3-sialyltransferase. *J. Biol. Chem.*, **267**, 21 004.
136. Gillespie, W., Paulson, J. C., Kelm, S., Pang, M., and Baum, L. G. (1993) Regulation of α-2,3-sialyltransferase expression correlates with conversion of peanut agglutinin (PNA)+ to PNA− phenotype in developing thymocytes. *J. Biol. Chem.*, **268**, 3801.
137. Wen, D. X., Livingston, B. D., Medzihradszky, K. F., Kelm, S., Burlingame, A. L., and Paulson, J. C. (1992) Primary structure of Galβ1,3(4)GlcNAc α-2,3-sialyltransferase determined by mass spectrometry sequence analysis and molecular cloning. *J. Biol. Chem.*, **267**, 21 011.
138. Drickamer, K. (1993) A conserved disulphide bond in sialyltransferases. *Glycobiology*, **3**, 2.
139. Livingston, B., Kitagawa, H., Wen, D., Medzihradsky, K., Burlingame, A., and Paulson, J. C. (1992) Identification of a sialylmotif in the sialyltransferase gene family. *Glycobiology*, **2**, 489.
140. Feizi, T. (1988) Oligosaccharides in molecular recognition. *Biochem. Soc. Trans.*, **16**, 930.
141. Feizi, T. (1990) Observations and thoughts on oligosaccharide involvement in malignancy. *Environ. Health Perspect.*, **88**, 231.
142. Hakomori, S. (1981) Glycosphingolipids in cellular interaction, differentiation and oncogenesis. *Annu. Rev. Biochem.*, **50**, 733.

143. Hakomori, S. I. (1989) Aberrant glycosylation in tumors and tumor-associated carbohydrate antigens. *Adv. Cancer. Res.*, **52**, 257.
144. Le Pendu, J., Cartron, J. P., Lemieux, R. U., and Oriol, R. (1985) The presence of at least two different H-blood-group-related β-D-Gal α-2-L-fucosyltransferases in human serum and the genetics of blood group H substances. *Am. J. Human Genetics*, **37**, 749.
145. Sarnesto, A., Köhlin, T., Hindsgaul, O., Thurin, J., and Blaszczyk-Thurin, M. (1992) Purification of the secretor-type β-galactoside α1→2-fucosyltransferase from human serum. *J. Biol. Chem.*, **267**, 2737.
146. Mollicone, R., Dalix, A. M., Jacobsson, A., Samuelsson, B. E., Gerard, G., Crainic, K. et al. (1988) Red cell H-deficient, salivary ABH secretor phenotype of Reunion Island. Genetic control of the expression of H antigen in the skin. *Glycoconjugate J.*, **5**, 499.
147. Ernst, L. K., Rajan, V. P., Larsen, R. D., Ruff, M. M., and Lowe, J. B. (1989) Stable expression of blood group H determinants and GDP-L-fucose:β-D-galactoside 2-α-L-fucosyltransferase in mouse cells after transfection with human DNA. *J. Biol. Chem.*, **264**, 3436.
148. Rajan, V. P., Larsen, R. D., Ajmera, S., Ernst, L. K., and Lowe, J. B. (1989) A cloned human DNA restriction fragment determines expression of a GDP-L-fucose:β-D-galactoside 2-α-L-fucosyltransferase in transfected cells. Evidence for isolation and transfer of the human H blood group locus. *J. Biol. Chem.*, **264**, 11 158.
149. Larsen, R. D., Ernst, L. K., Nair, R. P., and Lowe, J. B. (1990) Molecular cloning, sequence, and expression of a human GDP-L-fucose-β-D-galactoside 2-α-L-fucosyltransferase cDNA that can form the H blood group antigen. *Proc. Natl Acad. Sci. USA*, **87**, 6674.
150. Weston, B. W., Nair, R. P., Larsen, R. D., and Lowe, J. B. (1992) Isolation of a novel human α(1,3)fucosyltransferase gene and molecular comparison to the human Lewis blood group α(1,3/1,4)fucosyltransferase gene. Syntenic, homologous, nonallelic genes encoding enzymes with distinct acceptor substrate specificities. *J. Biol. Chem.*, **267**, 4152.
151. Weston, B. W., Smith, P. L., Kelly, R. J., and Lowe, J. B. (1992) Molecular cloning of a fourth member of a human α(1,3) fucosyltransferase gene family. Multiple homologous sequences that determine expression of the Lewis X, sialyl Lewis X, and difucosyl sialyl Lewis X epitopes. *J. Biol. Chem.*, **267**, 24 575.
152. Campbell, C. and Stanley, P. (1984) The Chinese hamster ovary glycosylation mutants LEC11 and LEC12 express two novel GDP-fucose:N-acetylglucosaminide 3-α-L-fucosyltransferase enzymes. *J. Biol. Chem.*, **259**, 11 208.
153. Howard, D. R., Fukuda M., Fukuda, M. N., and Stanley, P. (1987) The GDP-fucose:N-acetylglucosaminide 3-α-L-fucosyltransferases of LEC11 and LEC12 Chinese hamster ovary mutants exhibit novel specificities for glycolipid substrates. *J. Biol. Chem.*, **262**, 16 830.
154. Stanley, P. and Atkinson, P. H. (1988) The Lec11 Chinese hamster ovary mutant synthesizes N-linked carbohydrates containing sialylated, fucosylated lactosamine units. Analysis by one-and two-dimensional ^{1}NMR spectroscopy. *J. Biol. Chem.*, **263**, 11 374.
155. Shaw, D. and Eiberg, H. (1987) Report of the committee for chromosomes 17, 18, and 19. *Cytogenet. Cell Genet.*, **46**, 242.
156. Kukowska-Latallo, J. F., Larsen, R. D., Nair, R. P., and Lowe, J. B. (1990) A cloned human cDNA determines expression of a mouse stage-specific embryonic antigen and the Lewis blood group α(1,3/1,4)fucosyltransferase. *Genes Dev.*, **4**, 1288.

157. Johnson, P. H. and Watkins, W. M. (1992) Purification of the Lewis blood-group gene associated α-1,¾-fucosyltransferase from human milk: an enzyme transferring fucose primarily to Type 1 and lactose-based oligosaccharide chains. *Glycoconjugate J.*, **9**, 241.
158. Johnson, P. H., Donald, A., Feeney, J., and Watkins, W. M. (1992) Reassessment of the acceptor specificity and general properties of the Lewis blood-group gene associated α-1,¾-fucosyltransferase purified from human milk. *Glycoconjugate J.*, **9**, 251.
159. Lowe, J. B., Stoolman, L. M., Nair, R. P., Larsen, R. D., Berhend, T. L., and Marks, R. M. (1990) ELAM-1-dependent cell adhesion to vascular endothelium determined by a transfected human fucosyltransferase cDNA. *Cell*, **63**, 475.
160. de Heij, H. T., Tetteroo, P. A. T., van Kessel, A. H. M. G., Schoenmaker, E., Visser, F. J., and van den Eijnden, D. H. (1988) Specific expression of a myeloid-associated CMP-NeuAc:Galβ1-3GalNAcα-R α2-3-sialyltransferase and the sialyl-X determinant in myeloid human-mouse cell hybrids containing human chromosome 11. *Cancer Res.*, **48**, 1489.
161. Goelz, S. E., Hession, C., Goff, D., Griffiths, B., Tizard, R., Newman, B. *et al.* (1990) ELFT: a gene that directs the expression of an ELAM-1 ligand. *Cell*, **63**, 1349.
162. Potvin, B., Kumar, R., Howard, D. R., and Stanley, P. (1990) Transfection of a human α-(1,3)fucosyltransferase gene into Chinese hamster ovary cells. Complications arise from activation of endogenous α-(1,3)fucosyltransferases. *J. Biol. Chem.*, **265**, 1615.
163. Sarnesto, A., Köhlin, T., Hindsgaul, O., Vogele, K., Blaszczyk-Thurin, M., and Thurin, J. (1992) Purification of the β-N-acetylglucosaminide α1→3-fucosyltransferase from human serum. *J. Biol. Chem.*, **267**, 2745.
164. Koszdin, K. L. and Bowen, B. R. (1992) The cloning and expression of a human α-1,3 fucosyltransferase capable of forming the E-selectin ligand. *Biochem. Biophys. Res. Commun.*, **187**, 152.
165. Nishihara, S., Nakazato, M., Kudo, T., Kimura, H., Ando, T., and Narimatsu, H. (1993) Human α-1,3 fucosyltransferase (FucT-VI) gene is located at only 13 Kb 3' to the Lewis type fucosyltransferase (FucT-III) gene on chromosome 19. *Biochem. Biophys. Res. Commun.*, **190**, 42.
166. Yamamoto, F., McNeill, P. D., and Hakomori, S. (1992) Human histo-blood group A2 transferase coded by A2 allele, one of the A subtypes, is characterized by a single base deletion in the coding sequence, which results in an additional domain at the carboxyl terminal. *Biochem. Biophys. Res. Commun.*, **187**, 366.
167. Schachter, H., Michaels, M. A., Crookston, M. C., Tilley, C. A., and Crookston, J. H. (1971) A quantitative difference in the activity of blood group A-specific N-acetylgalactosaminyltransferase in serum from A1 and A2 human subjects. *Biochem. Biophys. Res. Commun.*, **45**, 1011.
168. Schachter, H., Michaels, M. A., Tilley, C. A., Crookston, M. C., and Crookston, J. H. (1973) Qualitative differences in the N-acetyl-D-galactosaminyltransferases produced by human A1 and A2 genes. *Proc. Natl Acad. Sci. USA*, **70**, 220.
169. Clausen, H., White, T., Takio, K., Titani, K., Stroud, M., Holmes, E. *et al.* (1990) Isolation to homogeneity and partial characterization of a histo-blood group A defined Fucα1→2Galα1→3-N-acetylgalactosaminyltransferase from human lung tissue. *J. Biol. Chem.*, **265**, 1139.
170. Yamamoto, F., Marken, J., Tsuji, T., White, T., Clausen, H., and Hakomori, S. (1990) Cloning and characterization of DNA complementary to human UDP-GalNAc:Fuc

α1→2Galα pha1→3GalNAc transferase (histo-blood group A transferase) mRNA. *J. Biol. Chem.*, **265**, 1146.
171. Nagata, Y., Yamashiro, S., Yodoi, J., Lloyd, K. O., Shiku, H., and Furukawa, K. (1992) Expression cloning of β1,4 N-acetylgalactosaminyltransferase cDNAs that determine the expression of GM2 and GD2 gangliosides. *J. Biol. Chem.*, **267**, 12 082.
172. Stanley, P. (1989) Chinese hamster ovary cell mutants with multiple glycosylation defects for production of glycoproteins with minimal carbohydrate heterogeneity. *Mol. Cell. Biol.*, **9**, 377.
173. Sarkar, M., Hull, E., Simpson, R. J., Moritz, R. L., Dunn, R., and Schachter, H. (1990) Rabbit liver UDP-N-acetylglucosamine:α-3-D-mannoside β-1,2-N-acetylglucosaminyltransferase I: characterization of a 2.5 kilobase cDNA clone. *Glycoconjugate J.*, **7**, 380.
174. Sarkar, M., Hull, E., Nishikawa, Y., Simpson, R. J., Moritz, R. L., Dunn, R. et al. (1991) Molecular cloning and expression of cDNA encoding the enzyme that controls conversion of high-mannose to hybrid and complex N-glycans: UDP-N-acetylglucosamine:α-3-D-mannoside β-1,2-N-acetylglucosaminyltransferase I. *Proc. Natl Acad. Sci. USA*, **88**, 234.
175. Schachter, H., Hull, E., Sarkar, M., Simpson, R. J., Moritz, R. L., Höppener, J. W. M. et al. (1991) Molecular cloning of human and rabbit UDP-N-acetylglucosamine: α-3-D-mannoside β-1,2-N-acetylglucosaminyltransferase I. *Biochem. Soc. Trans.*, **19**, 645.
176. Hull, E., Schachter, H., Sarkar, M., Spruijt, M. P. N., Höppener, J. W. M., Roovers, D. et al. (1990) Isolation of 13 and 15 kilobase human genomic DNA clones containing the gene for UDP-N-acetylglucosamine:α-3-D-mannoside β-1,2-N-acetylglucosaminyltransferase I. *Glycoconjugate J.*, **7**, 468.
177. Hull, E., Sarkar, M., Spruijt, M. P. N., Höppener, J. W. M., Dunn, R., and Schachter, H. (1991) Organization and localization to chromosome 5 of the human UDP-N-acetylglucosamine:α-3-D-mannoside β-1,2-N-acetylglucosaminyltransferase I gene. *Biochem. Biophys. Res. Commun.*, **176**, 608.
178. Kumar, R., Yang, J., Larsen, R. D., and Stanley, P. (1990) Cloning and expression of N-acetylglucosaminyltransferase I, the medial Golgi transferase that initiates complex N-linked carbohydrate formation. *Proc. Natl Acad. Sci. USA*, **87**, 9948.
179. Pownall, S., Kozak, C. A., Schappert, K., Sarkar, M., Hull, E., Schachter, H. et al. (1992) Molecular cloning and characterization of the mouse UDP-N-acetylglucosamine: α-3-D-mannoside β-1,2-N-acetylglucosaminyltransferase I gene. *Genomics*, **12**, 699.
180. Kumar, R., Yang, J., Eddy, R. L., Byers, M. G., Shows, T. B., and Stanley, P. (1992) Cloning and expression of the murine gene and chromosomal location of the human gene encoding N-acetylglucosaminyltransferase I. *Glycobiology*, **2**, 383.
181. Squire, J., Leong, P., Tan, J., D'Agostaro, G. A. F., Bendiak, B. K., Sarkar, M. et al. (1994) Mapping of the human UDP-N-acetylglucosamine:α-3-D-mannoside β,2-N-acetylglucosaminyltransferase I gene to band 5q35 and the human UDP-N-acetylglucosamine:α-6-D-mannoside β,2-N-acetylglucosaminyltransferase II gene to band 14q21 using fluorescent *in situ* hybridization. (In preparation.)
182. Burke, J., Pettitt, J. M., Schachter, H., Sarkar, M., and Gleeson, P. A. (1992) The transmembrane and flanking sequences of β-1,2-N-acetylglucosaminyltransferase I specify medial-Golgi localization. *J. Biol. Chem.*, **267**, 24 433.
183. Uhlén, M. and Moks, T. (1990) Gene fusions for purpose of expression: an introduction. *Meth. Enzymol.*, **185**, 129.
184. Sarkar, M. and Schachter, H. (1992) Expression of UDP-GlcNAc:α3-D-mannoside β-1,2-N-acetylglucosaminyltransferase I (GnT I) in insect cells. *Glycobiology*, **2**, 483.

185. Reck, F., Paulsen, H., Brockhausen, I., Sarkar, M., and Schachter, H. (1992) Synthesis of specific tetrasaccharide inhibitors for N-acetylglucosaminyltransferase II (GnT II) using a trisaccharide precursor and recombinant N-acetylglucosaminyltransferase I (GnT I). *Glycobiology*, **2**, 483.
186. Tang, B. L., Wong, S. H., Low, S. H., and Hong, W. (1992) The transmembrane domain of N-glucosaminyltransferase I contains a Golgi retention signal. *J. Biol. Chem.*, **267**, 10 122.
187. Nilsson, T., Pypaert, M., Hoe, M. H., Slusarewicz, P., Berger, E. G., and Warren, G. (1993) Overlapping distribution of two glycosyltransferases in the Golgi apparatus of HeLa cells. *J. Cell. Biol.*, **120**, 5.
188. D'Agostaro, G. A. F., Zingoni, A., Simpson, R. J., Moritz, R. L., Schachter, H., and Bendiak, B. K. (1993) Molecular cloning and expression of cDNA encoding rat UDP-N-acetylglucosamine:α-6-D-mannoside β-1,2-N-acetylglucosaminyltransferase II. *Glycoconjugate J.*, **10**, 234.
189. Petrarca, C., Bendiak, B. K., and D'Agostaro, G. A. F. (1993) Genomic mapping of the rat UDP-N-acetylglucosamine:α-6-D-mannoside β-1,2-N-acetylglucosaminyltransferase II gene. *Glycoconjugate J.*, **10**, 235.
190. Tan, J., D'Agostaro, G. A. F., Bendiak, B. K., Squire, J., and Schachter, H. (1993) Molecular cloning and localization to chromosome 14 of the human UDP-N-acetylglucosamine:α-6-D-mannoside β-1,2-N-acetylglucosaminyltransferase II gene (MGAT2). *Glycoconjugate J.*, **10**, 232.
191. Nishikawa, A., Ihara, Y., Hatakeyama, M., Kangawa, K., and Taniguchi, N. (1992) Purification, cDNA cloning, and expression of UDP-N-acetylglucosamine:β-D-mannoside β-1,4N-acetylglucosaminyltransferase III from rat kidney. *J. Biol. Chem.*, **267**, 18 199.
192. Piller, F., Piller, V., Fox, R. I., and Fukuda, M. (1988) Human T-lymphocyte activation is associated with changes in O-glycan biosynthesis. *J. Biol. Chem.*, **263**, 15 146.
193. Saitoh, O., Piller, F., Fox, R. I., and Fukuda, M. (1991) T-lymphocytic leukemia expresses complex, branched O-linked oligosaccharides on a major sialoglycoprotein, leukosialin. *Blood*, **77**, 1491.
194. Brockhausen, I., Kuhns, W., Schachter, H., Matta, K. L., Sutherland, D. R., and Baker, M. A. (1991) Biosynthesis of O-glycans in leukocytes from normal donors and from patients with leukemia: increase in O-glycan core 2 UDP-GlcNAc:Galβ3GalNAc α-R (GlcNAc to GalNAc) β(1-6)-N-acetylglucosaminyltransferase in leukemic cells. *Cancer Res.*, **51**, 1257.
195. Yousefi, S., Higgins, E., Daoling, Z., Pollex-Krüger, A., Hindsgaul, O., and Dennis, J. W. (1991) Increased UDP-GlcNAc:Galbeta1-3GalNAc-R (GlcNAc to GalNAc) β-1,6-N-acetylglucosaminyltransferase activity in metastatic murine tumor cell lines. Control of polylactosamine synthesis. *J. Biol. Chem.*, **266**, 1772.
196. Bierhuizen, M. F. A. and Fukuda, M. (1992) Expression cloning of a cDNA encoding UDP-GlcNAc:Galbeta1-3-GalNAc-R (GlcNAc to GalNAc) β1-6GlcNAc transferase by gene transfer into CHO cells expressing polyoma large tumor antigen. *Proc. Natl Acad. Sci. USA*, **89**, 9326.
197. Brockhausen, I., Matta, K. L., Orr, J., Schachter, H., Koenderman, A. H. L., and van den Eijnden, D. H. (1986) Mucin synthesis. VII. Conversion of R_1-β1-3Gal-R_2 to R_1-β1-3(GlcNAcβ1-6)Gal-R_2 and of R_1-β1-3GalNAc-R_2 to R_1-β1-3(GlcNAcβ1-6)GalNAc-R_2 by a β6-N-acetylglucosaminyltransferase in pig gastric mucosa. *Eur. J. Biochem.*, **157**, 463.

198. Ropp, P. A., Little, M. R., and Cheng, P. W. (1991) Mucin biosynthesis: purification and characterization of a mucin β6N-acetylglucosaminyltransferase. *J. Biol. Chem.*, **266**, 23 863.
199. Bierhuizen, M., Mattei, M. G., and Fukuda, M. (1993) Expression of the developmental I antigen by a cloned human cDNA encoding a member of a β-1,6-N-acetylglucosaminyltransferase gene family. *Genes Dev.*, **7**, 468.
200. Joziasse, D. H. (1992) Mammalian glycosyltransferases: genomic organization and protein structure. *Glycobiology*, **2**, 271.
201. Chatterjee, S. K. (1991) Molecular cloning of human beta1,4-galactosyltransferase and expression of catalytic activity of the fusion protein in *Escherichia coli*. *Int. J. Biochem.*, **23**, 695.

4 | Carbohydrate recognition in cell–cell interaction

JOHN B. LOWE

1. Introduction

Mammalian cell surfaces are decorated with complex carbohydrate molecules, also known as oligosaccharides or glycoconjugates. These molecules display a remarkable degree of structural diversity, that is in turn a function of the cell type, or tissue, or particular stage of development of the cell or tissue. Observations demonstrating that cell surface oligosaccharide structure varies with the differentiation state or lineage of a cell suggest that these molecules play important roles in mediating information transfer into and out of the cell. During mammalian development, for example, dynamic changes are observed in cell surface glycoconjugate expression. This work suggests that complex mechanisms have evolved to regulate these processes, and that they have important functional correlates. The existence of an ever-expanding group of mammalian carbohydrate binding proteins, or lectins (1), suggests that cell surface carbohydrate molecules may participate as ligands in adhesive processes mediated by mammalian lectins. Indeed, oligosaccharides have been shown to operate in processes whereby lysosomal enzymes are targeted to the lysosome (via the mannose-6-phosphate receptor; reference 2), in the clearance of desialylated serum glycoproteins (via the hepatic asialoglycoprotein receptor; reference 3), and in the clearance of circulating pituitary glycoprotein hormones like leutropin (4). Experimental studies during the past three years have implicated members of a family of mammalian lectins, known as the selectins, and a specific group of cell surface oligosaccharide determinants, in adhesive processes that mediate leukocyte adhesion to vascular endothelium. This chapter will discuss experimental approaches that have led to the discovery of these processes, and will discuss the enzymes, selectins, and other molecules that participate in carbohydrate-dependent leukocyte adhesion.

2. Oligosaccharides in lymphocyte homing

2.1 General aspects of lymphocyte homing

In mammals, lymphocytes circulate both in the intravascular compartment and in the lymphatic compartment. This is a directed process, in that vascular-born lymphocytes leave the vascular tree within lymph nodes. These cells then traverse the parenchyma of the lymph node or other lymphatic organ, and return to the vascular system by way of the lymphatic system (5). This process facilitates the exposure of lymphocytes to foreign antigens within the lymphatic tissues. Lymphocytes leave the vascular tree during their passage through the postcapillary venules within lymphoid organs. The initial event in this process involves adhesion between specific molecules expressed at the surface of the lymphocyte, and molecules displayed on the surfaces of the cells that line the postcapillary venular endothelium.

This endothelium within lymph nodes differs from other postcapillary endothelium in both function and morphology. The endothelial cells that line postcapillary venules in lymph nodes are plump, cuboidal cells, and the venules that contain them are known as high endothelial venules, or HEVs. Different lymphatic organs apparently maintain functionally distinct HEVs, since specific subsets of circulating lymphocytes have been shown to home specifically to one lymphatic organ, or set of organs, but not others. Immunochemical and cell trafficking studies have identified several distinct specificities for lymphocyte-HEV adhesive interactions (reviewed in reference 6). These interactions include those that are specific for HEV within chronically inflamed synovia, for HEV within lymph nodes that drain the pulmonary system, for HEV in lymphoid tissue within the gut (Peyer's patches), and for the postcapillary high endothelial venules within peripheral lymph nodes (PNHEV). These 'organ-specific' homing processes are apparently mediated by 'organ-specific' expression of distinct adhesive molecules, that in turn mediate organ-specific 'homing' of lymphocytes that display the corresponding counter-receptor. Lymphocyte homing to PNHEV has been shown to be mediated in part by carbohydrate-dependent mechanisms, and will be the focus of this section of the chapter.

The Stamper–Woodruff assay (7) has been used to demonstrate and study lymphocyte homing to PNHEV. This assay involves the application of viable lymphocytes to HEVs exposed in frozen sections of peripheral lymph notes. After an appropriate incubation period, the sections are washed; cells remaining attached to the sections are counted as a quantitative measure of adhesion between the applied lymphocytes and the exposed endothelium. This assay was used in the mouse to identify a lymphocyte cell surface determinant that mediates lymphocyte adhesion to PNHEV (8). This epitope is recognized by a monoclonal antibody, termed MEL-14, that was generated in an attempt to create antibodies that recognize cell surface antigens expressed by cultured lymphoma cell lines that bind to PNHEV (8). The MEL-14 monoclonal antibody binds to all murine peripheral lymphocytes, and, in the Stamper-Woodruff assay, blocks adhesion of these cells to PNHEV. MEL-14 can also prevent *in vivo* homing of lymphocytes to peripheral nodes, when the injected lymphocytes are first incubated with saturating amounts of the antibody. Homing to Peyer's patches is not altered by this treatment, however, nor is lymphocyte binding to Peyer's patch HEV.

2.2 Lymphocyte–PNHEV adhesive interactions are oligosaccharide-dependent

MEL-14 immunoprecipitates a 90 kDa lymphocyte surface protein (termed gp90^{MEL}) (8), whereas neutrophils express the molecule in a slightly larger form (9). Molecular cloning work indicates that the lymphocyte and neutrophil molecules differ in their posttranslational modification (by glycosylation and/or ubiquitination), but most certainly maintain co-linear polypeptide sequences (10). However, the initial studies of the function of this molecule took place prior to the cloning work. In these first studies, Stoolman and Rosen examined the possibility that carbohydrate determinants participated in lymphocyte-PNHEV adhesive interactions (11, 12). These studies were initiated after a consideration of others' morphological analysis of PNHEV. These studies had indicated that the surface of PNHEV consists of a rather prominent glycocalyx, which most probably represents the first surface encountered by the lymphocyte. The possibility that some carbohydrate component within this glycocalyx mediates adhesion was also consistent with the observation that glutaraldehyde fixation of exposed PNHEV does not diminish their ability to bind lymphocytes (7). Since glutaraldehyde fixation typically destroys peptide antigens, but leaves oligosaccharide epitopes relatively intact, these observations suggested the possibility that lymphocytes recognized carbohydrate determinants expressed on the glycocalyx of the PNHEV surface (11).

To examine this possibility, Stoolman and Rosen tested several different mono- and polysaccharides for their ability to block lymphocyte attachment to PNHEV (11). While most monosaccharides were found not to block adhesion, two did (D-mannose and L-fucose), although at relatively high concentrations (> 75 mM) and under conditions of high ionic strength. Amongst a group of polysaccharides tested that included heparin, dextran sulphate, chondroitin sulphate, and fucoidan (a sulphated high molecular weight polymer containing $\alpha(1\rightarrow2)$-and $\alpha(1\rightarrow3)$-linked fucose residues; reference 13), only fucoidan significantly inhibited adhesion, half maximally at low nanomolar concentrations.

These observations were extended in studies that examined several phosphorylated carbohydrates (12). Mannose-6-phosphate and fructose-1-phosphate, at physiological ionic strength, were found to block lymphocyte binding to PNHEV. Inhibition was half-maximal at concentrations between 2 and 5 mM, and complete inhibition was observed at 10 mM. A group of other phosphorylated monosaccharides did not inhibit binding. A polymeric polysaccharide known as PPME potently inhibited lymphocyte adherence, with half maximal activity at low nanomolar concentrations. PPME is a mannan-based substance enriched in mannose-6-phosphate groups, and is derived from the yeast *Hansunela holstii* (14). PPME-dependent inhibition of adhesion was observed when lymphocytes were pretreated with this compound, but pretreatment of PNHEV did not affect binding. Nonphosphorylated yeast mannans exhibited no effect on binding. These observations indicated that PPME interacts with a lymphocyte surface receptor responsible for lymphocyte-PNHEV interactions, and also suggested that this PPME and fucoidan maintain tertiary configurations similar to those displayed by counter-receptors expressed on the surface of PNHEVs.

This work was further extended in studies using fluorescent beads derivatized with PPME. These beads bind to lymphoid cells capable of adhering to PNHEV, but do not bind to nonadherent cell lines (15). Bead binding, like lymphocyte-PNHEV adhesion, is calcium-dependent, and can be blocked by the same carbohydrate-containing compounds that inhibit lymphocyte-PNHEV attachment. The relative inhibitory potencies observed in the rodent system were also maintained when examined in a human system, indicating that the receptor's carbohydrate binding specificity may have been conserved through evolution (16). Later studies demonstrated that the lymphocyte receptor for PPME is in fact the MEL-14 antigen (17). These experiments thus suggested that MEL-14 corresponds to a lymphocyte cell surface glycoprotein that recognizes a carbohydrate molecule displayed at the surface of PNHEV, and that this interaction yields an adhesive interaction between lymphocytes and PNHEV.

Subsequent studies revealed that sialic acid moieties are essential components of the putative carbohydrate counter-receptor for MEL-14. For example, sialidase treatment of paraformaldehyde-fixed PNHEV substantially diminishes lymphocyte adhesion (18). By contrast, sialidase treatment of Peyer's patch HEV has no effect on lymphocyte attachment, and only partially diminishes binding to treated mesenteric node HEV. These latter observations are consistent with previous experiments showing that mesenteric node HEV expresses both PNHEV ligands, whereas Peyer's patches HEV display an adhesive ligand entirely different from that found on Peyer's patches HEV (19).

Similar results have also been obtained in *in vivo* experiments. In a murine model of short term lymphocyte accumulation, for example, it has been shown that intravenous injection of *Vibrio* or *Clostridium* sialidase significantly diminishes short term accumulation of lymphocytes into peripheral lymph nodes, but does not alter lymphocyte accumulation in other lymphoid organs (Peyer's patches, spleen, kidney, and liver) (20). Histological studies in treated animals demonstrate that sialidase infusion effectively removes sialic acid residues from both peripheral node and Peyer's patches HEV. Moreover, PNHEV taken from sialidase-treated animals is not capable of supporting normal amounts of lymphocyte adherence, while Peyer's patches HEV from these animals exhibits normal binding efficiency. When considered together with the results obtained from *in vitro* sialidase treatment of PNHEV, this work implies that sialic acid moieties represent an essential component of the endogenous PNHEV for the MEL-14 molecule, and provides additional evidence that MEL-14 functions as a mammalian carbohydrate binding protein, or lectin.

2.3 The MEL-14 homing receptor contains a carbohydrate recognition domain

cDNAs encoding MEL-14 were cloned using protein sequence information (21, 22). The human equivalent (LAM-1) was cloned by expression cloning techniques (23,

Table 1 L-selectin nomenclature (27)

Previous names	Reference
LEC-CAM-1 (former consensus)	27
Mouse	
Mel-14	8
lymphocyte homing receptor	21
Ly-22	21
Human	
Leu8	28
TQ1	28
DREG-56	29
LAM-1	23

24) and by cross-hybridization with the murine probe (25, 26). While the consensus term L-selectin is now used to describe this molecule (27), it has had many other names (summarized in Table 1).

The sequence of L-selectin predicts a set of distinct domains (Figure 1). The amino-terminal signal sequence yields a type I transmembrane protein, oriented such that its amino terminus and the bulk of its mass lies outside of the cell. One report suggests the existence of alternatively spliced human L-selectin transcripts that could yield a form of L-selectin without a hydrophobic membrane-spanning domain (24). It has been proposed that this form of L-selectin is attached to the cell through a phospholipid anchor.

Sequence analysis of the 116 residues at the amino terminus of the mature protein segment disclose striking primary sequence similarity to a family of calcium-dependent carbohydrate binding proteins classified by Drickamer (1). This family of proteins are termed C-type lectins, and include the asialoglycoprotein receptor, the fucose binding protein of Kupffer cells, a flesh fly lectin, and a variety of mannose binding proteins. The carbohydrate recognition domains (CRDs) of these proteins are approximately 25–30% identical to the 116 amino terminal residues in L-selectin. Four cysteine residues that are conserved amongst these C-type lectins (1) are also conserved in L-selectin. The amino acid positions between these conserved residues varies substantially from the other C-type lectins, however, suggesting that the conserved cysteines serve to mould their tertiary conformation, whereas carbohydrate binding specificity may be dictated by intervening amino acids. The murine and human L-selectins CRDs share 83% sequence identity, consistent with previous observations that implied conservation of ligand recognition specificities across species (16).

The CRD of L-selectin is located next to a protein domain with primary sequence similarity to the epidermal growth factor precursor (EGF). This domain is 33–40 residues long, and corresponds to a protein sequence motif discovered in a variety of proteins, including the LDL receptor, factor X, and polypeptides encoded by the *Notch* locus of *Drosophila melanogaster*. The EGF-like segment is adjacent to a pair of

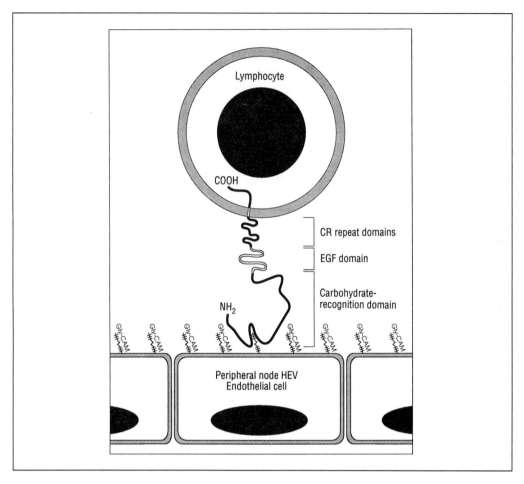

Fig. 1 Architecture of L-selectin. The domains of L-selectin defined by molecular cloning studies are shown approximately to scale. Their folded arrangements are displayed in a form proposed for the P-selectin molecule by Johnston *et al.* (30), using predicted intrachain disulphide bonds. L-selectin is shown here as a lymphocyte surface molecule, but is also found on other types of leukocytes, including neutrophils, as outlined in the text. The CRD of L-selectin mediates adhesion between circulating lymphocytes and oligosaccharide ligand(s) expressed by the cuboidal endothelial cells that line the 'high endothelial venules' within peripheral lymph nodes. Recent work indicates that some of these oligosaccharide ligands are displayed by *O*-linked oligosaccharide molecules attached to a mucin-like glycoprotein, termed GlyCAM, that is expressed by the endothelial cells

tandemly repeated 62 amino acid-long protein segments. Each of the two component segments exhibits primary sequence similarity to a peptide motif found in numerous other proteins including decay accelerating factor (DAF), complement receptor 1, the receptor for C3d, factor XIII, and the human IL-2 receptor. This peptide motif is known as a CR repeat, for its similarity to the complement receptor molecules (31). The CRD, EGF, and CR domains of L-selectin are anchored within the cell membrane via a hydrophobic transmembrane domain, that is in turn attached to a short cytoplasmic domain.

The genomic structures of human (32) and murine (33) L-selectin indicate that the genes from the two species share a similar organization. In each species, single distinct exons encode the signal peptide sequence, the CRD, the EGF domain, the unit CR repeats, and transmembrane segment. Two exons correspond to the cytosolic domain. *In situ* hybridization studies localize the human L-selectin gene to chromosome 1 between bands q22 and q25 (reference 23 and unpublished data cited therein and reference 34). Its position is distinct from the position of a series of genes localized to band 1q32 that encode several other CR-repeat-containing proteins (34, 35). As noted below, the genes that encode other selectin family members (E-selectin and P-selectin) each maintain an organization similar to the L-selectin gene, and are also found between 1q22 and 1q25 (34). These observations have prompted a suggestion that the three selectin genes correspond to an 'adhesion molecule locus' of very closely linked sequences that most probably evolved by gene duplication events (23).

2.4 Functions of the domains of L-selectin

The functional properties of the domains within L-selectin have been studied with recombinant L-selectin chimeras, and with antibody blocking methods. For example, Bowen *et al.* have explored the functional attributes of the CRD and EGF domains using recombinant chimeric mouse L-selectin molecules constructed from the mouse, intermixed with CRD segments, along with EGF and CR segments. These segments were in turn displayed by the hinge, CH2 and CH3 regions of human IgG1 (36). Studies with these molecules demonstrate that the MEL-14 epitope corresponds to a position within the amino terminal 53 residues of mouse L-selectin. Thus, since the MEL-14 antibody prevents binding of PPME to L-selectin (17), this work infers that L-selectin interacts with carbohydrate ligand(s) through its CRD. This work also showed that the EGF domain operates to maintain the CRD in a conformation competent to allow carbohydrate binding.

In a similar approach, chimeras constructed from L-selectin and P-selectin have been tested for PPME and fucoidan binding (6, 37). This work also showed that the L-selectin CRD interacts directly with its carbohydrate ligand(s), demonstrated that sequences within the COOH-terminal half of the CRD were critical for ligand binding specificity, and implied that the EGF domain participates in directing the proper folding of the CRD, but does not itself interact directly with ligand.

In an extension of this work, Spertini *et al.* have used monoclonal antibodies to define epitope position and correlate this with carbohydrate binding properties (38). These studies, and related ones (6, 37), have shown that many anti-L-selectin monoclonal antibodies that recognize the CRD also block attachment of lymphocytes to PNHEV, and inhibit binding of fucoidan or PPME (LAM1-3, MEL-14, DREG-56, TQ1). However, other antibodies specific for the CRD may block attachment yet not fucoidan or PPME binding (LAM1-6), or else do not perturb lymphocyte or carbohydrate binding (Leu8 and LAM1-10). Other monoclonal antibodies, directed against the EGF-like domain, have been shown to increase the affinity of the CRD for PPME (LAM1-1 and LAM-5, (6, 37, 38) or decrease its affinity (Ly-22, references 39 and 40). These studies thus also show that oligosaccharide recognition and lymphocyte binding properties are mediated through the CRD, and that the functional attributes of this domain are maintained by, and can be modulated by, the EGF domain.

Watson et al. reported the use of recombinant chimeric selectin molecules constructed from the lectin, EGF, and CR domains of mouse L-selectin, linked to the hinge, CH2 and CH3 domains of human IgG1 heavy chain Fc domain, to also explore the roles of the CR repeats in L-selectin (41). This work demonstrated that the two CR repeat segments within L-selectin are necessary for its proper conformation and function, as assessed by MEL-14 reactivity, and by PPME binding and PNHEV ligand recognition.

These chimeric molecules have also been used to explore the molecular nature of the endogenous PNHEV ligands for L-selectin. In this study, Watson et al. (42) demonstrated that binding of a recombinant L-selectin-IgG molecule to PNHEV is blocked by pretreatment of the PNHEV with *Clostridium perfringens* or *Vibrio cholera* sialidases. Likewise, pretreatment of PNHEV with *Limax flavus* agglutinin, a lectin specific for terminal sialic acid determinants, blocks binding of the recombinant chimera. Nonetheless, pretreatment with a different sialic acid-specific lectin (*Limulus polyphemus* agglutinin) did not diminish binding of the chimera. Thus, these results confirm, for the most part, previous observations that sialic acid moieties are necessary for L-selectin-dependent lymphocyte binding (43). The apparent discrepancy obtained with *Limulus polyphemus* agglutinin may perhaps be accounted for by differences in specificities exhibited by each lectin for different sialic acid linkages (44), for example, or by the wide variety of sialic acid linkages and sialic acid modifications (*O*-acetylation and *N*-glycolylation, for example) contained in mammalian oligosaccharides (45).

2.5 GlyCAM-1 is part of an endogenous ligand for L-selectin

Immunoprecipitation studies using a murine L-selectin-IgG chimera identify a 50 kDa radiolabelled glycoprotein from peripheral lymph nodes labelled with [^{35}S]sulphate, along with amounts of a 90 kDa radiolabelled molecule (41, 42, 46). Both proteins also immunoprecipitate with MECA-79 monoclonal antibody, shown previously to block lymphocyte attachment to PNHEV and thus recognize the L-selectin counter-receptor (47, 48). These molecules could be identified in mesenteric node HEV, but not on Peyer's patches HEV. They are trypsin-sensitive, and their immunoprecipitation is calcium-dependent and specifically inhibited by fucoidan and PPME. Radiolabelled fucose is also incorporated into these proteins, whereas radiolabelled mannose or glucose were not. These observations indicate that these molecules represent sulphated, fucosylated glycoproteins. Absence of labelling with mannose indirectly indicates that these proteins contain little, if any *N*-linked oligosaccharides, as these would label with mannose (49). Digestion of the proteins with N-glycanase did not alter the electrophoretic mobility of these proteins, providing further support for this notion. Imai et al. cite unpublished data that treatment of the molecules with reagents known to release O-linked sugars releases radiolabelled fucose and sulphate, suggesting, again indirectly, that the glycans on these two proteins are O-linked.

Imai et al. (46) have termed the 50-kDa immunoprecipitated molecule Sgp^{50}, and the 90 kDa molecule Sga^{90}, for sulphated glycoprotein of 50 kDa and 90 kDa, respectively. These candidate endogenous ligands for mouse L-selectin are most likely recognized by their glycan moieties, since the L-selectin-IgG chimera immunoprecipitates Sgp^{50} and Sgp^{90} even after they have been extracted from denaturing SDS polyacrylamide gels.

Sgp^{50} has been purified and cloned (50). The sequence of the cloned cDNA indicates that Sgp^{50} is 132 amino acids long, and is extremely rich in serine and threonine residues. There is a single potential N-linked glycosylation site. The abundance of serine and threonine residues, coupled with the large difference in size between the predicted (14, 154) and observed (~50 000 dalton) masses, indicates that the bulk of the mass of the protein in PNHEV is composed of O-linked carbohydrate, and suggest that this protein is a mucin-type glycoprotein. The protein is termed GlyCAM-1, for glycosylation-dependent cell adhesion molecule. It apparently does not maintain a transmembrane segment, suggesting that it maintains an association with the HEV surface through a COOH-terminal 21 residue segment predicted to be helical and amphipathic, or via interaction with another membrane-tethered molecule.

Northern blot analyses indicate that the GlyCAM-1 gene is transcribed in peripheral and mesenteric lymph nodes, and possibly in lung, but not in other organs. In situ hybridization and immunohistochemical analyses showed that GlyCAM-1 expression is restricted to the HEV within lymph nodes. Lasky et al. predict that GlyCAM-1 may assume a 'bottle brush' topology, wherein the bristles would be represented by O-linked oligosaccharide chains attached to the two serine/threonine-rich regions. They further propose that GlyCAM-1 functions largely to 'present' an oligosaccharide ligand to L-selectin.

2.6 Sialic acid and sulphate are essential to the adhesive activity of GlyCAM-1

The structures of the carbohydrates that make up the functional part(s) of the endogenous PNHEV L-selectin counter-receptor remain undefined at this time. As noted above, sialic acid residues are known to be essential for binding activity. To further address the nature of the glycosidic bond(s) (i.e. $\alpha 2 \rightarrow 3$, $\alpha 2 \rightarrow 6$, or $\alpha 2 \rightarrow 8$ linkages) that link sialic acid residues within the endogenous L-selectin ligand(s), Imai et al. tested two different sialidases in an assay that detects binding of PNHEV-derived, sulphate-labelled GlyCAM-1 to the L-selectin-IgG chimera (40). They found that a sialidase with specificity for $\alpha 2 \rightarrow 3$-linked sialic acid residues

(Newcastle disease virus neuraminidase) was nearly as effective at inactivating ligand activity as was a broad spectrum sialidase (*Arthrobacter ureafaciens* neuraminidase). Additional structural detail concerning the number and configurations of sialic acid moieties on GlyCAM-1 is not yet published.

A requirement for sulphate for functional activity has been demonstrated recently (51). In these studies, nonsulphated GlyCAM-1 was purified from PNHEVs treated with chlorate (an inhibitor of sulphation). This molecule did not sustain binding by the L-selectin-IgG chimera under conditions where a control GlyCAM-1 preparation did bind. Since the sulphate-depleted GlyCAM-1 molecule was apparently otherwise normally sialylated and fucosylated, the author concluded that sulphate, on carbohydrates, may form an essential part of the recognition motif for L-selectin. The chemical nature of such molecules remains to be defined. In this context, however, it is interesting to note that L-selectin has been shown to bind to both sialyl-Le^X and sialyl-Le^a, oligosaccharide ligands for E-selectin (described in detail below) (52, 53). Green *et al.*, have reported similar results, and have additionally shown that L-selectin binds with high affinity to sialyl-Le^X and sialyl-Le^a derivatives wherein the terminal sialic acid residue is replaced by a sulphate residue (54). It seems unlikely that either sialyl-Le^X or sialyl-Le^a actually represents an authentic endogenous ligand for L-selectin since, with one exception (55), these epitopes have not been detected on human HEV (cited in reference 52). It is not yet known if their sulphated analogues are expressed there; given the requirement for sialic acid in L-selectin interactions, it seems probable that they also will not represent the sole component of the oligosaccharide ligand for L-selectin.

2.7 L-selectin expression and function in neutrophil adhesion to endothelium

L-selectin is expressed on lymphoid cells, neutrophils, eosinophils, and monocytes. Lymphocyte L-selectin is approximately 78 kDa in size, whereas it is approximately 90–100 kDa as expressed on neutrophils. Lineage specific glycosylation accounts for the difference between these sizes, and the difference from the size predicted for its primary translation product (~40 kDa) (reviewed in reference 6).

L-selectin expression is regulated by at least two mechanisms. The amount of L-selectin expressed at the cell surface is affected by activation-dependent shedding (29, 56, 57, 58). Shedding occurs when lymphocytes are activated by phorbol 12-myristate 13-acetate (PMA) or by mitogens (28, 57). Similarly, treatment of granulocytes with PMA, GM-CSF, lipopolysaccharide, or calcium ionophores can cause release of L-selectin (59). Shedding is rapid (within minutes following activation) and may be inhibited by protein kinase C inhibitors (6, 57). Proteolytic cleavage of L-selectin immediately adjacent to its transmembrane segment is most probably responsible for this process. Soluble, functional L-selectin is found at surprisingly high levels in human plasma (1.6 µg/ml), but its role, if any, remains to be defined (60).

The affinity of L-selectin for its ligand(s) can also be regulated. Neutrophil activation by G-CSF, G/M-CSF, or tumour necrosis factor α results in an increase in the amount of PPME that is bound per cell (56). PPME binding under these circumstances is L-selectin dependent, and is also observed when lymphocytes are activated by T-cell receptor cross-linking manoeuvres. Since increased PPME binding is not associated with a change in the number of L-selectin molecules, these observations imply that these activational stimuli increase the intrinsic affinity of L-selectin for PPME, and, by inference, its natural ligand(s). This is analogous to the increase in intrinsic adhesive 'activity' of VLA and CD11/CD18 integrins observed upon leukocyte activation (61, 62). It has been speculated that the L-selectin-dependent, tissue-specific homing patterns of different cell types may be accounted for, in part at least, by cell-type responses to different activational stimuli.

L-selectin apparently functions to mediate adhesion of leukocytes to vascular endothelium outside the lymphoid system. For example, anti-L-selectin antibodies can inhibit emigration of neutrophils into inflamed peritoneum (63) or into inflammatory skin lesions (9). Likewise, neutrophil adhesion to cytokine-activated endothelium can be blocked if the neutrophils are first treated with anti-L-selectin antibodies (64, 65). L-selectin-dependent adhesion to extranodal endothelium occurs under conditions of flow, indicating that this adhesion is CD18-independent. Related observations indicate that monocyte adhesion to activated vascular endothelium is also L-selectin-dependent (66), as is attachment of neutrophils, monocytes, and lymphocytes to activated renal glomerular endothelium (67).

These observations are consistent with those made in *in vivo* systems capable of documenting the role of L-selectin in leukocyte 'rolling' (68–70). In these systems, intravital microscopy shows that leukocytes 'roll' along the endothelial cell lining of inflamed capillaries (68, 71, 72). Rolling apparently occurs in part because vasodilatation alters the vessel's flow characteristics, causing leukocytes to move away from their usual position at the centre of the vessel lumen where flow is most rapid (73). Dilatation of the vessel, along with increases in plasma viscosity cause the flow to slow, and in turn cause leukocytes to move to peripheral locations in the stream (74). Rotational forces are generated by differentials in flow rates between central positions (fast) and lateral positions (slow), and these cause leukocytes to roll along the wall of the dilated capillary. This effectively maximizes the interaction between the surface of the leukocyte and the endothelium. This interaction must include adhesion, since the velocities of rolling leukocytes have been shown to be substantially less than would be otherwise expected if these cells were tumbling freely (75). Experimental evidence implies that at least a portion of these adhesive interactions is L-selectin-dependent. For example, leukocyte rolling is blocked by intravascular administration of an L-selectin-IgG chimera, or anti-L-selectin antibodies in rabbit and rat models (68–70, 76, 77). Rolling is not blocked by anti-CD18 antibodies, however, although they block inhibit the firm attachment of rolling leukocytes (69, 70). These observations suggest that leukocyte recruitment occurs via a sequential process involving rolling first, followed by firm attachment, and that each of these two processes is mediated by a distinct class of adhesion molecules (L-selectin for rolling, and the integrins for firm attachment). This model (70, 78) will be discussed in additional detail at the end of this chapter. This model also predicts that neutrophil recruitment may be abrogated by blocking the first step in the process, namely L-selectin-dependent rolling. This has been successfully demonstrated in a murine model, where neutrophil recruitment to thioglycollate-inflamed peritoneum was inhibited by administering an intravascular dose of L-selectin-IgG chimera (77). Nonetheless, the counter-receptor for L-selectin-dependent neutrophil adhesion has not been characterized. It is possible, however, that some component of L-selectin-dependent neutrophil adhesion may involve E-selectin, since neutrophil L-selectin apparently displays oligosaccharide ligands for E-selectin (79). Other molecules that contribute to this process remain to be identified.

3. Oligosaccharide-dependent cell adhesion mediated by E-selectin

3.1 E-selectin-dependent adhesion of myeloid cells to activated endothelium

Cultured endothelial cell monolayers derived from human umbilical veins (HUVECs) have been used to study adhesion of leukocytes to vascular endothelium, and to identify molecules that mediate this process (80). HUVEC monolayers treated with the inflammatory cytokines interleukin 1β or tumour necrosis factor α, or with lipopolysaccharide, will support adhesion of neutrophils, monocytes, and myeloid-lineage cell lines like HL60 cells (80–84). Maximal adhesion occurs at roughly 6 h after cytokine exposure, and is protein synthesis-dependent, suggesting that cell surface adhesion molecules are expressed in response to cytokine-dependent endothelial cell activation. To identify such molecules, monoclonal antibodies were developed against cytokine-activated endothelium that specifically inhibit adhesion of myeloid cells to activated endothelial cell monolayers (82, 85). One pair of such antibodies immunoprecipitate a cytokine-dependent endothelial cell protein of M_r 115 000, whose temporal expression pattern correlates exactly with the temporal pattern of cytokine-dependent adhesion. This protein was initially named endothelial leukocyte adhesion molecule I, or ELAM-1, and later termed E-selectin (27, 82).

E-selectin is expressed by capillary endothelium in a variety of acute and chronic inflammatory conditions, including rheumatoid arthritis (86), sepsis (87, 88), immune complex-dependent acute lung injury (89), delayed hypersensitivity in the skin (90–93) and other skin conditions (94), and in human cardiac allografts (95). Counter-receptor(s) for E-selectin are found on neutrophils (80), monocytes (81, 96), eosinophils (97), memory T-cells (92, 98, 99), and NK cells (100). Since these cell types are also present in inflammatory sites where E-selectin is also expressed, it can be expected that E-selectin operates to promote the recruitment of these cells in both acute and chronic inflammation.

E-selectin turnover may occur via endocytosis (101), and by release from the cell surface (102, 103). Release may occur via a proteolytic process analogous to that responsible for generating soluble L-selectin. E-selectin has been reported to be phosphorylated at one or more serine residues on its extracellular domain (104). The function of this modification remains unknown.

3.2 E-selectin contains a carbohydrate recognition domain similar to the L-selectin CRD

Human E-selectin cDNAs have been cloned, using expression cloning methods (105, 106). The sequence of these cDNAs reveals that E-selectin exhibits substantial primary sequence similarity to L-selectin, with domains analogous to those found in L-selectin (Figure 2). There is an amino terminal domain sharing primary sequence similarity to C-type lectins and the L-selectin CRD, followed by an EGF-like domain and four CR repeat domains. These portions of the molecule associate with the cell surface via a hydrophobic transmembrane domain. A short COOH-terminal domain resides within the cytosol. E-selectin molecules cloned from rabbit (107) and mouse (108) are similarly organized.

The E-selectin lectin and EGF domains are approximately 61% identical in sequence to the corresponding domains in the L-selectin, whereas their transmembrane and cytosolic domains share less than 20% identity. The organization of the human E-selectin gene is similar to that of the L-selectin gene, in that the signal sequence, carbohydrate recognition domain, CR repeats, and the membrane-spanning segment, each correspond to a distinct exon, whereas the cytosolic segment is encoded by two exons (109). The E-selectin gene is found on human chromosome 1 (109) near the human L-selectin gene (23, 34).

3.3 Ligands for E-selectin—sialylated, fucosylated oligosaccharides as candidates

Primary sequence similarity between the E-selectin 'lectin' domain and the C-type lectins suggested that E-selectin mediated leukocyte adhesion via interaction with oligosaccharide counter-receptor(s) on these cells. This possibility was made especially tenable by the observations implicating oligosaccharide ligands for L-selectin. Given the apparent specificity of E-selectin for cells derived from the myeloid lineage, it was to be expected that E-selectin might interact with oligosaccharide determinants whose expression was restricted to these cell types. Furthermore, since the sequences of the amino terminal CRDs of L- and E-selectin were found to be so similar, and since sialic acid had been implicated as an essential component of the L-selectin ligand, it seemed reasonable to consider the possibility that sialic acid might also be part of an E-selectin ligand.

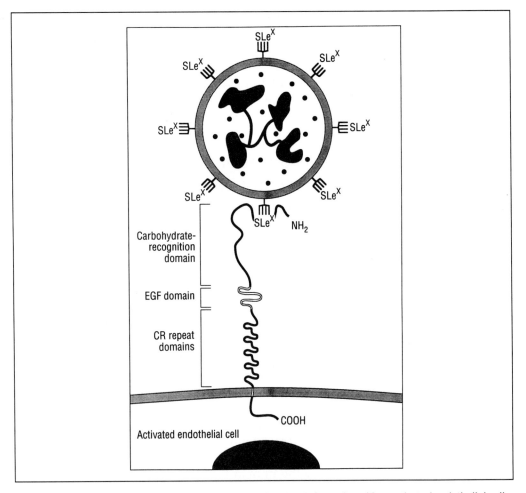

Fig. 2 Architecture of E-selectin. E-selectin is expressed at the surface of cytokine-activated endothelial cells. Its component domains are drawn approximately to scale, and displayed in an arrangement analogous to that proposed for P-selectin (30). The carbohydrate recognition domain of E-selectin mediates adhesion of neutrophils (shown here), monocytes, eosinophils, and some kinds of lymphocytes, via an interaction with carbohydrate counter-receptors (indicated by ⱨ) on these cells. For many of these cells, the sialyl-Le^x tetrasaccharide determinants is a critical component of these counter-receptors. E-selectin is displayed by activated capillary endothelium

Candidate ligands fulfilling these criteria could be identified among a set of sialylated, fucosylated oligosaccharides expressed on myeloid lineage cells and characterized by Fukuda and his collaborators (110–113), and by others (114). These molecules consist of protein-linked and lipid-linked oligosaccharides with cores represented by the lactosamine moiety [Galβ1→4GlcNAcβ1→3Gal], represented from one to several times as a linear chain. The polylactosaminoglycan structure is a common feature of O-linked glycans, N-linked glycans, and lipid-linked glycans on myeloid lineage cells (110–114). Its synthesis is catalysed by the sequential and alternating action of a pair of glycosyltransferases (Galβ1→4GlcN-Ac transferase and GlcNAcβ1→3Gal transferase, Figure 3). These polylactosamine chains can exist in an unmodified form, or may be modified by terminal α2→3- or α2→6-linked sialic acid or by α1→2-linked fucose residues. Such molecules are found on many types of blood cells, including those that do not bind to E-selectin (i.e. red cells, reference 119). Such glycoconjugates thus do not represent likely candidate ligands.

3.4 Ligands for E-selectin—biosynthesis of sialylated, fucosylated oligosaccharides

Polylactosamine chains are modified by α1→3-linked fucose residues in myeloid lineage cells, but not on other types of blood cells, thus suggesting a role for such molecules as E-selectin ligands. This type of fucose modification can occur on GlcNAc residues of lactosamine units (Figure 4). This modification is catalysed by a group of glycosyltransferases called α1→3fucosyltransferases (see Figure 4, reviewed in references 132 and 133). These studies indicate that this fucosylation is a terminal event in the biosynthesis of α1→3fucosylated molecules. Some, if not all α1→3fucosyltransferases can utilize α(2→3)sialylated polylactosamine moieties as substrates, while no known α2→3sialyltransferases can utilize an α1→3fucosylated glycan as a substrate (Figure 4).

Glycoconjugates constructed by these pathways include α(2→3) sialylated type II chains substituted with subterminal Fucα(1→3)GlcNAc linkages (known as the sialyl-Le^x moiety; Figure 4) and neutral type II fucosylated moieties (known as Le^x, SSEA-1, reference 134; or CD15; Figure 4). Singly fucosylated, α(2→3)sialylated molecules have also been described that contain an internally fucosylated lactosamine unit (VIM-2, reference 135; or CD65, Figure 4). The VIM-2 determinant may be synthesized by some α1→3fucosyltransferases that are unable to fucosylate the remaining unsubstituted GlcNAc on the terminal lactosamine unit (125). In other instances, the VIM-2 determinant may serve as a precursor to the difucosylated, sialylated species (136) in a pathway involving ordered fucosylation (122, 117) (Figure 4). These glycoconjugates, and multiply fucosylated variants, can be detected by a variety of monoclonal antibodies that specifically recognize the terminal tri- and tetrasaccharide determinants (136, 137). These considerations indicate that expression of these molecules may be controlled through regulating expression of α1→3fucosyltransferase genes. Four structurally distinct human α1→3fucosyltransferase genes have now been cloned (123–126, 128, 129). The acceptor substrate requirements of these enzymes indicate that each can construct a subset of the molecules shown in Figure 4. The expression patterns of these genes are not well understood, nor are mechanisms that regulate their expression.

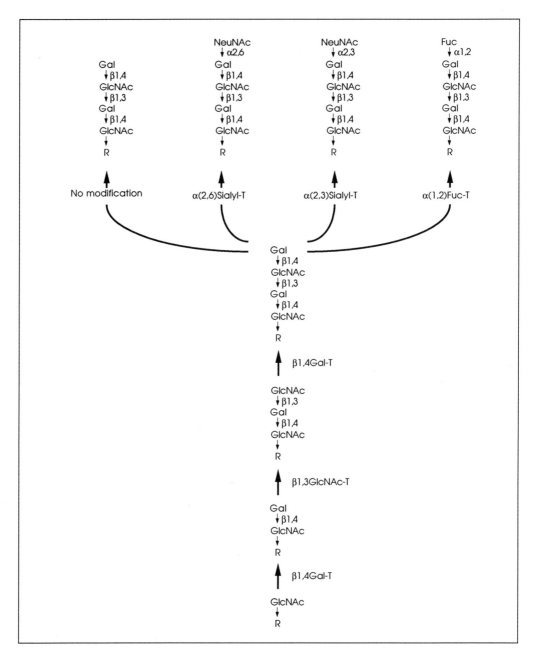

Fig. 3 Synthesis and chain termination of polylactosamines. Alternating galactose (Gal) and *N*-acetylglucosamine (GlcNAc) residues represent the component molecules of polylactosamine chains. These monosaccharides are linked together via alternating enzymatic reactions catalysed by a β1→4galactosyltransferase (β1,4Gal-T) and a β1→3*N*-acetylglucosaminyltransferase (β1,3GlcNAc-T) (111, 115). Branched forms arise from the linear backbone via the action of β1→6*N*-acetylglucosaminyltransferases (not shown). These linear or branched glycan chains may be underpinned by serine or threonine residues, in *O*-linked glycans (116), by asparagine residues (49), or by glycosphingolipids

3.5 Ligands for E-selectin—the sialyl-Le^X determinant as a myeloid cell counter-receptor for E-selectin

Several groups took advantage of existing information about the biosynthesis and expression patterns of these sialylated, fucosylated oligosaccharides to identify one of them, the sialyl-Le^X determinant, as a ligand for E-selectin. Lowe *et al.* (127) used cloned DNA segments encoding α(1→3)fucosyltransferases to alter the glycosylation phenotype of cultured cell lines (CHO and COS-1) that do not bind to E-selectin, and that are deficient in α(1→3)fucosyltransferase activity. Since these cell lines do construct oligosaccharide molecules that are precursors to Le^X and sialyl-Le^X determinants, transfection with α(1→3)fucosyltransferase genes (pCDNA1-α(1,3)FT; Fuc-TIV, reference 125; pCDM7-α(1,3/1,4)FT; Fuc-TIII, reference 123; Figure 4) yielded transfectants that expressed new surface-localized Le^X and/or sialyl-Le^a antigens. The expression patterns of these structures was then correlated with the E-selectin binding properties of each cell line. These studies found that sialyl-Le^X expression correlated with E-selectin binding, whereas cells that expressed Le^X, or VIM-2, in the absence of sialyl-Le^X, did not bind to E-selectin (125, 127). These studies indicated that one or more members of the family of α(2→3)sialylated, α(1→3)fucosylated polylactosamines represented by the sialyl-Le^X determinant function as endogenous ligands for E-selectin. This work further indicated that lineage-specific expression α(1→3)fucosyltransferase genes may control lineage-specific expression of these oligosaccharide ligands. This work also suggested that the oligosaccharide moieties themselves were primarily recognized by E-selectin, since E-selectin ligands were expressed on nonmyeloid cells, and were thus not components of myeloid-specific proteins and/or lipids.

Goelz *et al.* reached similar, if not identical conclusions (124). In this study, a monoclonal antibody (termed C2E5) was generated that was specific for a myeloid cell surface molecule, and that also blocked E-selectin-dependent cell adhesion. This antibody was used to expression clone a cDNA from HL-60 cells, that determined C2E5 reactivity when expressed in COS and CHO cells. E-selectin was shown to also bind to these transfectants. The sequence of the cDNA was similar to the human Lewis blood group α(1→3/1→4)fucosyltransferase cDNA (123), and cells transfected with this cDNA expressed α(1→3)fucosyltransferase activity. The enzyme encoded by the cDNA was termed ELFT, for ELAM-1 ligand fucosyltrans-

---

(117). The substructures are denoted by R. The terminal lactosamine repeat (Galβ1→4GlcNAcβ1→3) may exist unmodified (pathway at the *top left*), or the galactose residue may be subsequently modified by the action of α1→2fucosyltransferases, which attach an α1→2-linked fucose residue to form the H blood group determinant (pathway at the *top right*) (118). H determinants are expressed by cells of the erythroid lineage (119), but are not normally present on myeloid cells (111, 117). The terminal galactose may instead serve as an acceptor substrate for the action of sialyltransferases that can be attached by sialic acid (NeuNAc; N-acetyl neuraminic acid) in either α2→6- or in α2→3-linkage (top central two pathways) (120). N-acetylglucosamine moieties may also be modified by the action of α1→3fucosyltransferases that recognize N-acetylglucosamine moieties within polylactosamines (see Figure 4)

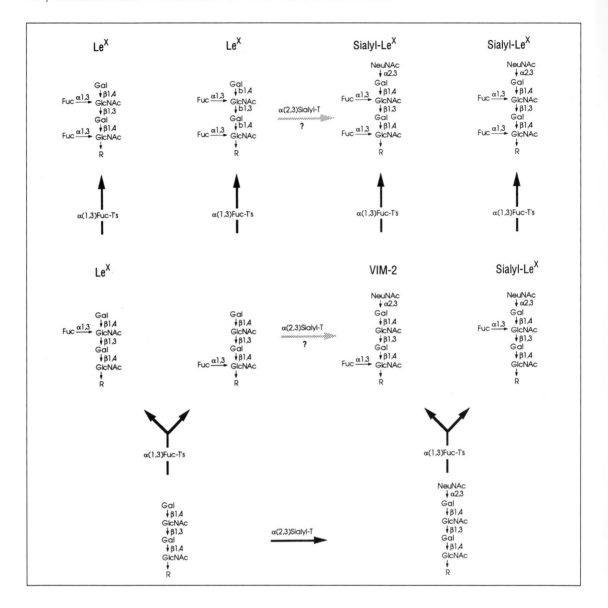

ferase. The authors did not report on the nature of the cell surface oligosaccharide antigens expressed by ELFT-transfected cells.

Kumar et al. independently used an expression cloning approach to isolate a human DNA segment identical to ELFT (clone D2.1, reference 126). These two DNA sequences are also identical to a gene termed Fuc-TIV, also cloned by Lowe et al. (125, 127; pCDNA1-α(1,3)FT). While these three isolates are identical in their sequences, different adhesion assay results have been obtained by the groups. While Goelz et al. found that this sequence determined E-selectin-dependent

Fig. 4 Synthesis of fucosylated oligosaccharides; the Le^X, sialyl-Le^X, and VIM-2 determinants. N-acetylglucosamine residues within neutral polylactosamines (*bottom left*) may be fucosylated by virtually any of the known α(1→3)fucosyltransferases (121–127). These monofucosylated oligosaccharides may then serve as substrates for additional fucosylations by α(1→3)fucosyltransferases, forming difucosylated species (vertical pathways shown at *left*), or (not shown) polylactosamine species containing more than two Fucα(1→)GlcNAc linkages (112). Fucosylation may proceed via an ordered process that is most probably enzyme-specific (121, 122). Neutral polylactosamine precursors may instead serve as substrates for the action of α(2→3)sialyltransferases (arrow at *bottom centre*), which form α(2→3)sialylated precursor molecules (*bottom right*). α(2→3)Sialylated oligosaccharides may then serve as acceptor substrates for the action of α(1→3)fucosyltransferases (vertical paths at *right*). Several α(1→3)fucosyltransferases (Fuc-TI (122), Fuc-TIII, and Fuc-TV (125–129)) efficiently fucosylate terminal N-acetylglucosamine moieties of sialylated precursors, forming monofucosylated sialyl-Le^X molecules (path at *far right*). Fucα(1→3)GlcNAc linkages may also be formed on an internal lactosamine unit (VIM-2 determinants) by these α(1→3)fucosyltransferases, and also by others that do not efficiently construct sialyl-Le^X molecules (Fuc-TII (122), Fuc-TIV or ELFT (124, 125, 127, 130)). Monofucosylated, α(2→3) sialylated molecules can also be fucosylated an additional time to form difucosylated sialyl-Le^X determinants (122, 131). Some α(1→3)fucosyltransferases add fucose first to internal lactosamine units (122), while others appear to add preferentially to terminal lactosamine units (122, 131). The known mammalian α(2→3)sialyltransferase activities apparently do not operate on polylactosamine molecules with a terminal Fucα(1→3)GlcNAc linkage (121, 131). The existence of α(2→3)sialyltransferases that can sialylate monofucosylated precursors to form VIM-2 determinants is also an open question. As discussed in the text, however, it will be necessary to pursue the possibility that there are α(2→3)sialyltransferase activities capable of completing these reactions (shaded arrows with question marks)

---

adhesion following transfection, Lowe *et al.* (125) and Kumar *et al.* (126) have been unable to demonstrate expression of E-selectin ligands on either COS cells (125) or CHO cells (125, 126) transfected with this gene. In addition, the enzyme encoded by this gene does not efficiently synthesize the sialyl-Le^X determinant *in vitro*, and does not determine its synthesis in transfected cells (125, 126). To account for this apparent discrepancy, Kumar *et al.* (126) discuss the possibility that endogenous CHO fucosyltransferase genes (like Fuc-TI, 122) capable of constructing sialyl-Le^X-based E-selectin ligands might have been 'activated' by the DNA transfection and amplification approaches, used in the experiments reported by Goelz *et al.* (124). Differences in assay methods (solution phase versus solid phase E-selectin presentation) or in the glycosylation phenotype of the CHO host cells may also be responsible for the observed differences. Final resolution of these discrepancies will probably await a biochemical analysis of the cell surface oligosaccharides expressed by the ELFT/Fuc-TIV transfectants, along with a more extensive characterization of the fucosyltransferase encoded by this gene.

In this context, it should be noted that the process whereby myeloid lineage cells synthesize sialyl-Le^X determinants is not well understood. As discussed above and diagrammed in Figure 4, current information suggests that fucosylation is the terminal event in the pathway. While ELFT/Fuc-TIV would seem to be the logical candidate fucosyltransferase for this reaction since it is expressed in the myeloid lineage, this enzyme does not efficiently create sialyl-Le^X determinants from the precursor substrate NeuNAcα(2→3)Galβ(1→4)GlcNAc *in vitro* (125, 126, 128, 129). Likewise, it cannot construct the sialyl-Le^X determinant when expressed in non-myeloid cell types (125, 126, 128, 129). Extracts prepared from neutrophils and other myeloid lineage cell types also do not contain an α1→3fucosyltransferase activity that can efficiently catalyse this reaction *in vitro*, however (132). These paradoxes may be explained by invoking a requirement by Fuc-TIV/ELFT for myeloid-specific molecules that facilitate its utilization of α(2→3)sialylated type II chains, in a manner where these factors operate in intact cells but not *in vitro*. Such factors might include a *trans*-acting molecules that modulate the acceptor substrate specificity of Fuc-TIV, similar to the way that α-lactalbumin modulates β1→4galactosyltransferase substrate specificity (reviewed in reference 138). Alternatively, such factors might consist of molecules that facilitate substrate utilization by presenting the oligosaccharide substrate acceptor to the enzyme in a preferential manner. This could include a myeloid-specific oligosaccharide, protein, or lipid, via a mechanism analogous to the way that the catalytic efficiency of a pituitary αGalNAc-transferase is regulated by the peptide that presents the oligosaccharide acceptor (139). It is also possible that an α(2→3)sialyltransferase exists in myeloid cells that can sialylate Galβ(1→4)[Fucα(1→3)]GlcNAc-terminated oligosaccharides (pathway indicated by the shaded arrow at the top of Figure 4). Answers to these questions will require additional biochemical and genetic studies.

Several other groups more or less simultaneously demonstrated that sialyl-Le^X containing oligosaccharides can operate as E-selectin ligands. Phillips *et al.* (140) employed a pair of CHO cell glycosylation mutants (LEC11, positive for sialyl-Le^X and Le^X; LEC12, positive for Le^X only; reference 122), and a HUVEC-based assay for E-selectin-dependent adhesion. The LEC11 cells adhered to activated HUVECs in an E-selectin-dependent manner, while the LEC12 cells did not. Sialidase treatment of the LEC11 cells, and pretreatment with anti-sialyl-Le^X antibody also blocked adhesion. Moreover, liposomes containing a purified sialyl-Le^X-containing glycolipid (difucosylated sialyl-Le^X) were shown to block E-selectin-dependent adhesion, while liposomes containing control glycolipids did not. This group has subsequently shown that a monofucosylated sialyl-Le^X molecule can also block binding as well as the difucosylated sialyl-Le^X glycolipid (141).

A recombinant E-selectin-IgG chimera was used by Walz *et al.* to probe for candidate E-selectin ligands on a panel of tumour cells and myeloid cell lines (142). An imperfect correlation was found between surface expression of the Le^X epitope and the cells' abilities to bind the recombinant E-selectin. Sialidase treatment of the cells eliminated binding by the chimera, however, pointing to the sialylated form of the Le^X determinant as the ligand. This was confirmed when a perfect correlation was found between sialyl-Le^X expression and E-selectin binding, and by studies where binding was blocked by pretreatment with anti-sialyl-Le^X antibody, but not by anti-Le^X or anti-VIM-2 antibodies. These results, and others reported in this paper, allowed the authors to conclude that E-selectin mediates leukocyte adhesion via the sialyl Le^X tetrasaccharide.

Another group approached the problem by directly testing fractionated myeloid cell glycolipids for their ability to support E-selectin-dependent cell adhesion (143). One purified acidic fraction that supported adhesion was subjected to structural analysis, and found to be a glycolipid consisting of 4 N-acetyllactosamine units linked to a ceramide group, terminated by α(2→3)linked sialic acid, and containing a single fucose in α(1→3)linkage on the GlcNAc residue of the subterminal

N-acetyllactosamine unit. This molecule is virtually identical to the VIM-2 determinant studied by Macher et al. (135; Figure 4). This result is in apparent conflict with results reported by Walz et al. (142) and Lowe et al. (125), which concluded that the VIM-2 determinant is neither necessary nor sufficient for E-selectin-dependent cell adhesion. Subsequent quantitative analyses indicate that E-selectin maintains an affinity for the VIM-2 determinant that is approximately one to two orders of magnitude less than its affinity for sialyl-Le^X-containing glycolipids (144).

3.6 Ligands for E-selectin—the sialyl-Le^a determinant as an E-selectin counter-receptor

The sialyl-Le^a moiety has also been shown to support E-selectin-dependent adhesion. The sialyl-Le^a moiety can be constructed from $\alpha(2\rightarrow3)$sialylated type I precursors through the action of $\alpha(1\rightarrow3/1\rightarrow4)$fucosyltransferase(s) (Figure 5). In studies reported by Berg et al. (146), a monoclonal antibody termed HECA-452 was used to screen a panel of neoglycoproteins for structures that would be distinct from the sialyl-Le^X determinant, but which would be candidates for ligands expressed by sialyl-Le^X-negative cell types (like skin homing memory T cells; refer-

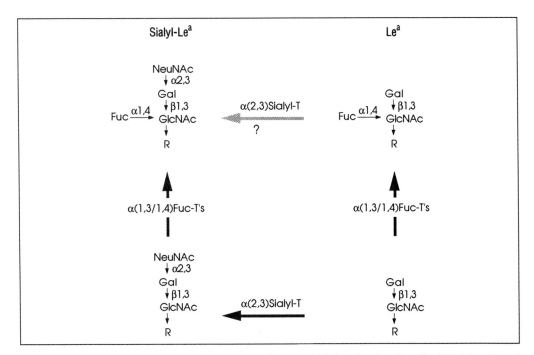

Fig. 5 Synthesis of type I Lewis determinants; the Le^a and sialyl-Le^a molecules. An $\alpha(2\rightarrow3)$sialylated type I precursor serves as a substrate for fucosylation to form the sialyl-Le^a molecule (145). This reaction may be catalysed by Fuc-TIII (123, 127, 129), which maintains the $\alpha(1\rightarrow4)$fucosyltransferase activity represented here, as well as an $\alpha(1\rightarrow3)$fucosyltransferase activity (outlined in Figure 4). Fuc-TIII is believed to be the product of the human Lewis blood group locus (123, 129)

ences 92 and 98) that bind to E-selectin. This approach identified a HECA-452-reactive neoglycoprotein molecule displaying the sialyl-Le^a moiety, and that maintained a slightly higher apparent affinity for E-selectin. Computer-generated molecular models presented in this paper indicated that the sialyl-Le^x and sialyl-Le^a molecules maintain the fucose and sialic acid residues in virtually identical positions in space on one face of each molecule. The nature of the E-selectin ligand on skin homing memory T-cells remains unknown at this time.

Takada et al. have independently demonstrated that sialyl-Le^a determinants mediate E-selectin-dependent cell adhesion (147). In these studies, the human colorectal carcinoma cell line Colo 201 showed E-selectin-dependent cell adhesion that was inhibitable by pretreatment with anti-sialyl Le^a antibodies, but not by anti-sialyl-Le^x antibodies. These cells were found to express large amounts of sialyl-Le^a determinants, and smaller amounts of sialyl-Le^x molecules. Liposomes containing sialyl-Le^a glycolipids, or sialyl-Le^x glycolipids, each effectively prevented sialyl-Le^a-dependent adhesion of Colo 201 cells to E-selectin, as well as sialyl-Le^x-dependent adhesion of HL-60 cells to E-selectin. These studies demonstrate that E-selectin can recognize both tetrasaccharides, and suggest that they interact with the same site(s) on E-selectin.

Tyrrel et al. have also shown that the sialyl-Le^a determinant supports E-selectin-dependent adhesion with an efficiency comparable to the sialyl-Le^x molecule (144). In studies also reported here, it was found that substitution of the N-acetylglucosamine at the reducing end of the sialyl-Le^x tetrasaccharide with glucose yields a molecule with an apparently enhanced affinity for E-selectin. Reduction of the glucose with sodium borohydride does not alter affinity significantly, indicating that the reducing sugar is not necessarily required to be maintained in a closed ring configuration for effective ligand activity. Likewise, sialyl-Le^x tetrasaccharides containing N-glycolyl and N-acetyl forms of sialic acid, or in which C-8 and C-9 of the sialic acid have been removed by mild periodate oxidation, exhibit affinities approximately equivalent to that maintained by a sialyl-Le^x tetrasaccharide with an unmodified sialic acid. However, isomers containing sialic acid in $\alpha(2\rightarrow6)$linkage are not bound by E-selectin to any significant degree. Molecular models of these isomeric compounds indicate that their fucose, sialic acid, and galactose residues occupy roughly similar tertiary positions, which presumably account for their similar binding affinities. More recent work has also shown that structural alterations at the reducing end of the molecule can substantially alter the binding affinity of the sialyl-Le^x and sialyl-Le^a determinants (148). For example, a sialyl-Le^a analogue in which the N-acetyl group was replaced with an amino group was found to be 36-fold more potent as a soluble inhibitor of E-selectin binding to sialyl-Le^x-containing neoglycoproteins.

Other recent studies indicate that E-selectin can also bind to analogues of the sialyl-Le^x and sialyl-Le^a determinants, in which the sialic acid moiety has been replaced by a sulphate group (149). Molecular models predict that the sulphate residue on each of these molecules is in a spatial position similar to the carboxyl portion of the sialic acid in the sialyl-Le^x and sialyl-Le^a determinants. This argues that productive E-selectin binding is mediated in part by charged residues displayed in space in a particular manner, in conjunction with appropriately located fucose residues. This notion is supported by mutational analysis of the E-selectin CRD, which has demonstrated that three positively charged residues (R97, K111, and K113) are critical for effective sialyl-Le^X binding (150), and may interact with the negatively charged portion of the sialic acid within this tetrasaccharide. In these latter studies, a model for the tertiary structure of E-selectin CRD has been generated using structural coordinates taken from the mannose binding protein. When coupled with functional analysis of mutants within the E-selectin CRD, this model predicts that the E-selectin CRD interacts with sialyl-Le^X at a position near a predicted antiparallel beta sheet composed of residues derived from the NH$_2$- and COOH-termini of the CRD, along with two loops derived from residues 43–48 and 94–100.

The sialyl-Le^X moiety is most certainly the glycan portion of the E-selectin ligand on myeloid cells, as it is expressed abundantly on the surfaces of neutrophils and monocytes. Sialyl-Le^a determinants are not found on these cells, however, and thus apparently do not play a physiological role in E-selectin-dependent myeloid cell adhesion. Sialyl-Le^a molecules are often expressed on carcinomas derived from pancreatic tissue, however, as well as on digestive tract epithelia cell malignancies (151). These malignancies, as well as those derived from the lung and the ovary, may also express sialyl-Le^X molecules. These observations have prompted speculation that these glycoconjugates may facilitate metastasis through E-selectin-dependent adhesion of malignancies to vascular endothelia (127, 142, 146, 147).

It will continue to be important to define the complete structures of the oligosaccharides that display these tetrasaccharide ligands, and to identify the proteins and lipids that display them at the cell surface. Biochemical studies have shown that sialyl-Le^X determinants are displayed by glycolipids (114) and glycoproteins (110), including some of the CD11/CD18 group of cell adhesion molecules (152). Display by these latter molecules may represent a mechanism that allows regulation of ligand affinity or functional display. For example, it has been shown that large numbers of sialyl-Le^X determinants on neutrophils are displayed by L-selectin (79). These sialyl-Le^X-containing L-selectin molecules are apparently displayed by microvillous processes at the neutrophil surface, in a manner that could facilitate presentation of sialyl-Le^X moieties to E-selectin.

3.7 Ligands for E-selectin — potential as anti-inflammatory pharmaceuticals

It seems probable that pharmacological manipulation of oligosaccharide-E-selectin adhesive interactions could be used to diminish or prevent unwanted leukocyte recruitment and leukocyte-dependent inflammation, especially through intravascular administration of oligosaccharide ligands. It has not been easily possible to synthesize oligosaccharide molecules of the size and complexity of the sialyl-Le^x tetrasaccharide (let alone higher order structures), through chemical methods, in amounts adequate even for animal testing (153, 154). These difficulties are being overcome through the development of enzyme-assisted synthetic procedures (155–158) that make use of cloned glycosyltransferases (reviewed in reference 133). These approaches can also produce 'unnatural' oligosaccharides, since some of the glycosyltransferases exhibit promiscuity in their substrate requirements (156). Structural characterization and functional testing of such 'mutant' oligosaccharides, along with their 'wild-type' counterparts, should ultimately allow the design of molecular mimics of the *bona fide* ligands with potent anti-inflammatory properties and desirable pharmacologic properties.

Alternate strategies for creating anti-inflammatory pharmaceuticals that manipulate sialyl-Le^x-E-selectin interactions include those focused on inhibiting the actions or expression of the glycosyltransferases and other enzymes that participate on the biosynthesis of the sialyl-Le^x determinant in myeloid cells. Since these are areas about which little is yet known, it can be expected that direct exploration of these approaches (as opposed to screening libraries of compounds for inhibitory activity) is several years away.

4. Oligosaccharide-dependent cell adhesion mediated by P-selectin

4.1 P-selectin in platelets and endothelial cells

P-selectin is a 140 kDa glycoprotein discovered as a component of platelet α-granules. Thrombin-dependent platelet activation releases P-selectin from the α-granule, and it is finally expressed within the platelet membrane (159, 160). P-selectin was originally termed granule membrane protein (GMP-140; 159), or platelet activation-dependent granule to external membrane protein (PADGEM, 160), and has been given the cluster designation CD62.

P-selectin is found in megakaryocytes as well as platelets (161, 162), in a human erythroleukaemia cell line (HEL, reference 163), and in Weibel–Palade bodies of vascular endothelial cells (164, 165). Endothelial cells express P-selectin following activation with thrombin, histamine, phorbol ester, and the calcium ionophore A23187 (166). These stimuli mobilize pre-formed P-selectin from Weibel–Palade bodies. Expression is maximal at about 10 min following stimulation, and returns to baseline levels after about one hour. P-selectin expression is most probably terminated through an endocytic mechanism (166). P-selectin is also expressed by endothelial cells exposed to micromolar concentrations of oxidants (167). Oxygen radical-dependent P-selectin expression is long lived (up to several hours), and is believed to be prolonged because of an oxygen radical-dependent defect in its endocytotic reinternalization (167). Oxygen radical-dependent, prolonged P-selectin expression may operate to allow neutrophils to amplify P-selectin-dependent leukocyte recruitment (163).

4.2 P-selectin contains a carbohydrate recognition domain similar to the L-selectin and P-selectin CRDs

Protein sequence determined from purified, platelet-derived human P-selectin has been used to clone P-selectin cDNAs (168). It maintains an amino terminal domain with primary sequence similarity to the L- and E-selectin CRDs, an adjacent domain similar to the EGF motif, nine CR domains, one membrane-spanning domain, and a short COOH-terminal cytosolic tail (Figure 6). The topology and domain structure of P-selectin is entirely analogous to those displayed by E- and L-selectin. Sequence analysis of P-selectin cDNAs, PCR experiments, and genomic structure analyses show that alternative splicing processes generate a total of at least three forms of this protein (168). One such alternately spliced transcript 'skips' exon 14, effectively deleting 40 amino acids encompassing the membrane spanning

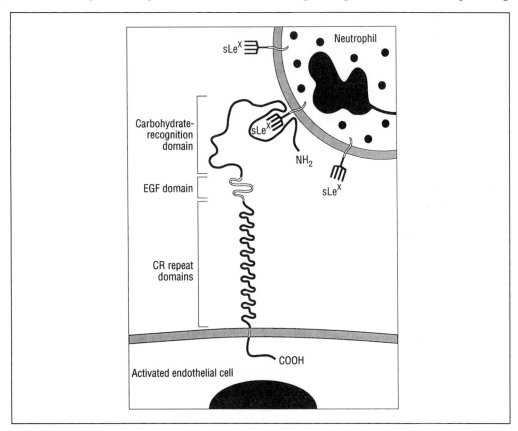

Fig. 6 Architecture of P-selectin. Domains are represented approximately to scale, and folded via predicted intrachain disulphide bond formation as illustrated by Johnston et al. (168). P-selectin is expressed by capillary endothelial cells within minutes after treatment with histamine or phorbol esters, via release of preformed molecules stored in Weibel–Palade bodies. The P-selectin CRD mediates adherence of myeloid lineage cells via interaction with surface-localized carbohydrate ligands found on glycoproteins (indicated by)

segment. This transcript exists in megakaryocytes, platelets, and endothelial cells, and generates a soluble P-selectin molecule. Another relatively minor transcript is expressed in endothelial cells, but not in platelets, and is deleted for sequences corresponding to exon 11. This transcript is predicted to yield a variant P-selectin molecule lacking the seventh CR repeat. The functional roles of these two variant P-selectins are not known, although some have suggested that the soluble form of P-selectin could bind to and cover up its counter-receptor on neutrophils, and thus modulate neutrophil adhesion during inflammation (see below).

4.3 P-selectin-dependent adhesion of myeloid cells to activated platelets and endothelium

A role for P-selectin in cell adhesion was suggested by the observation that activated platelets bind to a variety of myeloid cells, including neutrophils, monocytes, and HL-60 and U937 myeloid cell lines, whereas nonactivated platelets do not adhere (169, 170). Larsen *et al.* showed that this binding was P-selectin-dependent by using an assay where thrombin-activated platelets were allowed to form microscopically visible rosettes around myeloid cell types (171). Anti-P-selectin specific antisera blocked rosette formation around HL-60 cells, as did purified P-selectin or EDTA. Control antibodies against other platelet proteins (GPIIb/IIIa; thrombospondin; GPIV), or the proteins themselves, did not prevent rosette formation. Furthermore, fluorescently-labelled P-selectin-bearing phospholipid vesicles were shown to bind to the same panel of myeloid cell types rosetted by activated platelets, whereas nonsubstituted vesicles did not rosette. P-selectin-containing vesicles showed specific and saturable binding to U937 cells, and this binding was specifically inhibited with anti-P-selectin antibodies.

Hamburger *et al.* have also demonstrated that P-selectin mediates binding of thrombin-activated platelets to myeloid cells (172). In subsequent studies from this group, COS cells transfected with a P-selectin cDNA were shown to specifically bind HL-60 cells whereas control transfectants did not (173). Formalin fixation of HL-60 cells and neutrophils had no effect on binding efficiency, indicating that P-selectin-dependent adhesion can occur without metabolic activity. Furthermore, the resistance of the P-selectin counter-receptor to formalin fixation is consistent with later observations summarized below demonstrating that this counter-receptor is at least in part composed of a carbohydrate.

Geng *et al.* also demonstrated that neutrophils can adhere to endothelial cell monolayers in a P-selectin-dependent manner (173). P-selectin expression was induced by histamine or PMA treatment of the monolayers; binding of neutrophils under these circumstances was P-selectin-specific, and occurred in a temporal sequence that paralleled P-selectin expression. Activation-dependent neutrophil binding and P-selectin expression occurred even after pretreatment with RNA or protein synthesis inhibitors, demonstrating that preformed P-selectin is responsible for adhesion.

P-selectin-dependent myeloid cell adhesion is divalent cation dependent. Biochemical studies indicate that P-selectin maintains two high affinity sites for calcium, and indicate that calcium binding directs a conformational change in the P-selectin CRD that in turn allows it to bind its leukocyte ligand(s) (173, 174).

Calcium-dependent P-selectin-dependent adhesion of monocytes and neutrophils has also been reported by a third group (175, 176). This group has also published evidence that soluble P-selectin can inhibit CD18-dependent adhesion of TNFα-activated neutrophils to endothelium (175). This observation remains to be reconciled with the notion that CD18-dependent leukocyte adhesion occurs through interactions with endothelial cell ICAM-1 and ICAM-2 (reviewed in reference 177).

4.4 Ligands for P-selectin—sialylated, fucosylated oligosaccharides as candidates

The structural similarity of P-selectin, and its putative CRD, to the other two selectins suggested the possibility that P-selectin mediates cell adhesion via oligosaccharide ligand(s). Heparin apparently binds with high affinity to P-selectin, but this binding is calcium-independent, and may not reflect an interaction that mediates calcium-dependent cell adhesion (178). This group has subsequently shown that members of the same set of sulphate glycans examined previously (heparin, fucoidan, and high molecular weight dextran) also inhibit P-selectin binding to myeloid cells (176). By contrast, other sulphated glycans were substantially less able to block (lambda- and kappa-carrageenan, and low molecular dextran sulphate), and others (chondroitin 4-sulphate and chondroitin 6-sulphate) did not block at all. Since these experiments did not examine divalent cation requirements, their relationship to cation-dependent P-selectin-dependent cell adhesion is not yet clear.

In the first set of experiments that examined myeloid lineage-specific oligosaccharides as P-selectin ligands, Larsen *et al.* reported inhibition of P-selectin-dependent cell adhesion with a Le^X-active trisaccharide (179). This compound was tested after monoclonal antibody blocking studies implicated the CD15 determinant (Figure 4) as an essential component of a counter-receptor recognized by P-selectin. Purified oligosaccharides containing this determinant (lacto-N-fucopentaose III), or control oligosaccharides (lacto-N-fucopentaose I and lacto-N-fucopentaose II) were then tested for their ability to block P-selectin-dependent adhesion. LNF III inhibited half maximally at 50 µg/ml, while LNF I and LNF II did not significantly inhibit at concentrations up to 300 µg/ml. The authors concluded that the CD15 determinant was most probably only a part of the endogenous ligand since inhibition was observed only at very high concentrations of LNF III.

Several groups have shown that P-selectin-dependent cell adhesion requires sialic acid. For example, Corral *et al.* demonstrated that P-selectin-dependent adhesion of HL-60 cells, and of neutrophils, can be reduced by first treating these cells with any of three different broad-spectrum neuraminidases (*Clostridium perfringens*, *Vibrio cholera*, or *Arthrobacter ureafaciens*) (180). Since myeloid cells display glycans containing sialic acid in α2→3 and α2→6 linkage (110–113), then one or both types of sialic acid moieties may be essential components of P-selectin ligand(s). To further examine this issue, cells were subjected to binding assays after treatment with Newcastle disease virus (NDV) neuraminidase, which cleaves NeuNAcα2→3 or N-euNAcα2→8 linkages, but not NeuNAcα2→6 linkages (181). NDV neuraminidase pre-treatment did not decrease binding, under conditions where virtually all cell surface NeuNAcα2→3 linkages were removed. These observations indirectly implicate NeuNAcα2→6 linkages as important components of P-selectin-dependent cell adhesion, and suggest that NeuNAcα2→3 moieties are not required for P-selectin-dependent cell adhesion.

Moore et al. reached a different conclusion, however, using a technique involving binding of purified P-selectin to myeloid cells (182). In this assay, purified, soluble P-selectin bound specifically, and saturably, to myeloid cells, with half maximal binding at concentrations between 0.9 and 2.2 nM P-selectin. P-selectin binding was reduced by protease pretreatment, suggesting that the P-selectin ligand contains a protein component (182). Binding was reduced to 28% of control binding if cells were first treated with a broad spectrum sialidase (Vibrio cholera), while pretreatment with NDV neuraminidase reduced binding to 52% of control binding. Thus, these observations implicate NeuNAcα2→3 linkages as important components of P-selectin binding in contrast to the results of Corral et al. (180). Differences in assay procedures, or in neuraminidase preparations might account for the differences, although these have yet to be resolved.

Moore et al. also attempted to duplicate the inhibitory results reported by Larsen et al. (179), by attempting to inhibit P-selectin binding to neutrophils with an anti-CD15 antibody, and a multivalent Le^x-albumin conjugate. The anti-CD15 antibody tested by Moore et al. (82H5) did not inhibit binding, even when used at concentrations up to 400 μg/ml, whereas Larsen et al. reported inhibition to 35% of control binding with a different anti-CD15 antibody (80H5) used at a concentration of 20 μg/ml. Moore et al. (182) also found that the multivalent Le^x-albumin conjugate did not block P-selectin-dependent binding, while Larsen et al. (179) reported inhibition with a soluble Le^x-reactive oligosaccharide. It seems probable that the apparent discrepancy in results can be accounted for by differences in assay procedures, and by the fact that different reagents were used.

Polley et al. have shown that anti-sialyl-Le^x antibodies can inhibit P-selectin-dependent platelet rosetting, whereas anti-Lewis antibodies are only partially inhibitory (141). Furthermore, these investigators showed that a sialyl-Le^x-positive CHO cell mutant (LEC11 cells) bound activated platelets, while the Le^x-positive, sialyl-Le^x-negative CHO mutant LEC12 did not. Moreover, liposomes bearing sialyl-Le^x glycolipids effectively blocked P-selectin-dependent adhesion, as did a soluble hexasaccharide containing the sialyl-Le^x moiety (141). This latter reagent exhibited half maximal inhibitory activity at 2 μg/ml, whereas a Le^x-active pentasaccharide displayed half maximal inhibition at 54 μg/ml. These results implicate the sialyl-Le^x tetrasaccharide as an important component of endogenous myeloid cell ligand(s) for P-selectin.

4.5 Ligands for P-selectin—evidence for a protein component in addition to sialylated, fucosylated oligosaccharides

Similar observations have been reported by Zhou et al. (183). In these studies, an attempt was made to correlate P-selectin-dependent cell adhesion with glycosylation phenotype in a panel of cultured cell lines. An excellent, if imperfect, correlation between P-selectin-dependent binding and sialyl-Le^x expression emerged. Binding was decreased by pretreating cells with sialyl-Le^x antibodies, but not by using anti-Le^x antibodies. Le^x-positive, sialyl-Le^x-negative cell lines did not exhibit P-selectin-dependent adhesion. Quantitative determinations of binding site number and affinity was also reported for two different cell lines that exhibited P-selectin-dependent adhesion. HL-60 cells were found to express a low number of high affinity P-selectin binding sites, whereas a sialyl-Le^x-positive CHO cell line expressing a cloned fucosyltransferase cDNA (CHO-FT, reference 127) was found to express a large number of low affinity binding sites. These studies, when considered together with the others summarized above and reported elsewhere (184), indicate that P-selectin binds through glycan(s) containing the sialyl-Le^x determinant, and displayed by one or more protease-sensitive surface molecules. These studies also imply the existence of myeloid lineage-specific molecules that determine high affinity interactions with P-selectin.

This notion has been explored in studies that have identified a 120 kDa protein (reducing conditions) in HL-60 cells that binds in a calcium-dependent manner to P-selectin (185). This nature of this protein is not yet known, although Moore et al. have excluded it from being one of three other neutrophil membrane proteins (lamp-1, lamp-2, and leukosialin).

4.6 Ligands for P-selectin—evidence that P-selectin-dependent cell adhesion may also operate through sulphatides

Sulphatides (3-sulphated galactosylceramides) have also been reported to be ligands for P-selectin (186). In this study, HL-60 membrane glycolipids were fractionated by thin layer chromatography and stained with a recombinant P-selectin-IgG chimera in an attempt to identify P-selectin ligands Sulphatides were the only molecules bound by the chimera. Gangliosides, presumably including those that contain sialyl-Le^x determinants, did not bind the P-selectin chimera. Purified sulphatides inhibited P-selectin-dependent myeloid cell adhesion, while control glycolipids (lysosulphatide and galactosylceramide) did not. Subsequent studies have shown that 3-sulphated galactosylceramide inhibits P-selectin-dependent binding in a dose-dependent manner (187). While these results imply that sulphatides may represent a second class of P-selectin ligands, distinct from the protease-sensitive, sialyl-Le^x-containing P-selectin ligand, there is evidence to the contrary. Firstly, a variety of cells that express sulphatides, including erythrocytes and platelets, do not bind to P-selectin. Secondly, mutational analysis of the P-selectin CRD indicates that sulphatide binding by P-selectin mutants does not correlate at all with their cell binding properties (188). The physiological role for sulphatide binding by P-selectin thus remains open to question.

5. The sequential, multimolecule model for leukocyte rolling, adhesion, and transmigration

As alluded to above, leukocyte interactions with activated vascular endothelium under physiological flow conditions are first characterized by leukocyte rolling, followed by firm adhesion, spreading, and transmigration. Experiments described by Lawrence and Springer showed that P-selectin participates in this leukocyte 'rolling' process (189). In these experiments, leukocytes were allowed to flow over an artificial planar lipid bilayer under shear stress conditions analogous to those found in postcapillary venules. Videomicroscopy was used to quantitate the velocities of neutrophils as they flow through the chamber, and over the planar bilayer, and thus to determine if the leukocytes are rolling (adhesion-dependent) or tumbling (adhesion-independent). Planar bilayers reconstituted with P-selectin alone, with ICAM-1 alone, or with both adhesion molecules, were tested to allow a determination of the contributions to leukocyte rolling that are made by P-selectin, and by ICAM-1. Neutrophils 'rolled' on P-selectin-reconstituted bilayers, at relatively slow velocities, suggesting that the leukocytes were interacting with the P-selectin-substituted planar bilayer through an adhesive process. In contrast, leukocytes flowing over ICAM-1-substituted bilayers travelled at the higher velocities expected for tumbling particles that do not adhere to an adjacent surface. When leukocytes were made to flow over bilayers substituted with both P-selectin and ICAM-1, a cooperative interaction was observed. Neutrophils would roll along this surface, without stopping, unless they were activated with fMLP, a molecule that upregulates the adhesive avidity of the neutrophil integrin counter-receptors for ICAM-1. Under this circumstance, the rolling neutrophils would slow, stop, and then spread. These observations indicate that P-selectin and ICAM-1 operate in a sequentially cooperative manner; P-selectin brings passing leukocytes to the endothelial cell surface, where leukocyte integrins may become 'activated', bind to endothelial cell ICAM-1, and arrest the rolling cell.

These observations and related ones have been incorporated by several individuals into similar models that involve sequential adhesive events mediated by different classes of adhesion molecules, and modulated by a number of accessory molecules (65, 69, 70, 78, 189; Figure 7). In the earliest step in this model, pro-inflammatory molecules impinge on the microvasculature, and cause capillary dilatation and fluid leakage. This in turn causes the plasma to become more concentrated, which contributes to the slowing vascular flow that is a consequence of increased vessel diameter. Leukocytes flowing through such a vessel will tend to move away from a central location, to travel along a position near the perimeter of

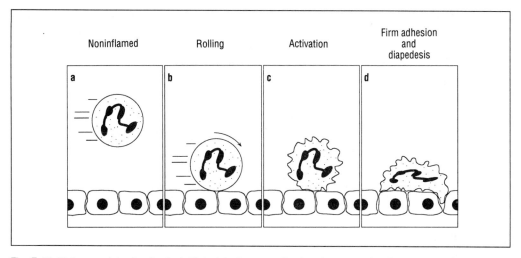

Fig. 7 Multistep model of selectin-initiated leukocyte adhesion (see text for details). (a) Under normal circumstance, leukocytes flow primarily through the central, rapidly flowing region of the vascular stream, without significant adhesive interaction with the noninflamed endothelial cell lining of the vascular wall. (b) Pro-inflammatory molecules impinge upon the vascular wall in venular capillaries, dilating the capillary, and promoting leakage of fluid from the vascular compartment. The resulting capillary dilatation and plasma protein concentration causes the flow rate to slow. This causes the leukocytes to travel at marginal positions within the stream, where they may participate in adhesive interactions with selectins displayed at the surface of the endothelial cell monolayer. This adhesion, in conjunction with forces directed at the lumenal side of the leukocytes, cause these cells to 'roll' along the endothelial lining. The leukocytes do not firmly attach to the vascular endothelium, however, because the leukocyte integrins remain in a low affinity, unactivated conformation, and thus cannot bind tightly to ICAM-1, their counter-receptor molecule. (c) Soluble and membrane-associated leukocyte-activating molecules emanating from endothelial cells and the subendothelial regions impinge upon the rolling leukocyte, and upregulate its integrin receptors. Leukocytes subsequently adhere tightly to ICAM-1 and stop. (d) Leukocytes firmly attached to the endothelium spread, and subsequently diapedese through the capillary wall

the lumen. The leukocyte will then be more likely to interact with E- and/or P-selectins expressed by endothelial cells activated by the pro-inflammatory cytokines. Selectin-dependent adhesion occurs where leukocytes interact with the vessel wall. When combined with flow-based forces applied to the free, lumenal side of the leukocyte, these events will cause the leukocyte to roll along the endothelial lining of the capillary. The cell will continue to roll since selectin-dependent adhesion has an inherently low affinity, and because the leukocyte integrins are in an unactivated conformation that exhibits a low affinity for endothelial ICAM-1. The leukocyte has been brought into close proximity to the vessel wall, however, and it is moving past at a relatively slower velocity. These circumstances thus expose the leukocyte to locally high concentrations of molecules that can upregulate its integrins. Such 'activating' substances include soluble molecules (interleukin-8, complement component C5a, N-formylated bacterial peptides, leukotriene B4), as well as membrane-associated molecules (platelet activating factor; PAF). Upregulation of integrin affinity may also occur through engagement of neutrophil ligands by E-selectin (190), or by PAF-dependent activation events that may be enhanced when P-selectin binds to its neutrophil ligands (191). The leukocyte slows and then stops, as its activated integrins take hold of endothelial cell surface ICAM-1 molecules. The leukocyte will then spread and transmigrate through the vascular wall. Given the large number of molecules that operate in this process, it can be imagined that there will be numerous opportunities to manipulate aberrant leukocyte recruitment by modifying their function or expression.

Acknowledgements

This work was supported in part by NIH grant 1P01AI33189. J.B.L. is an Associate Investigator of the Howard Hughes Medical Institute.

References

1. Drickamer, K. (1988) Two distinct classes of carbohydrate-recognition domains in animal lectins. *J. Biol. Chem.*, **263**, 9557.
2. Kornfeld, S. (1987) Trafficking of lysosomal enzymes. *FASEB J.*, **1**, 462.
3. Ashwell, G. and Harford, J. (1982) Carbohydrate-specific receptors of the liver. *Annu. Rev. Biochem.*, **51**, 531.
4. Fiete, D., Srivastava, V., Hindsgaul, O., and Baenziger, J. U. (1991) A hepatic reticuloendothelial cell receptor specific for SO4-4GalNAc beta 1,4GlcNAc beta 1,2Man alpha that mediates rapid clearance of lutropin. *Cell*, **67**, 1103.
5. Yednock, T. A. and Rosen, S. D. (1989) Lymphocyte homing. *Adv. Immunol.*, **44**, 313.
6. Kansas, G. S., Spertini, O., and Tedder, T. F. (1991) Leukocyte adhesion molecule-1 (LAM-1): structure, function, genetics, and evolution. In *Cellular and molecular mechanisms of inflammation* (ed. C. G. Cochrane and M. A. Gimbone, Jr), p. 31. Academic Press, San Diego.
7. Stamper, H. B., Jr and Woodruff, J. J. (1977) An *in vitro* model of lymphocyte homing. I. Characterization of the interaction between thoracic duct lymphocytes and specialized high-endothelial venules of lymph nodes. *J. Immunol.*, **119**, 772.
8. Gallatin, W. M., Weissman, I. L., and Butcher, E. C. (1983) A cell-surface molecule involved in organ-specific homing of lymphocytes. *Nature*, **304**, 30.
9. Lewinsohn, D. M., Bargatze, R. F., and Butcher, E. C. (1987) Leukocyte-endothelial cell recognition: evidence of a common molecular mechanism shared by neutrophils, lymphocytes, and other leukocytes. *J. Immunol.*, **138**, 4313.
10. Siegelman, M. H., Bond, M. W., Gallatin, W. M., St. John, T., Smith, H. T., Fried, V. A. et al. (1986) Cell surface molecules associated with lymphocyte homing is a ubiquitinated branched-chain glycoprotein. *Science*, **231**, 823.
11. Stoolman, L. M. and Rosen, S. D. (1983) Possible role for cell-surface carbohydrate-binding molecules in lymphocyte recirculation. *J. Cell. Biol.*, **96**, 722.
12. Stoolman, L. M., Tenforde, T. S., and Rosen, S. D. (1984) Phosphomannosyl receptors may participate in the adhesive interaction between lymphocytes and high endothelial venules. *J. Cell. Biol.*, **99**, 1535.
13. Mian, A. J. and Percival, E. (1973) Carbohydrates of the brown seaweed *Himanthalia lorea* and *Bifurcaria bifucata*. Part II. Structural studies of the 'fucans'. *Carbohyd. Res.*, **26**, 147.

14. San Blas, G. and Cunningham, W. L. (1974) Structure of cell wall and exocellular mannans from the yeast *Hansenula holstii*. I. Mannans produced in phosphate-containing medium. *Biochem. Biophys. Acta*, **354**, 233.
15. Yednock, T. A., Stoolman, L. M., and Rosen, S. D. (1987) Phosphomannosyl-derivatized beads detect a receptor involved in lymphocyte homing. *J. Cell Biol.*, **104**, 713.
16. Stoolman, L. M., Yednock, T. A., and Rosen, S. D. (1987) Homing receptors on human and rodent lymphocytes—evidence for a conserved carbohydrate-binding specificity. *Blood*, **70**, 1842.
17. Yednock, T. A., Butcher, E. C., Stoolman, L. M., and Rosen, S. D. (1987) Receptors involved in lymphocyte homing: relationship between a carbohydrate-binding receptor and the MEL-14 antigen. *J. Cell Biol.*, **104**, 725.
18. Rosen, S. D., Singer, M. S., Yednock, T. A., and Stoolman, L. M. (1985) Involvement of sialic acid on endothelial cells in organ-specific lymphocyte recirculation. *Science*, **228**, 1005.
19. Stevens, S. K., Weissman, I. L., and Butcher, E. C. (1982) Differences in the migration of B and T lymphocytes: organ-selective localization *in vivo* and the role of lymphocyte-endothelial cell recognition. *J. Immunol.*, **128**, 844.
20. Rosen, S. D., Ch, S.-I., True, D. D., Singer, M. S., and Yednock, T. A. (1989) Intravenously injected sialidase inactivates attachment sites for lymphocytes on high endothelial venules *J. Immunol.*, **142**, 1895.
21. Siegelman, M. H., van de Rijn, M., and Weissman, I. L. (1989) Mouse lymph node homing receptor cDNA clone encodes a glycoprotein revealing tandem interaction domains. *Science*, **243**, 1165.
22. Lasky, L. A., Singer, M. S., Yednock, T. A., Dowbenko, D., Fennie, C., Rodriguez, H. *et al.* (1989) Cloning of a lymphocyte homing receptor reveals a lectin domain. *Cell*, **56**, 1045.
23. Tedder, T. F., Isaacs, C. M., Ernst, T. J., Demetri, G. D., Adler, D. A., and Disteche, C. M. (1989) Isolation and chromosomal localization of cDNAs encoding a novel human lymphocyte cell surface molecule, LAM-1. *J. Exp. Med.*, **170**, 123.
24. Camerini, D., James, S. P., Stamenkovic, I., and Seed, B. (1989) Leu-8/TQ1 is the human equivalent of the Mel-14 lymph node homing receptor. *Nature*, **342**, 78.
25. Bowen, B. R., Nguyen, T., and Lasky, L. A. (1989). Characterization of a human homologue of the murine peripheral lymph node homing receptor. *J. Cell Biol.*, **109**, 421.
26. Siegelman, M. H. and Weissman, I. L. (1989) Human homologue of mouse lymph node homing receptor: evolutionary conservation at tandem cell interaction domain. *Proc. Natl Acad. Sci. USA*, **86**, 5562.
27. Bevilacqua, M., Butcher, E., Furie, B., Gallatin, M., Gimbrone, M. *et al.* (1991) Selectins: a family of adhesion receptors. *Cell*, **67**, 233.
28. Tedder, T. F., Penta, A. C., Levine, H. B., and Freedman, A. S. (1990) Expression of the human leukocyte adhesion molecule, LAM-1. Identity with the TQ1 and Leu-8 differential antigens. *J. Immunol.*, **144**, 532.
29. Kishimoto, T. K., Jutila, M. A., and Butcher, E. C. (1990) Identification of a human peripheral lymph node homing receptor: a rapidly down-regulated adhesion molecule. *Proc. Natl Acad. Sci. USA*, **87**, 224.
30. Johnston, G. I., Bliss, G. A., Newman, P. J., and McEver, R. P. (1990) Structure of the human gene encoding granule membrane protein-140, a member of the selectin family of adhesion receptors for leukocytes. *J. Biol. Chem.*, **265**, 21381.

31. Hourcade, D., Holers, V. M., and Atkinson, J. P. (1989) The regulators of complement activation (RCA) gene cluster. In *Advances in immunology* (ed. F. I. Dixon), Vol. 45, p. 381. Academic Press, New York.
32. Ord, D. C., Ernst, T. J., Zhou, L.-J., Rambaldi, A., Spertini, O., Griffin, J., and Tedder, T. F. (1990) Structure of the gene encoding the human leukocyte adhesion molecule-1 (TQ1, Leu-8) of lymphocytes and neutrophils. *J. Biol. Chem.*, **265**, 7760.
33. Dowbenko, D. J., Diep, A., Taylor, B. A., Lusis, A. J., and Lasky, L. A. (1991) Characterization of the murine homing receptor gene reveals correspondence between protein domains and coding exons. *Genomics*, **9**, 270.
34. Watson, M. L., Kingsmore, S. F., Johnston, G. I., Siegelman, M. H., Le Beau, M. M., Lemons, R. S. et al. (1990) Genomic organization of the selectin family of leukocyte adhesion molecules on human and mouse chromosome 1. *J. Exp. Med.*, **172**, 263.
35. Weis, J. H., Morton, C. C., Bruns, G. A., Weis, J. J., Klickstein, L. B., Wong, W. W. et al. (1987) A complement receptor locus: genes encoding C3b/C4b receptor and C3d/Epstein–Barr virus receptor map to 1q32. *J. Immunol.*, **138**, 312.
36. Bowen, B. R., Fennie, C., and Lasky, L. A. (1990) The Mel 14 antibody binds to the lectin domain of the murine peripheral lymph node homing receptor. *J. Cell. Biol.*, **110**, 147.
37. Kansas, G. S., Spertini, O., Stoolman, L. M., and Tedder, T. F. (1991) Molecular mapping of functional domains of the leukocyte receptor for endothelium, LAM-1. *J. Cell. Biol.*, **114**, 351.
38. Spertini, O., Kansas, G. S., Reimann, K. A., Mackay, C. R., and Tedder, T. F. (1991) Function and evolutionary conservation of distinct epitopes on the leukocyte adhesion molecule-1 (TQ-1, Leu-8) that regulate leukocyte migration. *J. Immunol.*, **147**, 942.
39. Siegelman, M. H., Cheng, I. C., Weissman, I. L., and Wakel, E. K. (1990) The mouse lymph node homing receptor is identical with the lymphocyte cell surface marker Ly-22: role of the EGF domain in endothelial binding. *Cell*, **61**, 611.
40. Imai, Y., Lasky, L. A., and Rosen, S. D. (1992) Further characterization of the interaction between L-selectin and its endothelial ligands. *Glycobiology*, **2**, 373.
41. Watson, S. R., Imai, Y., Fennie, C., Geoffrey, J., Singer, M., Rosen, S. D., and Lasky, L. A. (1991) The complement binding-like domains of the murine homing receptor facilitate lectin activity. *J. Cell. Biol.*, **115**, 234.
42. Watson, S. R., Imai, Y., Fennie, C., Geoffroy, J. S., Rosen, S. D., and Lasky, L. A. (1990) A homing receptor-IgG chimera as a probe for adhesive ligands of lymph node high endothelial venules. *J. Cell. Biol.*, **110**, 2221.
43. True, D. D., Singer, M. S., Lasky, L. A., and Rosen, S. D. (1990) Requirement for sialic acid on the endothelial ligand of a lymphocyte homing receptor. *J. Cell. Biol.*, **111**, 2757.
44. Liener, I. E., Sharon, N., and Goldstein, I. J. (1986) *The lectins. Properties, functions, and applications in biology and medicine*. Academic Press, Orlando, FL.
45. Schauer, R. (1985) Sialic acids and their roles as biological masks. *Trends Biochem. Sci.*, **10**, 357.
46. Imai, Y., Sinfer, M. S., Fennie, C., Lasky, L. A., and Rosen, S. D. (1991) Identification of a carbohydrate-based endothelial ligand for a lymphocyte homing receptor. *J. Cell. Biol.*, **113**, 1213.
47. Butcher, E. C. (1990) Cellular and molecular mechanisms that direct leukocyte traffic. *Am. J. Pathol.*, **136**, 3.
48. Streeter, P. R., Rouse, B. T. N., and Butcher, E. C. (1988) Immunohistologic and functional characterization of a vascular addressin involved in lymphocyte homing into peripheral lymphnodes. *J. Cell. Biol.*, **107**, 1853.

49. Kornfeld, R. and Kornfeld, S. (1985) Assembly of asparagine-linked oligosaccharides. *Annu. Rev. Biochem.*, **54**, 631.
50. Lasky, L. A., Singer, M. S., Dowbenko, D., Imai, Y., Henzel, W. J., Grimley C. *et al.* (1992) An endothelial ligand for L-selectin in a novel mucin-like molecule. *Cell*, **69**, 927.
51. Imai, Y., Lasky, L. A., and Rosen, S. D. (1993) Sulphation requirement for GlyCAM-1, an endothelial ligand for L-selectin. *Nature*, **361**, 555.
52. Berg, E. L., Magnani, J., Warnock, R. A., Robinson, M. K., and Butcher, E. C. (1992) Comparison of L-selectin and E-selectin ligand specificities: the L-selectin can bind to the E-selectin ligands sialyl Le^X and sialyl Le^a. *Biochem. Biophys. Res. Commun.*, **184**, 1048.
53. Foxall, C., Watson, S. R., Dowbenko, D., Fennie, C., Lasky, L. A., Kiso, M. *et al.* (1992) The three members of the selectin receptor family recognize a common carbohydrate epitope, the sialyl Lewis X oligosaccharide. *J. Cell Biol.*, **4**, 895.
54. Green, P. J., Tamatani, T., Watanabe, T., Miyasaka, M., Hasegawa, A., Kiso, M. *et al.* (1992) High affinity binding of leukocyte adhesion molecule L-selectin to 3'-sulphated-Le^a and Le^X oligosaccharides and the predominance of sulphate in this interaction demonstrated by binding studies with a series of lipid-linked oligosaccharides. *Biochem. Biophys. Res. Commun.*, **188**, 244.
55. Paavonen, T. and Renkonen, R. (1992) Selective expression of sialyl-Lewis X and Lewis A epitopes, putative ligands for L-selectin, on peripheral lympho-node high endothelial venules. *Am. J. Pathol.*, **141**, 1259.
56. Spertini, O., Kansas, G. S., Munro, J. M., Griffin, J. D., and Tedder, T. F. (1991) Regulation of leukocyte migration by activation of the leukocyte adhesion molecule-1 (LAM-1) selectin. *Nature*, **349**, 691.
57. Spertini, O., Freedman, A. S., Belvin, M. P., Penta, A. C., Griffin, J. D., and Tedder, T. F. (1991) Regulation of leukocyte adhesion molecule-1 (TQ1, Leu-8) expression and shedding by normal and malignant cells. *Leukemia*, **5**, 300.
58. Kishimoto, T. K., Jutila, M. A., Berg, E. L., and Butcher, E. C. (1989) Neutrophil Mac-1 and MEL-14 adhesion proteins inversely regulated by chemotactic factors. *Science*, **245**, 1238.
59. Griffin, J. D., Spertini, O., Ernst, T. J., Belvin, M. P., Levine, H. B., Kanakura, Y. *et al.* (1990) Granulocyte-macrophage colony-stimulating factor and other cytokines regulate surface expression of the leukocyte adhesion molecule-1 on human neutrophils, monocytes, and their precursors. *J. Immunol.*, **145**, 576.
60. Schleiffenbaum, B., Spertini, O., and Tedder, T. F. (1992) Soluble L-selectin is present in human plasma at high levels and retains functional activity. *J. Cell Biol.*, **119**, 229.
61. Dustin, M. L. and Springer, T. A. (1989) T-cell receptor cross-linking transiently stimulates adhesiveness through LFA-1. *Nature*, **341**, 619.
62. Shimizu, Y., Van Seventer, G. A., Horgan, K. J., and Shaw, S. (1990) Regulated expression and binding of three VLA(Beta 1) integrin receptors on T cells. *Nature*, **345**, 250.
63. Jutila, M. A., Rott, L., Berg, E. L., and Butcher, E. C. (1989) Function and regulation of the neutrophil MEL-14 antigen *in vivo*: comparison with LFA-1 and Mac-1. *J. Immunol.*, **143**, 3318.
64. Hallmann, R., Jutila, M. A., Smith, C. W., Anderson, D. C., Kishimoto, T. K., and Butcher, E. C. (1991) The peripheral lympho node homing receptor, L-selectin, is involved in CD18-independent adhesion of human neutrophils to endothelium. *Biochem. Biophys. Res. Commun.*, **174**, 236.

65. Smith, C. W., Kishimoto, T. K., Abbass, O., Hughes, B., Rothlein, R., McIntire, L. V. et al. (1991) Chemotactic factors regulate lectin adhesion molecule 1 (LECAM-1)-dependent neutrophil adhesion to cytokine-stimulated endothelial cells in vitro. *J. Clin. Invest.*, **87**, 609.
66. Spertini, O., Luscinskas, F. W., Gimbrone, M. A. Jr, and Tedder, T. F. (1992) Monocyte attachment to activated vascular endothelium *in vitro* is mediated by leukocyte adhesion molecule-1 (L-selectin) under non-static conditions. *J. Exp. Med.*, **175**, 1789.
67. Brady, H. R., Spertini, O., Jiminez, W., Brenner, B. M., Marsden, P. A., and Tedder, T. F. (1992) Neutrophils, monocytes, and lymphocytes bind to cytokine-activated kidney glomerular endothelial cells through L-selectin (LAM-1) *in vitro*. *J. Immunol.*, **49**, 2437.
68. Ley, K., Gaehtgens, P., Fennie, C., Singer, M. S., Lasky, L. A., and Rosen, S. D. (1991) Lectin-like cell adhesion molecule 1 mediates leukocyte rolling in mesenteric venules *in vivo*. *Blood*, **12**, 2553.
69. von Adrian, U. H., Chambers, J. D., McEvoy, L., Bargatze, R. F., Arfors, K. E., and Butcher, E. C. (1991) A two step model of leukocyte-endothelial cell interaction in inflammation: distinct roles for LECAM-1 and the leukocyte β2 integrins *in vivo*. *Proc. Natl Acad. Sci. USA*, **88**, 7538.
70. von Adrian, U. H., Hansell, P., Chambers, J. D., Berger, E. M., Torres, I., Butcher, E. et al. (1992) L-selectin function is required for $β_2$-integrin-mediated neutrophil adhesion at physiological shear rates *in vitro*. *Am. J. Physiol.*, **32**, H1034.
71. Atherton, A. and Born, G. V. (1973) Proceedings: effects of neuraminidase and N-acetyl neuraminic acid on the adhesion of circulating granulocytes and platelets in venules. *J. Physiol. (London)*, **234**, 66P.
72. Fiebig, E., Ley, K., and Arfors, K. E. (1991) Rapid leukocyte accumulation by 'spontaneous' rolling and adhesion in the exteriorized rabbit mesentery. *Int. J. Microcirc. Clin. Exp.*, **10**, 127.
73. Nobis, U., Pries, A. R., Cokelet, G. R., and Gaehtgens, P. (1985) Radial distribution of white cells during blood flow in small tubes. *Microvasc. Res.*, **29**, 295.
74. Chien, S. (1982) Rheology in the microcirculation in normal and low flow states. *Adv. Schock. Res.*, **8**, 71.
75. Atherton, A. and Born, G. V. (1973) Relationship between the velocity of rolling granulocytes and that of the blood flow in venules. *J. Physiol.*, **233**, 157.
76. Ley, K., Cerrito, M., and Arfors, K.-E. (1991) Sulphated polysaccharides inhibit leukocyte rolling in rabbit mesenteric venules. *Am. J. Physiol.*, **260**, H1667.
77. Watson, S. R., Fennie, C., and Lasky, L. A. (1991) Neutrophil influx into an inflammatory site inhibited by a soluble homing receptor-IgG chimaera. *Nature*, **349**, 164.
78. Butcher, E. C. (1991) Leukocyte-endothelial cell recognition: three (or more) steps to specificity and diversity. *Cell*, **67**, 1033.
79. Picker, L. J., Warnock, R. A., Burns, A. R., Doerschuk, C. M., Berg, E. L., and Butcher, E. C. (1991) The neutrophil LECAM-1 presents carbohydrate ligands to the vascular selectins ELAM-1 and GMP-140. *Cell*, **66**, 921.
80. Bevilacqua, M. P., Pober, J. S., Wheeler, M. E., Cotran, R. S., and Gimbrone, M. A., Jr (1985) Interleukin 1 acts on cultured human vascular endothelium to increase the adhesion of polymorphonuclear leukocytes, monocytes, and related leukocyte cell lines. *J. Clin. Invest.*, **76**, 2003.
81. Bevilacqua, M. P., Prober, J. S., Wheeler, M. E., Cotran, R. S., and Gimbrone, M. A., Jr (1985) Interleukin-1 activation of vascular endothelium. Effects on procoagulant activity and leukocyte adhesion. *Am. J. Pathol.*, **121**, 394.

82. Bevilacqua, M. P., Pober, J. S., Mendrick, D. L., Cotran, R. S., and Gimbrone, M. A. (1987) Identification of an inducible endothelial-leukocyte adhesion molecule. *Proc. Natl Acad. Sci. USA*, **84**, 9238.
83. Gamble, J. R., Harlan, J. M., Klebanoff, S. J., and Vada, M. A. (1985) Stimulation of the adherence of neutrophils to umbilical vein endothelium by human recombinant tumor necrosis factor. *Proc. Natl Acad. Sci. USA*, **82**, 8667.
84. Pober, J. S., Bevilacqua, M. P., Mendrick, D. L., Lapierre, L. A., Fiers, W., and Gimbrone, M. A. (1986) Two distinct monokines, interleukin 1 and tumor necrosis factor, each independently induce biosynthesis and transient expression of the same antigen on the surface of cultured human vascular endothelial cells. *J. Immunol.*, **136**, 1680.
85. Pohlman, T. H., Stanness, K. A., Beatty, P. G., Ochs, H. D., and Harlan, J. M. (1986) An endothelial cell surface factor(s) induced *in vitro* by lipopolysaccharide, interleukin 1, and tumor necrosis factor-alpha increases neutrophil adherence by a CDw18-dependent mechanism. *J. Immunol.*, **136**, 4548.
86. Koch, A. E., Burrows, J. C., Haines, G. K., Carlos, T. M., Harlan, J. M., and Leibovich, S. J. (1991) Immunolocalization of endothelial and leukocyte adhesion molecules in human rheumatoid and osteoarthritic synovial tissues. *Lab. Invest.*, **64**, 313.
87. Redl, H., Dinges, H. P., Buurman, W. A., van der Linden, C. J., Prober, J. S., Cotran, R. S. *et al.* (1991) Expression of endothelial leukocyte adhesion molecule-1 in septic but not traumatic/hypovolemic shock in the baboon. *Am. J. Pathol.*, **139**, 461.
88. Engelberts, I., Samyo, S. K., Leeuwenberg, J. F. M., van der Linden, C. J., and Buurman, W. A. (1992) A role for ELAM-1 in the pathogenesis of MOF during septic shock. *J. Surg. Res.*, **53**, 136.
89. Mulligan, M. S., Varani, J., Dame, M. K., Lane, C. L., Smith, C. W., Anderson, D. C., and Ward, P. A. (1991) Role of endothelial-leukocyte adhesion molecule 1 (ELAM-1) in neutrophil-mediated lung injury in rats. *J. Clin. Invest.*, **88**, 1396.
90. Cotran, R. S., Gimbrone, M. A. Jr, Bevilacqua, M. P., Mendrick, D. L., and Pober, J. S. (1989) Induction and detection of a human endothelial activation antigen *in vivo*. *J. Exp. Med.*, **164**, 661.
91. Groves, R. W., Allen, M. H., Barker, J. N., Haskard, D. O., and MacDonald, D. M. (1991) Endothelial leukocyte adhesion molecule-1 (E-selectin) expression in cutaneous inflammation. *Br. J. Dermatol.*, **124**, 117.
92. Picker, L. J., Kishimoto, T. K., Smith, C. W., Warnock, R. A., and Butcher, E. C. (1991) ELAM-1 is an adhesion molecule for skin-homing T cells. *Nature*, **349**, 796.
93. Griffiths, C. E., Barker, J. N., Kunkel, S., and Nickoloff, B. J. (1991) Modulation of leucocyte adhesion molecules, a T-cell chemotaxin (IL-8) and a regulatory cytokine (TNF-alpha) in allergic contact dermatitis (rhus dermatitis). *Br. J. Dermatol.*, **124**, 519.
94. Rohd, D., Schluter-Wigger, W., Mielke, V., von den Driesch, P., von Gaudecker, B., and Sterry, W. (1992) Infiltration of both T cells and neutrophils in the skin is accompanied by the expression of endothelial leukocyte adhesion molecule-1 (ELAM-1): an immunohistochemical and ultrastructural study. *J. Invest. Dermatol.*, **98**, 794.
95. Briscoe, D. M., Schoen, F. J., Rice, G. E., Bevilacqua, M. P., Ganz, P., and Pober, J. S. (1991) Induced expression of endothelial-leukocyte adhesion molecules in human cardiac allografts. *Transplantation*, **51**, 537.
96. Carlos, T. M., Dobrina, A., Ross, R., and Harlan, J. M. (1990) Multiple receptors on human monocytes are involved in adhesion to cultured human endothelial cells. *J. Leukocyte. Biol.*, **48**, 451.

97. Weller, P. F., Rand, T. H., Goelz, S. E., Chi-Rosso, G., and Lobb, R. R. (1991) Human eosinophil adherence to vascular endothelium mediated by binding to vascular cell adhesion molecule 1 and endothelial leukocyte adhesion molecule 1. *Proc. Natl Acad. Sci. USA*, **88**, 7430.
98. Shimizu, Y., Shaw, S., Graber, N., Gopal, T. V., Horgan, K. J., Van Seventer, G. A. et al. (1991) Activation-independent binding of human memory T cells to adhesion molecule ELAM-1. *Nature*, **349**, 799.
99. Graber, N., Gopal, T. V., Wilson, D., Beall, L. D., Polte, T., and Newman, W. (1990) T cells bind to cytokine-activated endothelial cells via a novel, inducible sialoglycoprotein and endothelial leukocyte adhesion molecule-1. *J. Immunol.*, **145**, 819.
100. Lobb, R. R., Chi-Rosso, G., Leone, D. R., Rosa, M. D., Bixler, S., Newman, B. M. et al. (1991) Expression and functional characterization of a soluble form of endothelial-leukocyte adhesion molecule 1. *J. Immunol.*, **147**, 124.
101. von Ausmith, E. J. U., Smeets, E. F., Ginsel, L. A., Onderwater, J. J. M., Leeuwenberg, J. F. M., and Buurman, W. A. (1992) Evidence for endocytosis of E-selectin in human endothelial cells. *Eur. J. Immunol.*, **22**, 2519.
102. Pigott, R., Dillon, L. P., Hemingway, L. H., and Gearing, A. J. H. (1992) Soluble forms of E-selectin, ICAM-1 and VCAM-1 are present in the supernatants of cytokine activated cultured endothelial cells. *Biochem. Biophys. Res. Commun.*, **187**, 584.
103. Newman, W., Beall, L. D., Carson, C. W., Hunder, G. G., Graben, N., Randhawa, Z. I. et al. (1993) Soluble E-selectin is found in supernatants of activated endothelial cells and is elevated in the serum of patients with septic shock. *J. Immunol.*, **150**, 644.
104. Smeets, E. F., de Vries, T., Leeuwenberg, J. F. M., van den Eijnden, D. H., Buurman, W. A., and Neefjes, J. J. (1993) Phosphorylation of surface E-selectin and the effect of soluble ligand (sialyl Lewis x) on the half-life of E-selectin. *Eur. J. Immunol.*, **23**, 147.
105. Bevilacqua, M. P., Stengelin, S., Gimbrone, M. A., and Seed, B. (1989) Endothelial leukocyte adhesion molecule 1: an inducible receptor for neutrophils related to complement regulatory proteins and lectins. *Science*, **243**, 1160.
106. Hession, C., Osborn, L., Goff, D., Chi-Rosso, G., Vassallo, C., Pasek, M. et al. (1990) Endothelial leukocyte adhesion molecule 1: direct expression cloning and functional interactions. *Proc. Natl Acad. Sci. USA*, **87**, 1673.
107. Larigan, J. D., Tsang, T. C., Rumberger, J. M., and Burns, D. K. (1992) Characterization of cDNA and genomic sequences encoding rabbit ELAM-1: conservation of structures and functional interactions with leukocytes. *DNA Cell Biol.*, **11**, 149.
108. Weller, A., Isenmann, S., and Vestweber, D. (1992) Cloning of the mouse endothelial selectins. Expression of both E- and P-selectin is inducible by tumor necrosis factor α. *J. Biol. Chem.*, **267**, 15 176.
109. Collins, T., Williams, A., Johnston, G. I., Kim, J., Eddy, R., Shows, T. et al. (1991) Structure and chromosomal location of the gene for endothelial-leukocyte adhesion molecule 1. *J. Biol. Chem.*, **266**, 2466.
110. Fukuda, M., Spooncer, E., Oates, J. E., Dell, A., and Klock, J. C. (1984) Structure of sialylated fucosyl lactosaminoglycan isolated from human granulocytes. *J. Biol. Chem.*, **259**, 10 925.
111. Fukuda, M. N., Dell, A., Oates, J. E., Wu, P., Klock, J. C., and Fukuda, M. (1985) Structures of glycosphingolipids isolated from human granulocytes. The presence of a series of linear poly-N-acetyllactosaminylceramide and its significance in glycolipids of whole blood cells. *J. Biol. Chem.*, **260**, 1067.
112. Spooncer, E., Fukuda, M., Klock, J. C., Oates, J. E., and Dell, A. (1984) Isolation and

characterization of polyfucosylated lactosaminoglycan from human granulocytes. *J. Biol. Chem.*, **259**, 4792.

113. Fukuda, M., Bothner, B., Ramsamooj, P., Dell, A., Tiller, P. R., Varki, A., and Klock, J. C. (1985) Structures of sialylated fucosyl polylactosaminoglycans isolated from chronic myelogenous leukemia cells. *J. Biol. Chem.*, **260**, 12957.

114. Symington, F. W., Hedges, D. L., and Hakomori, S.-I. (1985) Glycolipid antigens of human polymorphonuclear neutrophils and the inducible HL-60 myeloid leukemia line. *J. Immunol.*, **134**, 2498.

115. van den Eijnden, D. H., Koenderman, A. H., and Schiphorst, W. E. (1988) Biosynthesis of blood group i-active polylactosaminoglycans. Partial purification and properties of an UDP-GlcNAc:N-acetyllactosaminide beta 1----3-N-acetylglucosaminyltransferase from Novikoff tumor cell ascites fluid. *J. Biol. Chem.*, **263**, 12461.

116. Sadler, J. E. (1984) Biosynthesis of glycoproteins: formation of O-linked oligosaccharides. In *Biology of carbohydrates* (ed. V. Ginsburg and P. W. Robbins), p. 200. Wiley, New York.

117. Fukuda, M. N., Dell, A., Tiller, P. R., Varki, A., Klock, J. C., and Fukuda, M. (1986) Structure of a novel sialylated fucosyl lacto-N-norhexaosylceramide isolated from chronic myelogenous leukemia cells. *J. Biol. Chem.*, **261**, 2376.

118. Larsen, R. D., Ernst, L. K., Nair, R. P., and Lowe, J. B. (1990) Molecular cloning, sequence, and expression of human GDP-L-fucose:β-D-galactoside 2-α-L-fucosyltransferase cDNA that can form the H blood group antigen. *Proc. Natl Acad. Sci. USA*, **87**, 6674.

119. Fukuda, M., Dell, A., Oates, J. E., and Fukuda, M. N. (1984) Structure of branched lactosaminoglycan, the carbohydrate moiety of band 3 isolated from adult human erythrocytes. *J. Biol. Chem.*, **259**, 8260.

120. Weinstein, J., de Souza-e-Silva, U., and Paulson, J. C. (1982) Purification of a Gal β1→4GlcNAc α2→6sialyltransferase and a Gal β1→3(4)GlcNAc α2→3sialyltransferase to homogeneity from rat liver. *J. Biol. Chem.*, **257**, 13835.

121. Holmes, E. H., Ostrander, G. K., and Hakomori, S. (1985) Enzymatic basis for the accumulation of glycolipids with X and dimeric X determinants in human lung cancer cells (NCI-H69). *J. Biol. Chem.*, **260**, 7619.

122. Howard, D. R., Fukuda, M., Fukuda, M. N., and Stanley, P. (1987) The GDP-fucose: N-acetylglucosaminide 3-α-L-fucosyltransferases of LEC11 and LEC12 Chinese hamster ovary mutants exhibit novel specificities for glycolipid substrates. *J. Biol. Chem.*, **262**, 16830.

123. Kukowska-Latallo, J. F., Larsen, R. D., Nair, R. P., and Lowe, J. B. (1990) A cloned human cDNA determines expression of a mouse stage-specific embryonic antigen and the Lewis blood group α(1,3/1,4)fucosyltransferase. *Genes Dev.*, **4**, 1288.

124. Goelz, S. E., Hession, C., Goff, D., Griffiths, B., Tizard, R., Newman, B. *et al.* (1990) ELFT: a gene that directs the expression of an ELAM-1 ligand. *Cell*, **63**, 1349.

125. Lowe, J. B., Kukowska-Latallo, J. F., Nair, R. P., Larsen, R. D., Marks, R. M., Macher, B. A. *et al.* (1991) Molecular cloning of a human fucosyltransferase gene that determines expression of the Lewis X and VIM-2 epitopes but not ELAM-1-dependent cell adhesion. *J. Biol. Chem.*, **266**, 17467.

126. Kumar, R., Potvin, B., Muller, W. A., and Stanley, P. (1991) Cloning of a human alpha(1,3)-fucosyltransferase gene that encodes ELFT but does not confer ELAM-1 recognition on Chinese hamster ovary cell transfectants. *J. Biol. Chem.*, **266**, 21777.

127. Lowe, J. B., Stoolman, L. M., Nair, R. P., Larsen, R. D., Berhend, T. L., and Marks,

R. M. (1990) ELAM-1-dependent cell adhesion to vascular endothelium determined by a transfected human fucosyltransferase cDNA. *Cell*, **63**, 475.
128. Weston, B. W., Nair, R. P., Larsen, R. D., and Lowe, J. B. (1992) Isolation of a novel human α(1,3)fucosyltransferase gene and molecular comparison to the human Lewis blood group α(1,3/1,4)fucosyltransferase gene. Syntenic, homologous, nonallelic genes encoding enzymes with distinct acceptor substrate specificities. *J. Biol. Chem.*, **267**, 4152.
129. Weston, B. W., Smith, P. L., Kelly, R. J., and Lowe, J. B. (1992) Molecular cloning of a fourth member of a human α(1,3)fucosyltransferase gene family: multiple homologous sequences that determine expression of the Lewis X, sialyl Lewis X, VIM-2, and difucosyl sialyl Lewis x epitopes. *J. Biol. Chem.*, **267**, 24 575.
130. Potvin, B., Kumar, R., Howard, D. R., and Stanley, P. (1990) Transfection of a human α-(1,3)fucosyltransferase gene into Chinese hamster ovary cells. Complications arise from activation of endogenous α-(1,3)fucosyltransferases. *J. Biol. Chem.*, **265**, 1615.
131. Holmes, E. H., Ostrander, G. K., and Hakomori, S. (1986) Biosynthesis of the sialyl-Lex determinant carried by type 2 chain glycosphingolipids (IV3NeuAcIII3FucnLc4, VI3NeuAcV3FucnLc6, and VI3NeuAcIII3V3Fuc2nLc6) in human lung carcinoma PC9 cells *J. Biol. Chem.*, **261**, 3737.
132. Mollicone, R., Gibaud, A., Francois, A., Ratcliffe, M., and Oriol, R. (1990) Acceptor specificity and tissue distribution of three human alpha-3-fucosyltransferases. *Eur. J. Biochem.*, **191**, 169.
133. Lowe, J. B. (1992) Molecular cloning, expression, and uses of mammalian glycosyltransferases. *Seminars in Cell Biology*, **2**, 289.
134. Gooi, H. C., Feizi, T., Kapadia, A., Knowles, B. B., Solter, D., and Evans, M. J. (1981) Stage-specific embryonic antigen involves α1 to 3 fucosylated type 2 blood group chains. *Nature*, **292**, 156.
135. Macher, B. A., Buehler, J., Scudder, P., Knapp, W., and Feizi, T. (1988) A novel carbohydrate, differentiation antigen on fucogangliosides of human myeloid cells recognized by monoclonal antibody VIM-2. *J. Biol. Chem.*, **263**, 10 186.
136. Fukushi, Y., Hakomori, S., Nudelman, E., and Cochran, N. (1984) Novel fucolipids accumulating in human adenocarcinoma II. Selective isolation of hybridoma antibodies that differentially recognize mono-, di-, and trifucosylated type 2 chain. *J. Biol. Chem.*, **259**, 4681.
137. Fukushima, K., Hirota, M., Terasaki, P. I., Wakisaka, A., Togashi, H., Chia, D. *et al.* (1984) Characterization of sialosylated Lewis x as a new tumor-associated antigen. *Cancer Res.*, **44**, 5279.
138. Beyer, T. A., Sadler, J. E., Rearick, J. I., Paulson, J. C., and Hill, R. L. (1982) Glycosyltransferases and their use in assessing oligosaccharide structure and structure-function relationships. *Adv. Enzymol.*, **52**, 23.
139. Smith, P. L. and Baenziger, J. U. (1992) Molecular basis of recognition by the glycoprotein hormone specific N-acetylgalactosamine-transferase. *Proc. Natl Acad. Sci. USA*, **89**, 329.
140. Phillips, M. L., Nudelman, E., Gaeta, F. C., Perez, M., Singhal, A. K., Hakomori, S. *et al.* (1990) ELAM-1 mediates cell adhesion by recognition of a carbohydrate ligand, sialyl-Le^x. *Science*, **250**, 1130.
141. Polley, M. J., Phillips, M. L., Wayner, E., Nudelman, E., Singhal, A. K., Hakomori, S.-I. *et al.* (1991) CD62 and endothelial cell-leukocyte adhesion molecule 1 (ELAM-1) recognize the same carbohydrate ligand, sialyl-Lewis x. *Proc. Natl Acad. Sci. USA*, **88**, 6224.

142. Walz, G., Aruffo, A., Kolanus, W., Bevilacqua, M., and Seed, B. (1990) Recognition by ELAM-1 of the sialyl-Le^x determinant on myeloid and tumor cells. *Science*, **250**, 1132.
143. Tiemeyer, M., Sweidler, S. J., Ishihara, M., Moreland, M., Schweingruber, H., Hirtzer, P. et al. (1991) Carbohydrate ligands for endothelial-leukocyte adhesion molecule 1. *Proc. Natl Acad. Sci. USA*, **88**, 1138.
144. Tyrrel, D., Pames, P., Rao, N., Foxall, C., Abbas, S., Dasgupta, F. et al. (1991) Structural requirements for the carbohydrate ligand of E-selectin. *Proc. Natl Acad. Sci. USA*, **88**, 10 372.
145. Hansson, G. C. and Zopf, D. (1985) Biosynthesis of the cancer-associated sialyl-Le^a antigen. *J. Biol. Chem.*, **260**, 9388.
146. Berg, E. L., Robinson, M.K., Mansson, O., Butcher, E.C., and Magnani, J. L. (1991) A carbohydrate domain common to both sialyl Le^a and sialyl Le^x is recognized by the endothelial cell leukocyte adhesion molecule ELAM-1. *J. Biol. Chem.*, **266**, 14 869.
147. Takada, A., Ohmori, K., Takahashi, N., Tsuyuoka, K., Yago, A., Zenita, K. et al. (1991) Adhesion of human cancer cells to vascular endothelium mediated by a carbohydrate antigen, sialyl Lewis A. *Biochem. Biophys. Res. Commun.*, **179**, 713.
148. Nelson, R. M., Dolich, S., Aruffo, A., Cecconi, O., and Bevilacqua, M. P. (1993) Higher-affinity oligosaccharide ligands for E-selectin. *J. Clin. Invest.*, **91**, 1157.
149. Yuen, C.-T., Lawson, A. M., Chai, W., Larkin, M., Stoll, M. S., Stuart, A. C. et al. (1992) Novel sulphated ligands for the cell adhesion molecule E-selectin revealed by the neoglycolipid technology among O-linked oligosaccharides on an ovarian cystadenoma glycoprotein. *Biochemistry*, **31**, 9126.
150. Erbe, D. V., Wolitzky, B. A., Presta, L. G., Norton, C. R., Ramos, R. J., Burns, D. K. et al. (1992) Identification of an E-selectin region critical for carbohydrate recognition and cell adhesion. *J. Cell Biol.*, **119**, 215.
151. Kim, Y. S. and Itzkowitz, S. (1988) Carbohydrate antigen expression in the adenoma-carcinoma sequence. *Prog. Clin. Biol. Res.*, **279**, 241.
152. Asada, M., Furukawa, K., Kantor, C., Gahmberg, C. G., and Kobata, A. (1991) Structural study of the sugar chains of human leukocyte cell adhesion molecules CD11/CD18. *Biochemistry*, **30**, 1561.
153. Kameyama, A., Ishida, H., Kiso, M., and Hasegawa, A. (1991) Chemical synthesis of sialyl Lewis x. *Carbohydr. Res.*, **209**, C1.
154. Nicolaou, K. C., Hummel, C. W., Bockovich, N. J., and Wong, C. H. (1991) Stereocontrolled synthesis of sialyl Lex, the oligosaccharide binding ligand to ELAM-1 (Sialyl-N-acetylneuramin). *J. Chem. Soc.*, **13**, 870.
155. Wong, C. H. (1989) Enzymatic catalysts in organic synthesis. *Science*, **244**, 1145.
156. Dumas, D. P., Ichikawa, Y., Wong, C. H., Lowe, J. B., and Nair, R. P. (1991) Enzymatic synthesis of sialyl Le^x and derivatives based on a recombinant fucosyltransferase. *Bioorg. Med. Chem. Lett.*, **1**, 425.
157. Ball, G. E., O'Neill, R. A., Schultz, J. E., Lowe, J. B., Weston, B. W., Nagy, J. O. et al. (1992) Synthesis and structural analysis using 2-D NMR of sialyl Lewis x (SLe^X) and Lewis x (Le^X) oligosaccharides: ligands related to ELAM-1 binding. *J. Am. Chem. Soc.*, **114**, 5449.
158. Toone, E. J., Simon, E. S., Bednarski, M. D., and Whitesides, G. M. (1989) Enzyme-catalyzed synthesis of carbohydrates. *Tetrahedron Rep.*, **45**, 5365.
159. Stenberg, P. E., McEver, R. P., Shuman, M. A., Jacques, Y. V., and Bainton, D. F. (1985) A platelet alpha-granule membrane protein (GMP-140) is expressed on the plasma membrane after activation. *J. Cell Biol.*, **101**, 880.

160. Berman, C. L., Yeo, E. L., Wencel Drake, J. D., Furie, B. C., Ginsberg, M. H., and Furie, B. (1986) A platelet alpha granule membrane protein that is associated with the plasma membrane after activation. Characterization and subcellular localization of platelet activation-dependent granule-external membrane protein. *J. Clin. Invest.*, **78**, 130.
161. Beckstead, J. H., Stenberg, P. E., McEver, R. P., Shuman, M. A., and Bainton, D. F. (1986) Immunohistochemical localization of membrane and alpha-granule proteins in human megakaryocytes: application to plastic-embedded bone marrow biopsy specimens. *Blood*, **67**, 285.
162. McEver, R. P., Marshall-Carlson, L., and Beckstead, J. H. (1987) The platelet α-granule membrane protein GMP140 is also synthesized by human vascular endothelial cells and is present in blood vessels of diverse tissues. *Blood*, **70**, 355a.
163. Johnston, G. I., Kurosky, A., and McEver, R. P. (1989) Structural and biosynthetic studies of the granule membrane protein, GMP140, from human platelets and endothelial cells. *J. Biol. Chem.*, **264**, 1816.
164. Bonfanti, R., Furie, B. C., Furie, B., and Wagner, D. D. (1989) PADGEM is a component of Weibel-Palade bodies in endothelial cells. *Blood*, **73**, 1109.
165. McEver, R. P. (1989) GMP-140, a platelet α-granule membrane protein, is also synthesized by vascular endothelial cells and is localized in Weibel–Palade bodies. *J. Clin. Invest.*, **84**, 92.
166. Hattori, R., Hamilton, K. K., Fugate, R. D., McEver, R. D., and Sims, P. J. (1989) Stimulated secretion of endothelial von Willebrand factor is accompanied by rapid redistribution to the cell surface of the intracellular granule membrane protein GMP-140. *J. Biol. Chem.*, **264**, 7788.
167. Patel, K. D., Zimmerman, G. A., Prescott, S. M., McEver, R. P., and McIntyre, T. M. (1991) Oxygen radicals induce human endothelial cells to express GMP-140 and bind neutrophils. *J. Cell Biol.*, **112**, 749.
168. Johnston, G. I., Cook, R. G., and McEver, R. P. (1989) Cloning of GMP-140, a granule membrane protein of platelets and endothelium: sequence similarity to proteins involved in cell adhesion and inflammation. *Cell*, **56**, 1033.
169. Jungi, T. W., Spycher, M. O., Nydegger, U. E., and Barandun, S. (1986) Platelet-leukocyte interaction: selective binding of thrombin-stimulated platelets to human monocytes, polymorphonuclear leukocytes, and related cell lines. *Blood*, **67**, 629.
170. Silverstein, R. L. and Nachman, R. L. (1987) Thrombospondin binds to monocytes-macrophages and mediates platelet-monocyte adhesion. *J. Clin. Invest.*, **79**, 867.
171. Larsen, E., Celi, A., Gilbert, G. E., Furie, B. C., Erban, J. K., Bonfanti, R. et al. (1989) PADGEM protein: a receptor that mediates the interaction of activated platelets with neutrophils and monocytes. *Cell*, **59**, 305.
172. Hamburger, S. A. and McEver, R. P. (1990) GMP-140 mediates adhesion of stimulated platelets to neutrophils. *Blood*, **75**, 550.
173. Geng, J. G., Bevilacqua, M. P., Moore, K. L., McIntyre, T. M., Prescott, S. M., Kim, J. M. et al. (1990) Rapid neutrophil adhesion to activated endothelium mediated by GMP-140. *Nature*, **343**, 757.
174. Geng, J.-G., Moore, K. L., Johnson, A. E., and McEver, R. P. (1991) Neutrophil recognition requires a Ca^{2+}-induced conformational change in the lectin domain of GMP-140. *J. Biol. Chem.*, **266**, 22 313.
175. Gamble, J. R., Skinner, M. P., Berndt, M. C., and Vadas, M. A. (1990) Prevention of activated neutrophil adhesion to endothelium by soluble adhesion protein GMP140. *Science*, **249**, 414.

176. Skinner, M. P., Lucas, C. M., Burns, G. F., Chesterman, C. N., and Berndt, M. C. (1991) GMP-140 binding to neutrophils is inhibited by sulphated glycans. *J. Biol. Chem.*, **266**, 5371.
177. Springer, T. A. (1990) Adhesion receptors of the immune system. *Nature*, **346**, 425.
178. Skinner, M. P., Fournier, D. J., Andrews, R. K., Gorman, J. J., Chesterman, C. N., and Berndt, M. C. (1989) Characterization of human platelet GMP-140 as a heparin-binding protein. *Biochem. Biophys. Res. Comm.*, **164**, 1373.
179. Larsen, E., Palabrica, T., Sajer, S., Gilbert, G. E., Wagner, D. D., Furie, B. C. et al. (1990) PADGEM-dependent adhesion of platelets to monocytes and neutrophils is mediated by a lineage-specific carbohydrate, LNF III (CD15). *Cell*, **63**, 467.
180. Corral, L., Singer, M. S., Macher, B. A., and Rosen, S. D. (1990) Requirement for sialic acid on neutrophils in a GMP-140 (PADGEM) mediated adhesive interaction with activated platelets. *Biochem. Biophys, Res. Comm.*, **172**, 1349.
181. Paulson, J. C., Weinstein, J., Dorland, L., van Halbeek, H., and Vliegenthardt, J. F. (1982) Newcastle disease virus contains a linkage-specific glycoprotein sialidase. Application to the localization of sialic acid residues in N-linked oligosaccharides of alpha 1-acid glycoprotein. *J. Biol. Chem.*, **257**, 12 734.
182. Moore, K. L., Varki, A., and McEver, R. P. (1991) GMP-140 binds to a glycoprotein receptor on human neutrophils: evidence for a lectin-like interaction. *J. Cell Biol.*, **112**, 491.
183. Zhou, Q., Moore, K. L., Smith, D. F., Varki, A., McEver, R. P., and Cummings, R. D. (1991) The selectin GMP-140 binds to sialylated, fucosylated lactosaminoglycans on both myeloid and nonmyeloid cells. *J. Cell Biol.*, **115**, 557.
184. Larsen, G. R., Sako, D., Ahern, T. J., Shaffer, M., Erban, J., Sajer, S. A. et al. (1992) P-selectin and E-selectin. Distinct but overlapping ligand specificities. *J. Biol. Chem.*, **267**, 11 104.
185. Moore, K. L., Stults, N. L., Diaz, S., Smith, D. F., Cummings, R. D., Varki, A. et al. (1992) Identification of a specific glycoprotein ligand for P-selectin (CD62) on myeloid cells. *J. Cell Biol.*, **118**, 445.
186. Aruffo, A., Kolanus, W., Walz, G., Fredman, P., and Seed, B. (1991) CD62/P-selectin recognition of myeloid and tumor cell sulfatides. *Cell*, **67**, 35.
187. Todderud, G., Alford, J., Millsap, K. A., Aruffo, A., and Tramposch, K. M. (1992) PMN binding to P-selectin is inhibited by sulfatide. *J. Leukoc. Biol.*, **52**, 85.
188. Erbe, D. V., Watson, S. R., Presta, L. G., Wolitzky, B. A., Foxall, C., Brandley, B. K. et al. (1993) P- and E-selectin use common sites for carbohydrate ligand recognition and cell adhesion. *J. Cell Biol.*, **120**, 1227.
189. Lawrence, M. B. and Springer, T. A. (1991) Leukocytes roll on a selectin at physiologic flow rates: distinction from the prerequisite for adhesion through integrins. *Cell*, **65**, 859.
190. Lo, S. K., Lee, S., Ramos, R. A., Lobb, R., Rosa, M., Chi-Rosso, G. et al. (1991) Endothelial-leukocyte adhesion molecule 1 stimulates the adhesive activity of leukocyte integrin CR3 (CD11b/CD18, Mac-1 amb2) on human neutrophils. *J. Exp. Med.*, **173**, 1493.
191. Lorant, D. E., Patel, K. D., McIntyre, T. M., McEver, R. P., Prescott, S. M., and Zimmerman, G. A. (1991) Coexpression of GMP-140 and PAF by endothelium stimulated by histamine or thrombin. A juxtracine system for adhesion and activation of neutrophils. *J. Cell Biol.*, **115**, 223.

5 | Chemical synthesis of oligosaccharides

SHAHEER H. KHAN and OLE HINDSGAUL

1. Introduction

That oligosaccharides can function as ligands in biological recognition, usually through their binding to protein receptor sites, is now beyond dispute. The active structures have been identified in cases ranging in diversity from bacterial and viral adhesion to cell–cell recognition in the development of whole organisms. But the potential range of activities seems almost unlimited as noted in a recent, comprehensive and unusually open-minded review article entitled 'Biological function of oligosaccharides: all of the theories are correct' (1).

As in all cases where relatively small molecules of molecular weight near 1000 Da have been found to possess biological activity, the potential for the design of synthetic analogues is apparent. The possibility of inhibiting the adhesion of bacteria (2–4) or viruses (5, 6) to human cell surfaces has in itself spawned the birth of a small biotechnology industry. Even more promising with regard to commercialization has been the goal of preventing both inflammation and metastasis using oligosaccharide analogues, derivatives of the naturally occurring cell surface glycoproteins and glycolipid tetrasaccharide sequence termed sialyl-Le^x (7–10). Oligosaccharide analogues are also sought after as reagents for assaying the enzymes which regulate the biosynthesis of glycoconjugates: the glycosyltransferases (11–17) and glycosidases (18–20).

Fortunately, but perhaps not accidentally, at the same time that the interest in cell surface carbohydrate ligands is growing so is an interest in the challenge of synthesizing such structures in organic chemical laboratories. The natural biologically active ligands in question are invariably conjugated to proteins or lipids and presented, especially in cell–cell recognition and adhesion, as polyvalent arrays. A major challenge to synthetic carbohydrate chemists at this point in time is to produce relatively small molecules, amenable to facile structural manipulation, which can effectively compete with the natural ligands. This is the case for both recognition by receptors (selectins) or by biosynthetic enzymes like glycosidases and glycosyltransferases. It would be even more impressive if such synthetic inhibitors were fully active when monovalent. The objective of this chapter is to put into perspective, for a biologically motivated readership, the current state of synthetic oligosaccharide chemistry as it tackles the production of such biologically active molecules.

1.1 Our current capability in oligosaccharide synthesis

The field of synthetic carbohydrate chemistry has grown almost exponentially in the last decade. A number of biologically significant oligosaccharides containing 10 or more sugars, with different residues and different types of anomeric configuration, have been synthesized during that time whereas only 15 years ago the synthesis of a disaccharide was considered a major accomplishment. Most such synthetic oligosaccharides have been prepared for specific purposes which include their use in antibody production, screening of antibody, lectin and selectin specificity, interaction studies with virus and bacterial receptors, substrates for glycosidases and glycosyltransferases as well as other enzymes, and probes in molecular recognition studies including conformational analysis. Most synthetic oligosaccharides contain two to five sugar residues and their syntheses are now frequently described as 'routine'. 'Routine', in that context, means that in the hands of a well-trained and motivated chemist such compounds can with reasonable confidence be prepared in under a year. The advances made in this sphere have been achieved as a result of the introduction of new reagents and improved methods for separation and characterization of products, and also because of the concerted efforts of a considerable number of synthetic laboratories engaged in this endeavour. A detailed discussion of synthetic methodologies for oligosaccharide synthesis is beyond the scope of this chapter; however, the reader is referred to review articles (21–45) which have thoroughly covered the different aspects of oligosaccharide structure and synthesis.

2. The problems in oligosaccharide synthesis

The synthesis of oligosaccharides is much more complicated than peptides or nucleotides — mostly because of inherent properties of the molecules themselves but also due to the lack of a single set of 'optimum' reaction conditions for stereospecific formation of glycosidic bonds. To put the problem into a common perspective, improvements in peptide synthesis a dozen years ago resulted in amino-acid coupling yields increasing from 90 to over 99% with less than 1% racemization, i.e. without detectable formation of diastereomers. Today, in 'modern' oligosaccharide synthesis, addition of one monosaccharide to a preformed dimer is generally proudly accomplished in a yield of 60–80% and, in favourable instances, with stereochemical isomers formed in as little as 5% yield.

The challenges of glycosidic bond formation in oligosaccharide synthesis are summarized in Table 1 which lists the molecular structures of the most commonly occurring sugar residues of glycoproteins and glycolipids along with their relative degree of difficulty for chemical synthesis. All of the sugars involved are potentially polyvalent, having at least three hydroxyl groups, and the anomeric carbon can

Table 1 Common mammalian glycosidic linkages and their relative ease of formation by chemical synthesis*

Sugar residue	Linkage	Easy ←——— Degree of difficulty ———→ Difficult
Glc	β / α	
Gal	β / α	
Man	β / α	
Fuc	α	
Xyl	β	
GlcNAc	β / α	
GalNAc	β / α	
NANA	α	

* For simplicity, OH-groups have been deleted and replaced by lines.

have two different configurations (α and β) either as the free reducing sugar or when in a glycosidic linkage. As a consequence, in oligosaccharides not only can the sequence of individual sugar residues vary but also the site of coupling, the configuration of the glycosidic linkage, and the branching pattern. This extraordinary ability to provide structural diversity is summarized in Table 2, adapted from Schmidt (25), where it is compared to the corresponding ability of peptides.

Table 2 Comparison of isomeric possibilities for sequences of oligopeptides and oligosaccharides

Oligomer	Composition	Number of possible isomers	
		Oligopeptide	Oligosaccharide
Dimer	AA/AB	1/2	11/20
Trimer	AAA/ABC	1/6	120/720
Tetramer	AAAA/ABCD	1/24	1424/34560
Pentamer	AAAAA/ABCDE	1/120	17872/2144640

Oligosaccharides can be seen to have an unmatched potential as carriers of unique specificity for biological recognition.

2.1 Protection–deprotection in oligosaccharide synthesis

The presence of three or more hydroxyl groups in each sugar residue necessitates the protection of those hydroxyl groups which are not to become glycosylated in order to accomplish an unambiguous synthesis. In carbohydrate chemistry, protecting groups are generally distinguished as either 'temporary' or 'persistent'. Some hydroxyl groups need to be selectively protected in a different manner (temporary) than others (persistent) so that they can be deprotected in a desired intermediate and made available for subsequent glycosylation. Once the desired protected oligosaccharide is obtained, then the persistent protecting groups can be removed to yield the final product. This sequence of steps is illustrated in Figure 1 for the synthesis of a hypothetical trisaccharide with sequence C-B-A. A unique protecting group ($R_{aglycon}$) for the reducing end of the product oligosaccharide is frequently included to permit its covalent attachment to other molecules such as proteins. The products are often referred to as neoglycocoproteins (46).

The diverse protecting groups and strategies required for their manipulation have been largely defined (40, 43, 47–50). The most frequently used protecting groups, along with standard methods for their attachment and removal, are presented in Table 3. As a general rule, about five steps are required to selectively protect a monosaccharide for use in a synthetic scheme such as that delineated in Figure 1.

2.2 Glycosylation methods

The most challenging obstacle in carbohydrate chemistry involves the stereospecific coupling of two sugar residues. In an oligosaccharide synthesis this is usually accomplished by reaction of the anomeric carbon (C-1) of one completely protected sugar (glycosyl donor, Figure 1) with a free hydroxyl group on another appropriately protected sugar (glycosyl acceptor). In most coupling steps, mixtures

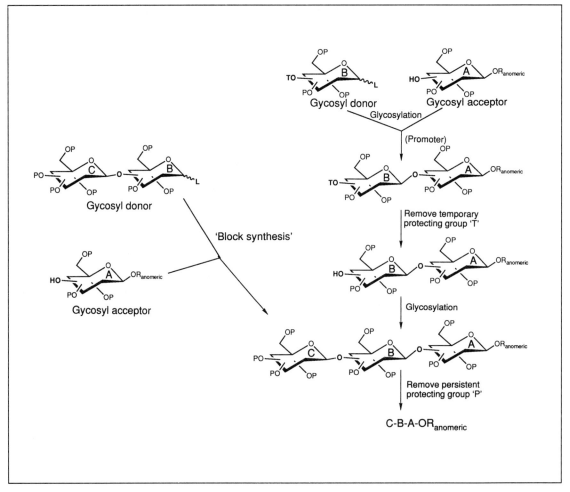

Fig. 1 Generalized schemes for both stepwise (*right*) and block synthesis (*left*) of a hypothetical trisaccharide C-B-A-OR. P = persistent protecting group, T = temporary protecting group, R = anomeric aglycon, L = leaving group on the glycosyl donor

of α and β-anomers are formed, *even though they are often not reported*, and these must be separated by chromatography.

In planning the synthesis of a required oligosaccharide the glycosidic linkages between individual sugar residues are classified into the following groups, each of which requires a different strategy for preparation. The common monosaccharides present in most of the glycoconjugates can have either α or β configuration. The α-linkage in the case of D-Glc, D-GlcNAc, D-Gal, D-GalNAc, D-Xyl, and L-Fuc corresponds to the 1,2-*cis* type whereas the β-linkage corresponds to 1,2-*trans* type. In contrast, the 1,2-*cis* in D-Man corresponds to the β-anomer while the 1,2-*trans* linkage corresponds to the α anomer. Successful glycosylation reactions require:

- *Regioselectivity.* This is normally controlled by selective protection of OH groups as discussed above. Under special circumstances, and in well-studied cases, only one of several free OH groups can on occasion become selectively glycosylated.
- *Stereoselectivity.* Broadly speaking, there are two general classes of glycosylation reactions: those proceeding with neighbouring group participation to give 1,2-*trans* stereochemistry (βGlc, βGal, βGlcNAc, βGalNAc, βXyl, and αMan) and those yielding 1,2-*cis* stereochemistry without neighbouring group participation (αGlc, αGal, αGalNAc, βMan, and αFuc). Sialic acids are linked at C-2 and special conditions are required for the synthesis of sialosides.

The most frequently used strategies for stereospecific glycosyl-bond formation are summarized in Figures 2 and 3. Figure 2 (a and b) show how neighbouring group participation from an ester (acetate, benzoate, or pivalate) at C-2 gives rise to βGlc, βGal, or βXyl-linkages. Both αGal and αGlc linkages are formed as shown in Figure 2 (c or d) with a nonparticipating group at O-2, usually the persistent benzyl ether. This method termed '*in situ* anomerization' was originally introduced by Lemieux (51) but many new modifications (23) have since been introduced. Stereochemistry can also be influenced by a participating solvent as noted in Figure 2 (d) where use of

Table 3 Common protecting groups and methods for their attachment and removal

Protecting group	Method of attachment	Method of removal
Ester		
Acetyl [Ac]	Ac_2O or AcCl/pyridine	NaOMe/MeOH
Chloroacetyl [Cl-Ac]	Chloroacetyl chloride/pyridine	Thiourea/EtOH/pyridine
Benzoyl [Bz]	BzCl/pyridine	NaOMe/MeOH
Pivaloyl [Piv]	Pivoloyl chloride/pyridine	NaOMe/MeOH
Ether		
Benzyl [Bn]	BnBr/DMF, NaH	$Pd-C/H_2$
pMethoxybenzyl [CH_3O-Bn]	$pCH_3O-BnCl$/DMF, NaH	$TFA/CHCl_3/H_2O$ or DDQ/CH_2Cl_2 or CAN
Allyl [All]	$CH_2=CH-CH_2Cl$/DMF, NaH	Bu^tOK/DMSO or $(Ph_3P)_3RhCl$
Trityl [$C(C_6H_5)_3$]	$(C_6H_5)_3CCl$/pyridine	$AcOH-H_2O$
Silyl [$Si(CH_3)_2Bu^t$]	$Bu^t(CH_3)_2SiCl$/pyridine	Pyridine—H_2O or Bu_4NF
[$Si(C_6H_5)_2Bu^t$]	$Bu^t(C_6H_5)_2SiCl$/imidazole	Bu_4NF
Acetal		
Isopropylidine [$C(CH_3)_2$]	$(CH_3)_2C(OCH_3)_2$/acetone, H^+	$AcOH-H_2O$ or $TFA/CHCl_3/H_2O$
Benzylidine [CHC_6H_5]	PhCHO or $PhCH(OCH_3)_2$/DMF, $ZnCl_2$ or H^+	$AcOH-H_2O$ or $TFA/CHCl_3/H_2O$ or $Pd-C/H_2O$
p-Methoxybenzylidine	$pCH_3O-C_6H_4CHO/ZnCl_2$	$AcOH-H_2O$ or $TFA/CHCl_3/H_2O$
For nitrogen protection		
N-Phthalimido (Phth)	Phthalic anhydride/pyridine	$NH_2-NH_2 \cdot H_2O$/EtOH
Azido (N_3)	Multistep synthesis	H_2/Pd or H_2S aq. pyridine

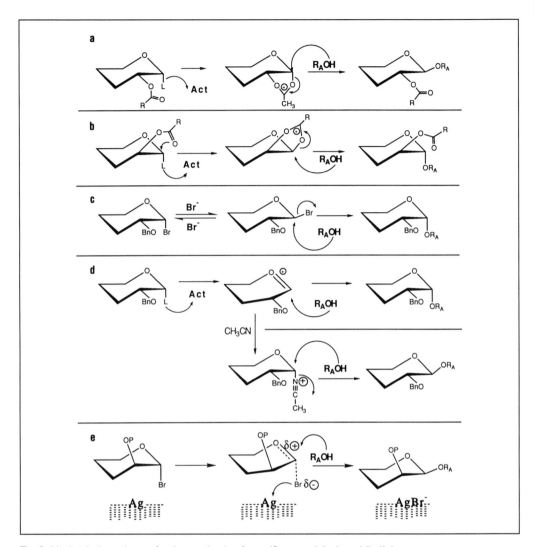

Fig. 2 Mechanistic pathways for the synthesis of specific α- and β-glycosidic linkages

acetonitrile changes the selectivity of glycosidic bond formation from α to β, presumably by the mechanism shown (37). Finally, in Figure 2(e), special conditions utilizing insoluble silver catalysts (52, 53) are required for β-mannoside formation and this area is still in great need of improvement (42, 54–56).

Figure 3 shows the corresponding approaches for the synthesis of 2-acetamido sugars. The 2-phthalimido group behaves as a participating neighbouring group (57), or it is sterically shielding, and gives rise to the β-anomers (Figure 3a) while a 2-azido group effectively protects the 2-position as a nonparticipating group giving rise to α-linked structures (Figure 3b). As in Figure 2(d), inclusion of acetonitrile increases the proportion of β-glycosides. Sialic acid derivatives (Figure 3c) are now

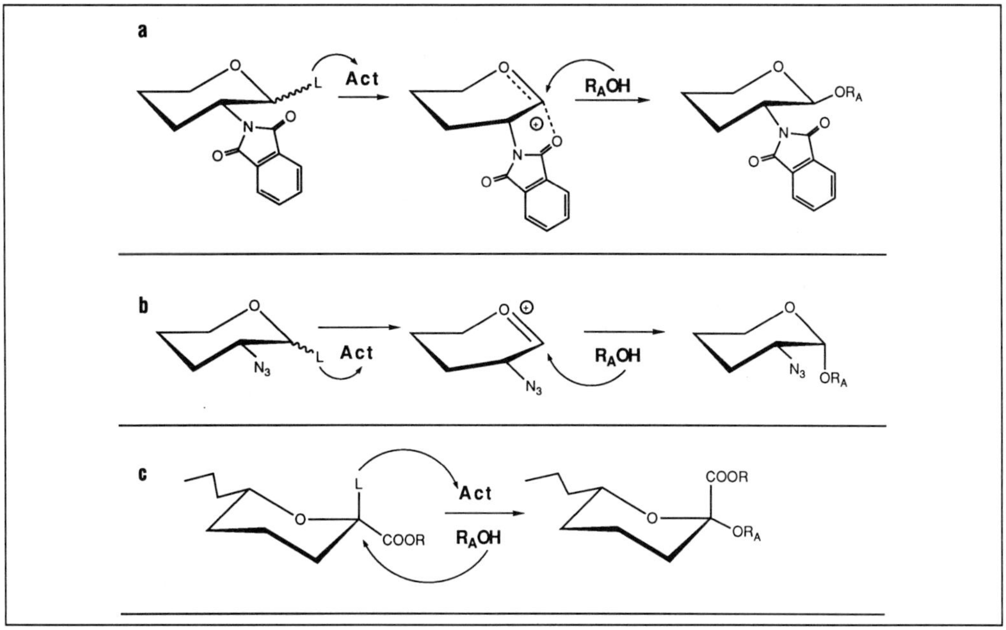

Fig. 3 Mechanistic pathways for the synthesis of specific α- and β-glycosides of 2-amino sugars and α-sialosides

generally prepared under a special set of experimental conditions (31, 33, 42) which proceed without neighbouring group participation, though in earlier work a stereocontrolling auxillary was temporarily introduced at C-3 (58–60).

The 'generic' glycosylation reactions summarized in Figures 1–3 all use a glycosyl donor where the reactive substituent at C-1 has been designated 'L' (for Leaving group). 'L' is usually selectively activated by addition of a specific reagent designated here as 'Act'. Selection of the correct 'L and Act' pairs is absolutely critical to the success of glycosylation (23) and the selection must take into account also the reactivity of the alcohol termed the glycosyl acceptor. The most commonly used leaving groups, and their activators, are described in Table 4.

An extraordinary variation of 'L and Act' pairs are used in the literature for glycosylation by the many different research groups engaged in oligosaccharide synthesis. The reason for the apparent lack of consensus reactions is due in part to different research groups promoting their own methodology. But a major consideration for such varied glycosylation protocols arises from the complexity of the two molecules, the donor and the acceptor, which must meet with precise timing for an effective glycosylation reaction to occur. Our understanding of the transition state structure for these reactions is very poor. The net result, as true today as when first noted by Paulsen 10 years ago (23), is that *'each oligosaccharide synthesis remains an independent problem whose resolution requires considerable systematic research and a good deal of know how. There are no universal reaction conditions for oligosaccharide synthesis'*.

Table 4 Some commonly used gycosyl donors and their activators

Glycosyl donor	Activator	Comments	Reference
L = Br	Br⁻, HgCN$_2$, HgBr$_2$, AgOTf	Most commonly used donor. Rate of reaction can be modulated by choice of activator	23, 24, 29, 51
L = Cl	HgCN$_2$, HgBr$_2$, AgOTf	More stable and less flexible than the bromide	23, 24, 29
L = F	SnCl$_2$—AgClO$_4$, AgOTf—HfCp$_2$Cl$_2$, AgOTf—SnCl$_2$, Tf$_2$O—HfCp$_2$Cl$_2$, Tf$_2$O	Efficient donors under mild conditions, react with thioglycosides in excellent yields	36, 61–64
L = SR	DMTST, TfOH—NIS, IDCP	Rapid, high yielding, and stereoselective	27, 40, 65, 66
L = OAc	BF$_3$, SnCl$_4$, TMSOTf	Mild conditions	67
L = (CH$_2$)$_3$CH=CH$_2$	IDCP, TfOH—NIS, AgOTf—NIS, TESOTf—NIS	Instantaneous reaction, reactivity controlled by appropriate promoter	41, 65, 68
L = C(=NH)—CCl$_3$	BF$_3$—Et$_2$O	Mild reaction conditions and good yields	22, 25, 29, 69
L = SOPh	Tf$_2$O	Rapid glycosylation of unreactive acceptors	70–72
L = SCN	TPMP, TMSOTf	Stereoselective, favouring 1,2-*cis* coupling	73, 74
L = S—C(=S)—OEt	DMTST, Cu(OTf)$_2$	Gives α/β mixtures in good yields	37
(N-methyl-pyridinium ether structure)	PTS, TMSOTf	Stereoselective, favouring 1,2-*trans* coupling	39

2.3 Convergent block synthesis

The synthesis of disaccharides is usually straightforward and involves the coupling of two monosaccharide units to give the product. There are few surprises. However, it is the synthesis of oligosaccharides with three or more sugar residues which requires more strategic planning and a difficult choice has to be made. The synthesis can be carried out by the stepwise method (Figure 1, *right*), which means that the chemistry involved must be planned in such a way as to couple two sugar residues which are susceptible to further chemical manipulations and stepwise condensation. Alternatively, 'blocks' (23) of two or more residues are prepared and coupled together in a convergent manner (Figure 1, *left*). Both of these strategies (42, 75–78) have been successfully used in the synthesis of complex oligosaccharides. In stepwise synthesis, the yields of individual steps are more critical and so is the steric control. If a particular glycosylation step gives an α,β mixture, the anomers must be separated before the next step can begin. In larger saccharide chains, certain glycosidic linkages impose limitations on the type of chemical manipulations that can be performed during subsequent protection–deprotection and glycosylation. On the other hand, most such transformations are performed on smaller oligosaccharides in the case of block synthesis which makes the chemistry, purification, and characterization of intermediates easier. However, the preparation of the required 'blocks' necessitates many steps and usually the yield of adding a more complex structure is substantially lower than adding a single sugar residue. The classical glycosylating agents (for example glycosyl bromides or chlorides, Table 4) have proven to be very successful donors in the stepwise approach (23, 24) whereas in block synthesis it can sometimes be difficult to convert a protected building block into the glycosyl-halide. Furthermore, glycosyl-halides of complex oligosaccharides decompose rapidly under most reaction conditions. Thioglycosides (42, 77, 78) and trichloroacetimidates (29, 42, 69) are therefore currently among the most useful 'blocks' in convergent block synthesis. The block approach is particularly efficient when a series of compounds having a common saccharide unit has to be prepared (79–82).

3. Synthesis of acceptor substrates for glycosyltransferases

The state of synthetic oligosaccharide chemistry, while remaining comparatively immature as described above, is nevertheless sufficiently advanced that most target molecules can be assembled with reasonable predictability. Among published synthetic targets, partial sequences of naturally occurring carbohydrate chains have been the most sought after (24, 29, 34). These are often used to probe carbohydrate–protein interactions. The rest of this chapter will focus on one particular aspect of the application of synthetic oligosaccharides — namely their use as acceptors in the assay of glycosyltransferases and as inhibitors of this class of enzyme. We will specifically concentrate here on the so-called branching enzymes involved in the biosynthesis of N-linked oligosaccharides, and on the glycosyltransferases controlling terminal glycosylation patterns.

Glycosyltransferases (11–17) catalyse the transfer of single sugar residues from sugar nucleotides to growing oligosaccharide chains (Figure 4). The binding site of these mostly Golgi-localized enzymes must therefore recognize the oligosaccharide portion of the acceptor sugar chains and, in particular, the OH group to which the new glycosyl residue is transferred. Almost universally, glycosyltransferase assays involve the quantitation of the rate of transfer of ^{3}H or ^{14}C-labelled sugar residues from sugar nucleotides (termed the glycosyl donors as in the chemical reactions) to form labelled products which are quantitated by liquid scintillation counting. There are two well appreciated difficulties in assaying glycosyltransferases in this manner. First, there is the problem of isolating sufficient quantities of *well-characterized* oligosaccharide acceptor — a minimum requirement for such assays. In addition, many naturally occurring acceptor sequences are substrates for more

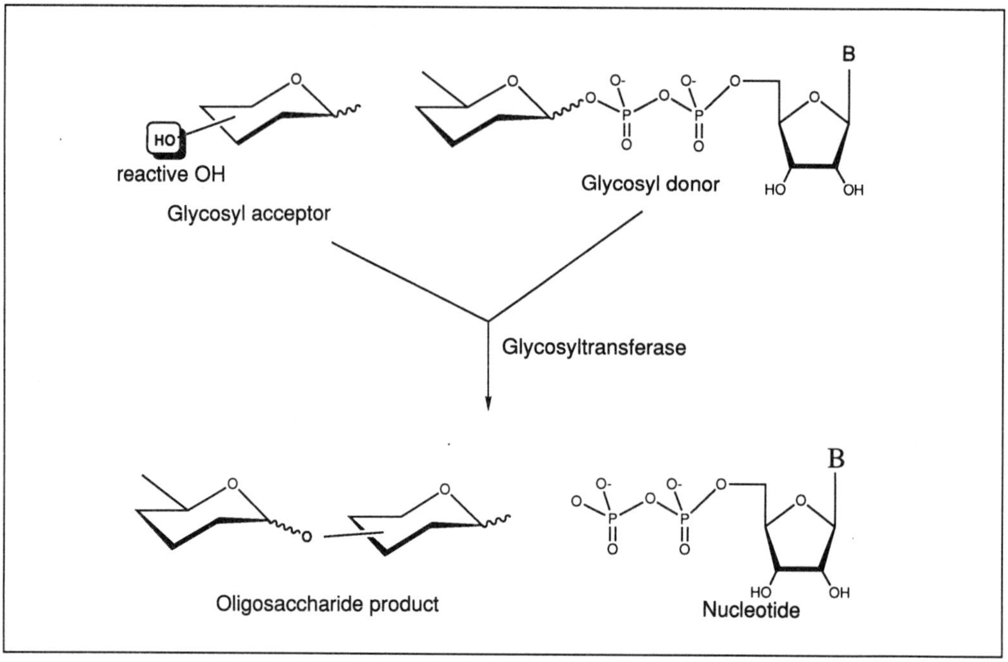

Fig. 4 A generalized glycosyltransferase reaction showing the donor sugar nucleotide, the acceptor oligosaccharide, and the hypothetical product formed

than one enzyme utilizing the same sugar nucleotide. This means that radiolabelled products have to be separated not only from unreacted sugar nucleotides, and their degradation products, but also from isomeric products. This situation is particularly difficult in the case of the branching N-acetylglucosaminyltransferases (GlcNAcTs).

Our contributions to this area have involved defining minimum oligosaccharide sequences which can serve as efficient acceptors for a specific enzyme. Essentially, we have been 'carving out' fragments of complex N-linked oligosaccharides and these have been prepared through total chemical synthesis using the general procedures and strategies described above (82–89). Several groups have been involved in similar efforts, the most notable being those of Paulsen (90, 91) and Matta (91–94). Also, Venot et al. (95, 96) have worked extensively on developing substrates for fucosyl and sialyl-transferases. We will restrict ourselves here mostly to a discussion of our own results where we are directly familiar with the synthetic chemistry involved.

All substrates we have developed are attached to hydrophobic linking arms — usually the 'Lemieux arm' (28, 97) (8-methoxycarbonyloctyl) or the simpler and preferred octyl group. When such hydrophobic glycosides are used as acceptor substrate, the radiolabelled products adsorb onto reverse-phase C-18 resins, from which they can be eluted with methanol, affording a very simple yet rapid procedure termed the 'SepPak Assay' (98). Table 5 lists the structures of 11 synthetic

Table 5 Acceptor sequences for glycosyltransferases, number of steps required for their synthesis, and products formed

Entry	Enzyme	Acceptor	No. of steps	Donor	Product	Reference
A	β(1→2)GlcNAc-T-I	α-Man-(1→6)⎤ 　　　　　　　⎬β-Man-OR α-Man-(1→3)⎦	16	UDP-GlcNAc	α-Man-(1→6)⎤ 　　　　　　　　　　　　⎬β-Man-OR β-GlcNAc-(1→2)-α-Man-(1→3)⎦	83, 84
B	β(1→2)GlcNAc-T-II	α-Man-(1→6)⎤ 　　　　　　　　　　　⎬β-Man-OR β-GlcNAc-(1→2)-α-Man-(1→3)⎦	17*	UDP-GlcNAc	β-GlcNAc-(1→2)-α-Man-(1→6)⎤ 　　　　　　　　　　　　　　⎬β-Man-OR β-GlcNAc-(1→2)-α-Man-(1→3)⎦	83
C	β(1→3)GlcNAc-T-III	β-GlcNAc-(1→2)-α-Man-(1→6)⎤ 　　　　　　　　　　　　　　⎬β-Man-OR β-GlcNAc-(1→2)-α-Man-(1→3)⎦	18*	UDP-GlcNAc	β-GlcNAc-(1→2)-α-Man-(1→6)⎤ 　　　　　　　　　　β-GlcNAc-(1→4)⎬β-Man-OR β-GlcNAc-(1→2)-α-Man-(1→3)⎦	83
D	β(1→4)GlcNAc-T-IV	β-GlcNAc-(1→2)-α-Man-(1→3)-β-Man-OR	20	UDP-GlcNAc	β-GlcNAc-(1→4)⎤ 　　　　　　　　⎬α-Man-(1→3)-β-Man-OR β-GlcNAc-(1→2)⎦	91, 99
E	β(1→6)GlcNAc-T-V	β-GlcNAc-(1→2)-α-Man-(1→6)-β-Glc-OR	19	UDP-GlcNAc	β-GlcNAc-(1→6)⎤ 　　　　　　　　⎬α-Man-(1→6)-β-Glc-OR β-GlcNAc-(1→2)⎦	82, 85
F	β(1→4)GlcNAc-T-VI	β-GlcNAc-(1→6)⎤ 　　　　　　　　⎬α-Man(1→6)-β-Man-OR β-GlcNAc-(1→2)⎦	27	UDP-GlcNAc	β-GlcNAc-(1→6)⎤ β-GlcNAc-(1→4)⎬α-Man-(1→6)-β-Man-OR β-GlcNAc-(1→2)⎦	86, 87
G	α(1→2)Fuc-T	β-Gal-(1→3)-β-GlcNAc-OR	10	GDP-Fuc	α-Fuc-(1→2)-β-Gal-(1→3)-β-GlcNAc-OR	88, 97
H	α(1→3)Fuc-T	β-Gal-(1→4)-β-GlcNAc-OR	12	GDP-Fuc	β-Gal-(1→4)⎤ 　　　　　　⎬β-GlcNAc-OR α-Fuc-(1→3)⎦	88
I	α(1→4)Fuc-T	β-Gal-(1→3)-β-GlcNAc-OR	10	GDP-Fuc	α-Fuc-(1→4)⎤ 　　　　　　⎬β-GlcNAc-OR β-Gal-(1→3)⎦	88, 97
J	β(1→6)-GlcNAc-T (Mucin core-2)	β-Gal-(1→3)-α-GalNAc-OR	10	UDP-GlcNAc	β-GlcNAc-(1→6)⎤ 　　　　　　　　⎬α-GalNAc-OR β-Gal-(1→3)⎦	88, 100
K	α(2-3)-sialyl-T	β-Gal-(1→4)-β-GlcNAc-OR	12	CMP-NeuNAc	α-NeuNAc-(2→3)-β-Gal-(1→4)-β-GlcNAc-OR	88
L	β-(1→3)-GlcNAc-T-i	β-Gal-(1→4)-β-GlcNAc-(1→6)-α-Man-(1→6)-β-Man-OR	29	UDP-GlcNAc	β-GlcNAc-(1→3)-β-Gal-(1→4)-β-GlcNAc-(1→6)-α-Man-(1→6)-β-Man-OR	89

* One enzymatic step was used in the synthetic scheme; R = $(CH_2)_8CO_2CH_3$ or $(CH_2)_7CH_3$.

oligosaccharides which have been shown to be acceptors for 12 different glycosyltransferases, the sugar nucleotides used, and the products formed by the action of these enzymes. The kinetic parameters for these compounds can be found in the indicated references. They are substrates essentially equivalent to the natural glycopeptides, sometimes better. Also listed in Table 5 are the approximate number of synthetic steps required for the preparation of the acceptors beginning with commercial unprotected monosaccharides. This averages over five steps per monosaccharide residue. Of practical importance is that each step requires about one week of work for a trained individual. This includes the consideration that each chemical step is usually carried out twice—first on a small trial scale to make sure the selected reaction conditions are effective. Then on a preparative scale. This means also that two chromatographic purification procedures must be carried out and on the average (in our hands) this requires six high field (> 250 MHz) NMR spectra per compound (^{1}H and ^{13}C) to be recorded and analysed. The net result is that oligosaccharide synthesis is very expensive—usually about \$15 000 for a trisaccharide produced on a normal synthetic scale of 15 mg.

3.1 Synthesis of acceptor analogues

Frequently, 'carved-out' structures such as those in Table 5, consisting of the natural sugars in the sequence normally recognized by a given glycosyltransferase, are also substrates for other enzymes as noted above. In that case, to avoid tedious product separation, acceptor analogues can be synthesized where the cross-reactive OH group is either removed or masked. The resulting deoxy or O-methylated structures (Table 6) can then in principle be monospecific acceptors for the target enzyme *provided they are still recognized by the enzyme*. A particular case in point is the biantennary pentasaccharide in entry C (Table 5). This compound was shown to be a substrate for GlcNAcT-III, yielding the indicated bisected hexasaccharide, but GlcNAcTs-IV and -V can also act on this structure making it of much lesser value in glycosyltransferase assays (13). This problem could be solved by either deoxygenating *both* cross-reactive OH groups (101, 102) or O-methylating (103) them. In both instances, the resulting pentasaccharide analogue fortunately remained fully active. The other seven analogues in Table 6 also proved extremely useful to specifically assay the indicated enzyme activity.

The purpose for presenting Table 6 is not only to reiterate the power of organic synthesis to yield unnatural analogues with unique and desirable properties, here as enzyme substrates. The main reason is to alert the reader to the fact that the synthesis of oligosaccharide analogues containing modified sugars is a much more complex task than to prepare the 'natural' compounds shown in Table 5. In fact, as can be seen from the entries noting the number of synthetic steps, almost twice as many steps are required—about 10 steps per monosaccharide residue. Also of importance, and not obvious from Table 6, is that the steps are much more difficult since most chemical protocols in the literature have been optimized for the natural unmodified sugar residues. This means that the syntheses of analogues take more

Table 6 Modified acceptor substrates for glycosyltransferases, number of steps required for their synthesis, and products formed

Entry	Enzyme	Modified acceptor	No. of steps	Donor	Product	Reference
A	β(1→2)-GlcNAc-T-II	β-GlcNAc-(1→2)-α-Man-(1→6)[6-deoxy]\α-Man-(1→3)[3-deoxy]/β-Man-OR	38	UDP-GlcNAc	β-GlcNAc-(1→2)-α-Man-(1→6)[6-deoxy]\α-Man/β-Man-OR · β-GlcNAc-(1→2)-α-Man-(1→3)[4-deoxy]\β-Man-OR	101; 101, 102
B	β-(1→3)-GlcNAc-T-III	β-GlcNAc-(1→2)-α-Man-(1→6)[6-deoxy]\β-GlcNAc-(1→2)-α-Man-(1→3)[4-deoxy]/β-Man-OR	39	UDP-GlcNAc	β-GlcNAc-(1→2)-α-Man-(1→6)[6-deoxy], β-GlcNAc-(1→4), β-GlcNAc-(1→2)-α-Man-(1→3)[4-deoxy] → β-Man-OR	103
C	β-(1→3)-GlcNAc-T-III	β-GlcNAc-(1→2)-α-Man-(1→6)[6-OMe]\β-GlcNAc-(1→2)-α-Man-(1→3)[4-OMe]/β-Man-OR	41	UDP-GlcNAc	β-GlcNAc-(1→2)-α-Man-(1→6)[6-OMe], β-GlcNAc-(1→4), β-GlcNAc-(1→2)-α-Man-(1→3)[4-OMe] → β-Man-OR	104
D	β-(1→6)-GlcNAc-T-V	β-GlcNAc-(1→2)-α-Man-(1→6)-β-Glc-OR [3-OMe]	19	UDP-GlcNAc	β-GlcNAc-1→6, β-GlcNAc-(1→2)-α-Man-(1→6)-β-Glc-OR [3-OMe]	104
E	β-(1→6)-GlcNAc-T-V	β-GlcNAc-(1→2)-α-Man-(1→6)-β-Glc-OR [2,3,4-tri-OMe]	18	UDP-GlcNAc	β-GlcNAc-(1→6), β-GlcNAc-(1→2)-α-Man-(1→6)-β-Glc-OR [2,3,4-tri-OMe]	105
F	α-(1→3)Fuc-T	β-Gal-(1→4)-β-GlcNAc-OH [2-OMe]	15	GDP-Fuc	2-OMe, β-Gal-(1→4), α-Fuc-(1→3)-β-GlcNAc-OH	92, 94, 106
G	α-(1→3)Fuc-T	β-Gal-(1→4)-β-GlcNAc-OBn [3-SO₃Na]	12	GDP-Fuc	3-SO$_3$Na, β-Gal-(1→4), α-Fuc-(1→3)β-GlcNAc-OBn	94
H	α-(1→4)Fuc-T	β-Gal-(1→3)-β-GlcNAc-OBn [2-OMe]	14	GDP-Fuc	α-Fuc-(1→4), β-Gal-(1→3)-β-GlcNAc-OBn [2-OMe]	93, 94, 107

One enzymatic step was used in the synthetic scheme; R = (CH$_2$)$_8$CO$_2$CH$_3$ or (CH$_2$)$_7$CH$_3$.

than twice as long and, too frequently, key reactions fail completely required a very expensive and frustrating restart of the synthesis using a new route.

4. Inhibitors for glycosyltransferases

The interest in the generation of specific inhibitors for glycosyltransferases arises from the expectation that such compounds, when they are finally prepared in a membrane-permeable form, should act as specific modifiers of protein and lipid glycosylation, and thereby also as specific modifiers of cell-surface glycosylation (108, 109). Thus, they should be invaluable as tools for the study of the function of glycosylation, and also potential as drugs where specific carbohydrate sequences potentially control a biological phenomenon such as cell–cell, bacterial, or viral adhesion.

One of the most promising approaches for generating *specific* glycosyltransferase-inhibitors postulates that the specificity of this class of enzyme most likely resides in the recognition of acceptor sequences, and not donor sequences, since over 10 glycosyltransferases utilize any given sugar nucleotide. To date, several inhibitors (Table 7) have been prepared by taking known acceptor sequences and modifying either the reactive OH group to which a glycosyltransferase transfers (usually by either deoxygenation or O-methylation) or a site immediately adjacent. In Table 7, several examples are presented where important but limited success has been achieved in this regard. As seen in this table, in most instances the inhibitor thus produced is competitive with the acceptor oligosaccharide as expected. In only two cases (Table 7, entries D and G) are the K_is better than the acceptor K_m (110, 112) and in one of these the mode of inhibition is mechanistically unclear. This latter example, entry G in Table 7, represents the best acceptor-based inhibitor for a glycosyltransferase reported to date (112).

Modification of acceptors has become a common approach for development of glycosyltransferase inhibitors, but too many times the analogues are inactive. This is especially troublesome since the synthesis of analogues, as discussed in section 3.1, is the most time consuming and expensive area of oligosaccharide synthesis. In the original paper (88) which surveyed the possibility of using deoxygenated acceptor analogues as glycosyltransferase inhibitors, only half of the compounds synthesized showed any activity. It is hoped that other approaches, including the development of covalently inactivating inhibitors will prove even more successful.

5. Enzyme-assisted oligosaccharide synthesis

While the purpose of this chapter is to summarize the state of *chemical* oligosaccharide synthesis, glycosyltransferases and glycosidases are increasingly being used to assist in the preparation of target molecules. Over 20 glycosyltransferases have been used in this regard, and many glycosidases. The area is the subject of a recent excellent review (116) and we will comment here only on the practicality of this approach.

Table 7 Synthetic glycosyltransferase inhibitors

Entry	Enzyme	Donor	Acceptor	Inhibitor	Acceptor K_M (μM)	Inhibitor K_i (μM)	Inhibitor mode	Reference
A	α(1→2)FucT	GDP-Fuc	βGal(1→2)βGlcNAc-OR$_1$	βGal(1→3)βGlcNAc-OR$_1$, 2-deoxy	200	800	Competitive	88, 113
B	β(1→4)GalT	UDP-Gal	βGlcNAc(1→3)βGal-OCH$_3$	βGlcNAc(1→3)βGal-OCH$_3$, 6-thio	1000	1000	Competitive	114
C	β(1→6)GlcNAc-T-V	UDP-GlcNAc	βGlcNAc(1→2)αMan(1→6)βGlc-OR$_2$	βGlcNAc(1→2)αMan(1→6)βGlc-OR$_2$, 6-deoxy, 4-O-Methyl	23	30	Competitive	88, 110, 115
D	β(1→6)GlcNAc-T-V	UDP-GlcNAc	βGlcNAc(1→2)αMan(1→6)βGlc-OR$_2$	βGlcNAc(1→2)αMan(1→6)βGlc-OR$_2$, 6-deoxy	23	14	Competitive	110
E	β(1→6)GlcNAc-T-i	UDP-GlcNAc	βGal(1→3)αGalNAc-OR$_1$	βGal(1→3)αGalNAc-OR$_1$, 3-deoxy	80	560	Competitive	88
F	α(1→3)-'A'-GalNAc-T	UDP-GalNAc	αFuc(1→2)βGal-OR$_2$	αFuc(1→2)βGal-OR$_2$, 3-amino	1.5	68	Competitive	111
G	α(1→3)-'A'-GalNAc-T	UDP-GalNAc	αFuc(1→2)βGal-OR$_2$	αFuc(1→2)βGal-OR$_2$	1.5	0.2	Mixed-mode	112

R$_1$ = (CH$_2$)$_8$CO$_2$CH$_3$; R$_2$ = (CH$_2$)$_7$CH$_3$.

As shown in Figure 4, and as summarized in Tables 5 and 6, a very large number of chemical steps is required to yield a final tri-to pentasaccharide. Also from these tables, it is clear that glycosyltransferases, although they biosynthetically act on very large and complex glycolipids and glycoproteins, often act equally well on small synthetic oligosaccharides of a size more amenable to relatively facile preparations through chemical synthesis. The limitation to the use of glycosyltransferases in synthesis is the availability of the enzyme required for a particular step, since the sugar nucleotide, though expensive, is commercially available. Only a few of these enzymes are commercially available, but an increasing number of simple partial purification procedures, requiring under a week of work, have been reported (116).

The most efficient use of glycosyltransferases in the combined chemical-enzymatic synthesis of oligosaccharides is to add terminal sugar residues enzymatically on to presynthesized 'primers' (34, 116–120). In Table 5, two partially purified N-acetylglucosaminyltransferases (GlcNAcT-I and II) were used (83) in this manner to convert the synthetic trimannoside (entry A) to the tetrasaccharide (entry B) and on to the pentasaccharide (entry C). Similarly in Table 6 (entry B), the second enzyme was used in the preparation of an analogue (101). These conversions were carried out on a 50–100 mg scale. Had these two enzymes not been used to 'top off' existing compounds, the chemical syntheses would have had to be completely redesigned at great expense in time. Many similar examples have been reported, the most extensive one being the use of four sequential glycosyltransferases on a synthetic disaccharide to produce an O-linked sialyl-Le^x hexasaccharide (121). As discussed in recent reviews (122, 123), glycosyltransferases will also transfer unnatural sugar residues from their synthetic sugar nucleotides although such reactions are generally much less efficient than the natural reaction. Simple deoxygenated analogues are generally the most effective.

6. Conclusion

The chemical synthesis of oligosaccharides remains a major laboratory exercise requiring many months for a single new compound. The preparation of analogues, such as deoxy, O-methyl, amino, fluoro etc., takes approximately twice as long as the preparation of unmodified oligosaccharides. The use of enzymes, especially glycosyltransferases, can greatly simplify the procedure since the requirement for the chemistry is then diminished to the synthesis of more readily accessible 'primer' oligosaccharides which can be elongated in a controlled manner. In all instances, however, the work required to yield 10 mg samples of compounds remains a serious commitment and target molecules should therefore be carefully chosen for their ability to contribute to structure-activity relationships in glycobiology.

Acknowledgements

The authors are grateful to the Alberta Heritage Foundation for Medical Research for a Fellowship (S.H.K.) and Research grant support (O.H.) from the Natural Sciences and Engineering Research Council of Canada.

References

1. Varki, A. (1993) Biological roles of oligosaccharides: all of the theories are correct, *Glycobiology*, **3**, 97.
2. Bock, K., Breimer, M. E., Brignole, A., Hasson, G. C., Karlsson, K.-A., Larson, G. et al. (1985) Specificity of binding of a strain of uropathogenic *Escherichia coli* to Galα1→4Gal containing glycosphingolipids. *J. Biol. Chem.*, **260**, 8545.
3. Krivan, H. C., Roberts, D. D., and Ginsburg, V. (1988) Many pulmonary pathogenic bacteria bind specifically to the carbohydrate sequence GalNAcβ1→4Gal found on some glycolipids. *Proc. Natl Acad. Sci.*, **85**, 6157.
4. Karlsson, K.-A. (1989) Animal glycosphingolipids as membrane attachment sites for bacteria. *Annu. Rev. Biochem.*, **58**, 309.
5. Pritchett, T. J., Brossmer, R., Rose, U., and Paulson, J. C. (1987) Recognition of monovalent sialosides by influenza virus H3 hemagglutinin. *Virology*, **160**, 502.
6. Gruters, R. A., Neefjes, J. J., Tersmette, M., de Goede, R. E. Y., Tulp, A., Huisman, G. et al. (1987) Interference with HIV-induced syncytium formation and viral infectivity by inhibitors of trimming glucosidase. *Nature*, **330**, 74.
7. Phillips, M. L., Nudelman, E., Gaeta, F. C. A., Perez, M., Singhal, A. K., Hakamori, S-I. et al. (1990) ELAM-1 mediates cell adhesion by recognition of a carbohydrate ligand, sialyl-Le[x]. *Science*, **250**, 1130.
8. Brandley, B. K. (1991) Cell surface carbohydrates in cell adhesion. *Semin. Cell Biology*, **2**, 281.
9. Lasky, L. A. (1992) Selectins: interpreters of cell-specific carbohydrate information during inflammation. *Science*, **258**, 964.
10. Bevilacqua, M. (1993) Endothelial-leukocyte adhesion molecules. *Annu. Rev. Immunol.*, **11**, 767.
11. Beyer, T. A., Sadler, J. E., Rearick, J. I., Paulson, J. C., and Hill, R. L. (1981) Glycosyltransferases and their use in assessing oligosaccharide structure and structure–function relationships. *Adv. Enzymol.*, **52**, 23.
12. Schachter, H. (1984) Glycoproteins: Their structure, biosynthesis and possible clinical inplications. *Clinical Biochem.*, **17**, 3.
13. Schachter, H. (1986) Biosynthetic controls that determine the branching and microheterogenity of protein-bound oligosaccharides. *Biochem. Cell Biol.*, **64**, 163.
14. Schachter, H. (1991) Enzymes associated with glycosylation. *Curr. Opin. Struct. Biol.*, **1**, 755.
15. Schachter, H. (1991) The 'yellow brick road' to branched complex N-glycans. *Glycobiology*, **1**, 453.
16. Schachter, H. and Brockhausen, I. (1992) The biosynthesis of serine (Threonine)-N-acetylgalactosamine-linked carbohydrate moieties. In *Glycoconjugates: composition, structure, and function* (ed. H. J. Allen and E. C. Kisailus), p. 263. Marcel Dekker, New York.
17. Schachter, H. (1992) Branching of N- and O-glycans: biosynthetic controls and functions. *Trends Glycosci. Glycotechnol.*, **4**, 241.
18. DiCioccio, R. A., Klock, P. J., Barlow, J. J., and Matta, K. L. (1980) Rapid procedures for determination of endo-N-acetyl-α-D-galactosaminidase in *Clostridium perfringens*, and of the substrate specificity of exo-β-D-galactosidases. *Carbohydr. Res.*, **81**, 315.
19. Skyes, D. E., Abbas, S. A., Barlow, J. J., and Matta, K. L. (1983) Substrate specificity and other properties of the β-D-galactosidase from *Aspergillus niger*. *Carbohydr. Res.*, **116**, 127.
20. Yazawa, S., Madiyalakan, R., Chawda, R. P., and Matta, K. L. (1986) α-L-Fucosidase

from *Aspergillus niger*: demonstration of a novel α-L-(1→6)-fucosidase acting on glycopeptides. *Biochem. Biophys. Res. Commun.*, **136**, 563.
21. Wulff, G. and Röhle, G. (1974) Results and problems of *O*-glycoside synthesis. *Angew. Chem. Int. Ed. Engl.*, **13**, 157.
22. Sinaÿ, P. (1978) Recent advances in glycosylation reactions. *Pure Appl. Chem.*, **50**, 1437.
23. Paulsen, H. (1982) Advances in selective chemical synthesis of complex oligosaccharides. *Agnew. Chem. Int. Ed. Engl.*, **21**, 155.
24. Paulsen, H. (1983) Synthesis of complex oligosaccharide chains of glycoproteins. *Chem. Soc. Rev.*, **13**, 15.
25. Schmidt, R. (1986) New methods for the synthesis of glycosides and oligosaccharides — Are there alternatives to the Koenings–Knorr method? *Angew. Chem. Int. Ed. Engl.*, **25**, 212.
26. Flowers, H. M. (1987) Chemical synthesis of oligosaccharides. *Methods Enzymol.*, **138**, 359.
27. Fügedi, P., Garegg, P. J., Lönn, H., and Norberg, T. (1987) Thioglycosides as glycosylating agents in oligosaccharide synthesis. *Glycoconjugate J.*, **4**, 97.
28. Lemieux, R. U. (1989) The origin of the specificity in the recognition of oligosaccharides by proteins. *Chem. Soc. Rev.*, **18**, 347.
29. Paulsen, H. (1990) Synthesis, conformations and X-ray structure analyses of the saccharide chains from the core regions of glycoproteins. *Angew. Chem. Int. Ed. Engl.*, **29**, 823.
30. Malik, A., Bauer, H., Tschakert, J., and Voelter, W. (1990) Solid phase synthesis of carbohydrates. *Chemiker-Zeitung*, **114**, 371.
31. Okamoto, K. and Goto, T. (1990) Glycosylation of sialic acid. *Tetrahedron*, **46**, 5835.
32. Bundle, D. R. (1990) Synthesis of oligosaccharides related to bacterial *O*-antigens. *Topics Curr. Chem.*, **154**, 1.
33. DeNinno, M. P. (1991) The synthesis and glycosidation of *N*-acetylneuraminic acid. *Synthesis*, 583.
34. Hindsgaul, O. (1991) Synthesis of carbohydrates for applications in glycobiology. *Semin. Cell Biol.*, **2**, 319.
35. Jung, K-H. and Schmidt, R. R. (1991) Structure and synthesis of biologically active glycopeptides and glycolipids. *Curr. Opin. Struct. Biol.*, **1**, 721.
36. Nicolaou, K. C., Caulfield, T. J., and Groneberg, R. D. (1991) Synthesis of novel oligosaccharides. *Pure Appl. Chem*, **63**, 555.
37. Sinaÿ, P. (1991) Recent advances in glycosylation reactions. *Pure Appl. Chem.*, **63**, 519.
38. Glaudemans, C. P. J. (1991) Mapping of subsites of monoclonal, anti-carbohydrate antibodies using deoxy and deoxyfluoro sugars. *Chem. Rev.*, **91**, 25.
39. Banoub, J., Boullanger, P., and Lafont, D. (1992) Synthesis of oligosaccharides of 2-amino-2-deoxy sugars. *Chem. Rev.*, **92**, 1167.
40. Garegg, P. J. (1992) Saccharides of biological importance: challenges and opportunities for organic synthesis. *Acc. Chem. Res.*, **25**, 575.
41. Fraser-Reid, B., Udodong, U. E., Wu, Z., Ottosson, H., Merritt, J. R., Rao, C. S., Roberts, C., and Madsen, R. (1992) *n*-Pentenyl glycosides in organic chemistry: A contemporary example of serendipity. *Synlett.*, **12**, 927.
42. Kanie, O. and Hindsgaul, O. (1992) Synthesis of oligosaccharides, glycolipids and glycopeptides. *Curr. Opin. Struct. Biol.*, **2**, 674.
43. Lee, Y. C. and Lee, R. T. (1992) Synthetic glycoconjugates. In *Glyconjugates: composition, structure, and function* (ed. H. J. Allen and E. C. Kisailus), p. 121. Marcel Dekker, New York.

44. Györgydeák, Z., Szilágyi, L., and Paulsen, H. (1993) Synthesis, structure and reactions of glycosyl azides. *J. Carbohydr. Chem.*, **12**, 139.
45. Kaji, E. and Lichtenthaler, F. W. (1993) 2-Oxo- and 2-oximinoglycosyl halides: versatile glycosyl donors for the construction of β-D-mannose and β-D-mannosamine-containing oligosaccharides. *Trends Glycosci. Glycotechnol.*, **5**, 121.
46. Stowell, C. P. and Lee, Y. C. (1980) Neoglycoproteins: the preparation and application of synthetic glycoproteins. *Adv. Carbohydr. Chem. Biochem.*, **37**, 225.
47. Hanies, A. H. (1981) The selective removal of protecting groups in carbohydrate chemistry. *Adv. Carbohydr. Chem. Biochem.*, **39**, 13.
48. David, S. and Hanessian, S. (1985) Regioselective manipulation of hydroxyl groups via organotin derivatives. *Tetrahedron*, **47**, 643.
49. Zehavi, U. (1988) Application of photosensitive protecting groups in carbohydrate chemistry. *Adv. Carbohydr. Chem. Biochem.*, **46**, 179.
50. Binkley, R. W. (1988) Protective groups. In *Modern carbohydrate chemistry*, p. 113. Marcel Dekker, New York.
51. Lemieux, R. U., Hendriks, K. B., Stick, R. V., and James, K. (1975) Halide ion catalyzed glycosidation reactions. Synthesis of α-linked disaccharides. *J. Am. Chem. Soc.*, **97**, 4056.
52. Paulsen, H. and Lockhoff, O. (1981) Neue effective β-glycosidsynthese für mannose-glycoside synthesen von mannose-haltigen oligosacchariden. *Chem. Ber.*, **114**, 3102.
53. Garegg, P. J. and Ossowski, P. (1983) Silver zeolite as promoter in glycoside synthesis. The synthesis of β-D-mannopyranosides. *Acta Chem. Scan. (B)*, **37**, 249.
54. Barresi, F. and Hindsgaul, O. (1991) Synthesis of β-mannopyanosides by intramolecular aglycon delivery. *J. Am. Chem. Soc.*, **113**, 9376.
55. Barresi, F. and Hindsgaul, O. (1992) Improved synthesis of β-mannopyanosides by intramolecular aglycon delivery. *Synlett.*, **9**, 759.
56. Stork, G. and Kim, G. (1992) Stereocontrolled synthesis of disaccharides via the temporary silicon connection. *J. Am. Chem. Soc.*, **114**, 1087.
57. Lemieux, R. U., Takeda, T., and Chung, B. Y. (1976) Synthesis of 2-acetamido-2-deoxy-β-D-glucopyranosides. *ACS Symp. Ser.*, **39**, 90.
58. Ito, Y., Numata, M., Sugimoto, M., and Ogawa, T. (1989) Highly stereoselective synthesis of ganglioside GD_3. *J. Am. Chem. Soc.*, **111**, 8508.
59. Ito, Y. and Ogawa, T. (1990) Highly stereoselective glycosylation of sialic acid aided by stereocontrolling auxiliaries. *Tetrahedron*, **46**, 89.
60. Nicolaou, K. C., Hummel, C. W., Bockovich, N. J., and Wong, C.-H. (1991) Stereocontrolled synthesis of sialyl Le^x, the oligosaccharide binding ligand to ELAM-1 (Sialyl = N-acetylneuramin). *J. Chem. Soc., Chem. Commun.*, 870.
61. Mukaiyama, T., Murai, Y., and Shoda, S-I. (1981) An efficient method for glycosylation of hydroxy compounds using glucopyranosyl fluoride. *Chem. Lett.*, 431.
62. Nicolaou, K. C., Caulfield, T. J., Kataoka, H., and Stylianides, N. A. (1990) Total synthesis of the tumor-associated Le^x family of glycosphingolipids. *J. Am. Soc.*, **112**, 3693.
63. Wessel, H. P. and Ruiz, N. (1991) α-Glycosylation reactions with 2,3,4,6-tetra-O-benzyl-β-D-glycopyranosyl fluoride and triflic anhydride as promoter. *J. Carbohydr. Chem.*, **10**, 901.
64. Matsuzaki, Ito, Y., Nakahara, Y., and Ogawa, T. (1993) Synthesis of branched poly-N-acetyl-lactosamine type pentaantennary pentacosasaccharide: glycan part of a glycosyl ceramide from rabbit erythrocyte membrane. *Tetrahedron Lett.*, **34**, 1061.

65. Konradsson, P., Motto, D. R., McDevitt, R. E., and Fraser-Reid, B. (1990) Iodonium ion generated *in situ* from N-iodosuccinimide and trifluoromethanesulphonic acid promotes direct linkage of 'Disarmed' pent-4-enyl glycosides. *J. Chem. Soc., Chem. Commun.*, 270.
66. Veeneman, V. H., van Leeuwen, S. H., and Van Boom, J. H. (1990) Iodonium ion promoted reactions at the anomeric centre. II. An efficient thioglycoside mediated approach toward the formation of 1,2-*trans* linked glycosides and glycosidic esters. *Tetrahedron Lett.*, **31**, 1331.
67. Ogawa, T., Beppu, K., and Nakabayashi, S. (1981) Trimethylsilyl trifluoromethanesulfonate as an effective catalyst for glycoside synthesis. *Carbohydr. Res.*, **93**, C6.
68. Merritt, R. J. and Fraser-Reid, B. (1992) n-Pentenyl glycoside methodology for rapid assembly of homoglycans exemplified with the nonasaccharide component of a high-mannose glycoprotein. *J. Am. Chem. Soc.*, **114**, 8334.
69. Ogawa, T., Sugimoto, M., Kitajama, T., Sadozai, K. K., and Nukada, T. (1986) Total synthesis of a undecasaccharide: a typical carbohydrate sequence for the complex type of glycan chains of a glycoprotein. *Tetrahedron Lett.*, **27**, 5739.
70. Kahne, D., Walker S., Cheng, Y., and Engen, D. V. (1989) Glycosylation of unreactive substrates. *J. Am. Soc.*, **111**, 6881.
71. Sarkar, A. K. and Matta, K. L. (1992) 2,3,4,6-Tetra-O-benzyl-β-D-galactopyranosyl phenyl sulfoxide as a glycosyl donor. Synthesis of some oligosaccharides containing an α-D-galactopyranosyl group. *Carbohyd. Res.*, **233**, 245.
72. Raghvan, S. and Kahne, D. (1993) A one step synthesis of the Ciclamycin trisaccharide. *J. Am. Chem. Soc.*, **115**, 1580.
73. Kochetkov, N. K., Klimov, E. M., and Malysheva, N. N. (1989) Novel highly stereoselective method of 1,2-*cis*-glycosylation. Synthesis of α-D-glucosyl-D-glucoses. *Tetrahedron Lett.*, **30**, 5459.
74. Kochetkov, N. K., Klimov, E. M., Malysheva, N. N., and Demchenko, A. V. (1992) Stereospecific 1,2-*cis*-glycosylation: a modified thiocyanate method. *Carbohydr. Res.*, **232**, C1.
75. Kováč, P. (1986) Efficient chemical synthesis of methyl β-glycosides of β-(1→6)-linked D-galacto-oligosaccharides by a stepwise and a blockwise approach. *Carbohydr. Res.*, **153**, 237.
76. Paulsen, H., Heume, M., and Nurnberger, H. (1990) Synthese der verzweigten nonasaccharid-sequenz der 'bisected' struktur von N-glycoproteinen. *Carbohydr. Res.*, **200**, 127.
77. Kameyama, A., Ishida, H., Kiso, M., and Hasegawa, A. (1991) Synthetic studies on sialoglycoconjugates 22: total synthesis of tumor-associated ganglioside, sialyl lewis X. *J. Carbohydr. Chem.*, **10**, 549.
78. Lorentzen, J. P., Halpap, B., and Lockhoff, O. (1991) Synthesis of an elicitor-active heptaglucan saccharide for investigation of defense mechanisms of plants. *Angew. Chem. Int. Ed. Eng.*, **30**, 1681.
79. Thomas, R. L., Dubey, R., Abbas, S. A., and Matta, K. L. (1987) Synthesis of O-(2-acetamido-2-deoxy-β-D-glucopyranosyl)-(1→3)-O-β-D-galactopyranosyl-(1→4)-2-acetamido-2-deoxy-D-glucopyranose and two related trisaccharides containing the 'lacto-N-biose II' unit. *Carbohydr. Res.*, **169**, 201.
80. Dubey, R., Abbas, S. A., and Matta, K. L. (1988) Further use of O-(2,3,4-tri-O-acetyl-α-L-fucopyranosyl)-(1→3)-O-(2-acetamido-4,6-di-O-acetyl-2-deoxy-β-D-glucopyranosyl-(1→3)-2,4,6-tri-O-acetyl-α-D-galactopyranosyl bromide as a glycosyl donor. Synthesis of two mucin-type tetrasaccharides. *Carbohydr. Res.*, **181**, 236.

81. Jain, R. K., Kohata, K., Abbas, S. A., and Matta, K. L. (1988) Synthetic mucin fragments. Synthesis of a tetra- and two penta-saccharides containing the O-β-D-galactopyranosyl-(1→3)-O-(2-acetamido-2-deoxy-β-D-glucopyranosyl-(1→3)-D-galactopyranose ('lacto-N-triose 1') unit. *Carbohydr. Res.*, **182**, 290.
82. Khan, S. H., Abbas, S. A., and Matta, K. L. (1989) Synthesis of some oligosaccharides containing the O-(2-acetamido-2-deoxy-β-D-glucopyranosyl)-(1→2)-O-α-D-mannopyranosyl unit. Potential substrates for UDP-GlcNAc: α-D-mannopyranosyl-(1→6)-N-acetyl-β-D-glucosaminyltransferase (GnT-V). *Carbohydr. Res.*, **193**, 125.
83. Kaur, K. J., Alton, G., and Hindsgaul, O. (1991) Use of N-acetylglucosaminyltransferase I and II in the preparative synthesis of oligosaccharides. *Carbohydr. Res.*, **210**, 145.
84. Kaur, K. J. and Hindsgaul, O. (1991) A simple synthesis of octyl 3,6-di-O-(α-mannopyranosyl)-β-D-mannopyranoside and its use as an acceptor for the assay of N-acetylglucosaminyltransferase I activity. *Glycoconjugate J.*, **8**, 90.
85. Srivastava, O. P., Hindsgaul, O., Shoreibah, M., and Pierce, M. (1988) Recognition of oligosaccharide substrates by N-acetylglucosaminyl transferase V. *Carbohydr. Res.*, **179**, 137.
86. Hindsgaul, O., Tahir, S. H., Srivastava, O. P., and Pierce, M. (1988) The trisaccharide β-D-GlcpNAc-(1→2)-α-D-Manp-(1→6)-β-D-Manp, as its 8-methoxycarbonyloctyl glycoside, is an acceptor selective for N-acetylglucosaminyltransferase V. *Carbohydr. Res.*, **173**, 263.
87. Brockhausen, I., Hull, E., Hindsgaul, O., Schachter, H., Shah, R. N., Michnick, S. W. et al. (1989) Control of glycoprotein synthesis. *J. Biol. Chem.*, **264**, 11 211.
88. Hindsgaul, O., Kaur, K. J., Srivastava, G., Blaszczyk-Thurin, M., Crawley, S. C., Heerze, L. D. et al. (1991) Evaluation of deoxygenated oligosaccharid acceptor analogues as specific inhibitors of glycosyltransferases. *J. Biol. Chem.*, **266**, 17 858.
89. Srivastava, G. and Hindsgaul, O. (1992) Synthesis of a tetrasaccharide acceptor for use in the assay of UDP-GlcpNAc:β-D-Galp-(1→4)-β-D-GlcpNAc(GlcNAc to Gal) β(1→3)-N-acetylglucosaminyltransferase activity and the pentasaccharide product that would be formed by its enzymic glycosylation. *Carbohydr. Res.*, **224**, 83.
90. Möller, G., Reck, F., Paulsen, H., Kaur, K. K., Sarkar, M., Schachter, H. et al. (1992) Control of glycoprotein synthesis: substrate specificity of rat liver UDP-GlcNAc:Manα3R β2-N-acetylglucosaminyl-transferase I using synthetic substrate analogs. *Glyconjugate J.*, **9**, 180.
91. Brockhausen, I., Möller, G., Yang, J.-M., Khan, S. H., Matta, K. L., Paulsen, H. et al. (1992) Control of glycoprotein synthesis. Characterization of (1→4)-N-acetyl-β-D-glucosaminyltransferase acting on the α-D-(1→3)- and α-(1→6)-linked arms of N-linked oligosaccharides. *Carbohydr. Res.*, **238**, 281.
92. Madiyalakan, R., Yazawa, S., Abbas, S. A., Barlow, J. J., and Matta, K. L. (1986) Use of N-acetyl-2'-O-methyllactosamine as a substrate acceptor for the determination of α-L-(1→3)-fucosyltransferase in human serum. *Anal. Biochem.*, **152**, 22.
93. Yazawa, S., Madiyalakan, R., Jain, R. K., Shimoda, N., and Matta, K. L. (1990) Use of benzyl 2-acetamido-2-deoxy-3-O-(2-O-methyl-β-D-galactosyl)-β-D-glucopyranoside [2'-O-methyllacto-N-biose I β Bn] as a specific acceptor for GDP-Fucose: N-acetylglucosaminide α-(1→4)-L-fucosyltransferase. *Anal. Biochem.*, **187**, 374.
94. Chandrasekaran, E. V., Jain, R. K., and Matta, K. L. (1992) Ovarian cancer α-1,3-L-fucosyltransferase. *J. Biol. Chem.*, **267**, 23 806.
95. Mollicone, P., Candelier, J. J., Mennesson, B., Couillin, P., Venot, A., and Oriol, R.

(1992) Five specificity patterns of (1→3)-α-L-fucosyltransferase activity defined by use of synthetic oligosaccharide acceptors. Differential expression of the enzymes during human embryonic development and in adult tissues. *Carbohydr. Res.*, **228**, 265.

96. Wlasichuk, K. B., Kashem, M. A., Nikrad, P. V., and Venot, A. P. (1993) Determination of the specificities of rat liver Gal(β1→4)GlcNAc α2,6-sialyltransferase and Gal(β1→3/4)GlcNAc α2,3-sialyltransferase using synthetic modified acceptors. *J. Biol. Chem.*, **268**, 13971.

97. Lemieux, R. U., Bundle, D. R., and Baker, D. A. (1975) The properties of a 'Synthetic' antigen related to the human blood-group Lewis a. *J. Am. Chem. Soc.*, **97**, 4076.

98. Palcic, M. M., Heerze, L. D., Pierce, M., and Hindsgaul, O. (1988) The use of hydrophobic synthetic glycosides as acceptors in glycosyltransferase assays. *Glycoconjugate J.*, **5**, 49.

99. Paulsen, H. and Helpap, B. (1991) Synthes von teilislukturen der N-glycoproteine des komlexen typs. *Carbohdyr. Res.*, **216**, 289.

100. Ratcliffe, R. M., Baker, D. A., and Lemieux, R. U. (1981) Synthesis of the T [β-D-Gal-(1→3)-α-D-GalNAc]-antigenic determinant in a form useful for the preparation of an effective artificial antigen and the corresponding immunoabsorbent. *Carbohydr. Res.*, **93**, 35.

101. Kaur, K. J. and Hindsgaul, O. (1992) Combined chemical-enzymic synthesis of a dideoxypentasaccharide for use in a study of the specificity of N-acetylglucosaminyl-transferase-III. *Carbohydr. Res.*, **226**, 219.

102. Alton, G., Kanie, Y., and Hindsgaul, O. (1993) The use of a synthetic dideoxygenated pentasaccharide as a specific acceptor for N-acetylglucosaminyltransferase-III. *Carbohydr. Res.*, **228**, 339.

103. Khan, S. H. and Hindsgaul, O. (1992) Synthesis of a dimethylated pentasaccharide acceptor for N-acetylglucosaminyl-transferase-III. *Abstracts, 16th International Carbohydrate Symposium, Paris,* p. 67.

104. Crawley, S. C. (1993) ELISA Assay, substrate specificity and inhibitors of N-acetylglucosaminyltransferase V. Ph.D. dissertation, University of Alberta, Edmonton.

105. Linker, T., Crawley, S. C., and Hindsgaul, O. (1993) Recognition of the acceptor β-D-GlcpNAc-(1→2)-α-D-Manp-(1→6)β-D-GlcpOR by N-acetylglucosaminyltransferase-V. *Carbohydyr. Res.*, **245**, 323.

106. Abbas, S. A., Piskorz, C. P., and Matta, K. L. (1987) Synthesis of 2-acetamido-2-deoxy-4-O-(2-O-methyl-β-D-galactopyranosyl)-D-glucopyranose (N-acetyl-2'-O-methyllactosamine). *Carbohydr. Res.*, **167**, 131.

107. Sarkar, A. K., Jain, R. K., and Matta, K. L. (1990) Synthesis of benzyl O-(2-O-methyl-β-D-galactopyranosyl)-(1→3)-2-acetamido-2-deoxy-β-D-glucopyranoside [benzyl 2'-O-methyllacto-N-bioside I], and its higher saccharide containing an O-(2-O-methyl-β-D-galactopyranosyl)-(1→3)-2-acetamido-2-deoxy-β-D-glucopyranosyl group as a potential substrate for (1→4)-α-L-fucosyltransferase. *Carbohydr. Res.*, **203**, 33.

108. Khan, S. H. and Matta, K. L. (1992) Recent advances in the development of potential inhibitors of glycosyltransferases. In *Glyconjugates: composition, structure, and function* (ed. H. J. Allen and E. C. Kisailus), p. 361. Marcel Dekker, New York.

109. Palcic, M. M., Heerze, L. D., Srivastava, O. P., and Hindsgaul, O. (1989) A bisubstrate analog inhibitor for α-(1→2)-fucosyltransferase. *J. Biol. Chem.*, **264**, 17174.

110. Khan, S. H., Crawley, O., Kanie, O., and Hindsgaul, O. (1993) A trisaccharide acceptor analog for N-acetylglucosaminyltransferase V which binds to the enzyme but sterically precludes the transfer reaction. *J. Biol. Chem.*, **268**, 2468.

111. Lowary, T. L. and Hindsgaul, O. (1993) Recognition of synthetic deoxy and deoxyfluoro analogs of the acceptor α-L-Fucp-(1→2)-β-D-Galp-OR by the blood group A and B glycosyltransferases. *Carbohydr. Res.*, **249**, 163.
112. Lowary, T. L. and Hindsgaul, O. (1994) Recognition of synthetic O-methyl, epimeric, and amino analogs of the acceptor α-L-Fucp-(1→2)-β-D-Galp-OR by the blood group A and B gene-specified glycosyltransferases. *Carbohydr. Res.* (In press.)
113. Khare, D. P., Hindsgaul, O., and Lemieux, R. U. (1985) The synthesis of monodeoxy derivatives of lacto-N-biose I and N-acetyl-lactosamine to serve as substrates for the differentiation of α-L-fucosyltransferases. *Carbohydr. Res.*, **136**, 285.
114. Kajihara, Y., Hashimoto, H., and Kodama, H. (1992) Methyl-3-O-(2-acetamido-2-deoxy-6-thio-β-D-glucopyranosyl)-β-D-galactopyranoside: a slow reacting acceptor-analogue which inhibits glycosylation by UDP-D-galactose-N-acetyl-D-glucosamine-(1→4)-β-D-galactosyltransferase. *Carbohydr. Res.*, **229**, C5.
115. Palcic, M. M., Ripka, J., Kaur, K. K., Shoreibah, M., Hindsgaul, O., and Pierce, M. (1990) Regulation of N-acetylglucosaminyltransferase V activity. *J. Biol. Chem.*, **265**, 6759.
116. Ichikawa, Y., Look, G. C., and Wong, C-H. (1992) Enzyme-catalysed oligosaccharide synthesis. *Anal. Biochem.*, **202**, 215.
117. Kashem, M. A., Jiang, C., Venot, A., and Alton, G. (1992) Combined chemical-enzymic synthesis of an internally monofucosylated hexasaccharide corresponding to the CD-65/VIM-2 epitope: use of a terminal α2-6-linked N-acetylneuraminic acid as a temporary blocking group. *Carbohydr. Res.*, **230**, C7.
118. Ball, G. E., O'Neill, R. A., Schultz, J. E., Lowe, J. B., Weston, B. W., Nagy, J. O. *et al.* (1992) Synthesis and structural analysis using 2-D-NMR of sialyl lewis X (SLe^x) and lewis X (Le^x) oligosaccharides: Ligands related to E-selectin [ELAM-1] binding. *J. Am. Chem. Soc.*, **114**, 5449.
119. Compston, C. A., Condon, C., Hanna, H. R., and Mazid, M. A. (1993) Rapid production of a panel of a blood group A-active oligosaccharides using chemically synthesized di- and trisaccharide primers and an easily prepared porcine (1→3)-α-N-acetyl-D-galactosaminyltransferase. *Carbohydr. Res.*, **239**, 176.
120. Ito, Y. and Paulson, J. C. (1993) A novel strategy for the synthesis of a ganglioside GM3 using an enzymatically produced sialoside glycosyl donor. *J. Am. Chem. Soc.*, **115**, 1603.
121. Oehrlein, R., Hindsgaul, O., and Palcic, M. M. (1993) Use of 'core-2'-N-acetylglucosaminyltransferase in the chemical-enzymatic synthesis of a sialyl-Le^x containing hexasaccharide found on O-linked glycoproteins. *Carbohydr. Res.*, **244**, 149.
122. Hindsgaul, O., Kaur, K. J., Gokhale, U. B., Srivastava, G., Alton, G., and Palcic, M. M. (1991) Use of glycosyltransferases in synthesis of unnatural oligosaccharide analogs. *Am. Chem. Soc. Symp. Ser.*, **466**, 38.
123. Heidlas, J. E., Williams, K. W., and Whitesides, G. M. (1992) Nucleoside phosphate sugars: synthesis on practical scales for use as reagent in the enzymatic preparation of oligosaccharides and glycoconjugates. *Acc. Chem. Res.*, **25**, 307.

6 | Conformational studies on oligosaccharides

STEVEN W. HOMANS

1. Introduction

Since the initial demonstration by Burnet that the influenza virus requires a specific carbohydrate signal on the erythrocyte for attachment and infectivity (1), oligosaccharides have been implicated as recognition signals in a variety of biological processes. Indeed, in the field of cell adhesion, we now have knowledge of the oligosaccharide ligands and a detailed knowledge of their protein receptors has given rise to the term 'selectins' for this class of membrane proteins (reviewed in references 2, 3, see also Chapter 4). As techniques for the purification and sequence analysis of oligosaccharides improve in efficiency and sensitivity, we can anticipate the characterization of the precise oligosaccharide ligand and its receptor in other important biological processes. In parallel with these studies is a requirement to understand the molecular mechanism of oligosaccharide–protein interactions and, in particular, the maintenance of its specificity. In this chapter I shall review the contribution of high-resolution NMR, X-ray crystallography, and molecular modelling to our knowledge of the conformational properties of oligosaccharides, followed by a discussion of our current understanding of the molecular basis of oligosaccharide-mediated recognition. I shall not discuss in detail the various experimental and theoretical techniques for the determination of oligosaccharide conformations, since these have recently been reviewed (4). In addition, I shall primarily focus attention upon the N-linked class of oligosaccharides, not from the implication that these are functionally more relevant than other oligosaccharide or polysaccharide classes, but rather since most research in this area has been focused on compounds of this class, and it is impossible to review all aspects of oligosaccharide conformations in this chapter.

2. Conformational parameters of oligosaccharides

At the outset it is convenient to define the various parameters which will be used throughout this article to describe the conformational properties of oligosaccharides. The absolute configuration of the majority of monosaccharide residues which will be discussed are in the $^4C_1$ D-configuration, except fucose which exists in the

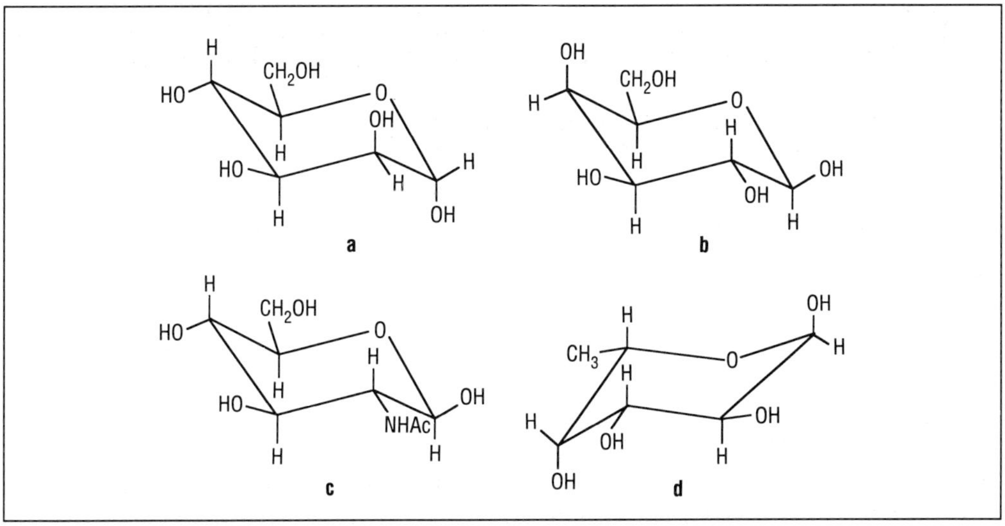

Fig. 1 Sterical structures of the four neutral monosaccharides found in N-linked glycans. (a) α-D-mannopyranose; (b) β-D-galactopyranose; (c) β-D-N-acetylglucosamine; (d) α-L-fucopyranose

$^1C_4$ L-configuration. This gives rise to the sterical structures shown in Figure 1. These configurations should be assumed unless otherwise stated. When two monosaccharide residues are glycosidically linked, then their relative spatial dispositions are defined by the two torsion angles φ and ψ about the glycosidic linkage, or additionally by a third torsion angle ω in the case of 1–6 glycosidic linkages (Figure 2). When discussing NMR-derived conformations, these angles are generally measured with respect to protons, and are defined as follows; $φ_H$ = H1-C1-O1-CX, $ψ_H$ = C1-O1-CX-HX (C1-O1-C6-C5 in the case of 1–6 linkages), and $ω_H$ = O6-C6-C5-H5, where the subscripts distinguish the 'NMR definitions' of these angles from the conventional IUPAC definitions, and CX and HX refer to the aglyconic atoms.

3. Sterical structures of oligosaccharides

The high resolution three-dimensional structures of oligosaccharides have been investigated in the main by three techniques, namely conformational energy calculations, high-resolution NMR, and X-ray crystallography. By their nature, X-ray measurements present a rather static picture of conformation. Early theoretical predictions and NMR measurements were likewise interpreted where possible in terms of essentially fixed geometry about glycosidic linkages. However, it is now known that oligosaccharides exist with rather more flexibility than was first supposed, and hence we briefly summarize these early results together with the results of X-ray measurements, before considering in detail the nature of oligosaccharide dynamics in section 4.

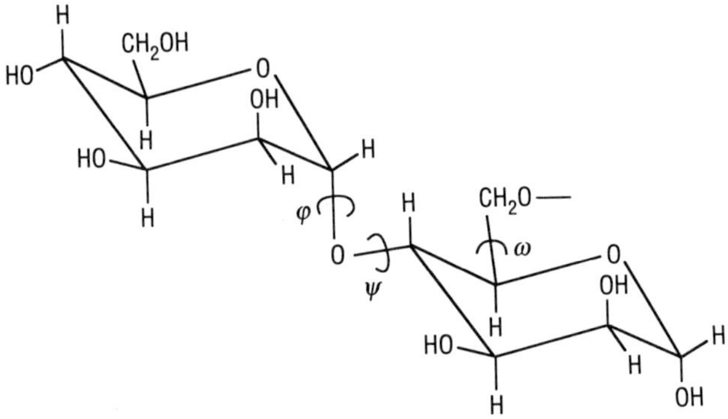

Fig. 2 Definitions of the torsion angles φ (H1-C1-O1-CX), ψ (C1-O1-CX-HX), and ω (H5-C5-C6-O6). The example illustrates a (1-4) linkage, and hence $X = 4$

3.1 Theoretical predictions

Many early studies on oligosaccharide conformations relied in the main on molecular mechanical energy calculations (5–9), from which useful initial conclusions could be drawn. Indeed, the nature of high-resolution NMR data is such that energy calculations of some form are essential for their interpretation (see section 3.2).

The computation of the internal energy is computationally intensive, even for a small molecule such as a disaccharide. Hence, full *ab initio* quantum mechanical calculations are effectively out of reach with presently available computing power, and computations on oligosaccharides have almost invariably been molecular mechanical in nature. The differences between methods in application to oligosaccharides have been reviewed elsewhere (4, 10), but briefly the molecular mechanical approach treats the molecule as a series of point masses and charges for the atoms, together with a series of springs for the bonds. In this manner it is possible to treat the molecule as a classical object, in the sense that classical mechanics can be used to derive the energy and its derivative. Despite the inherent approximations in this approach, molecular mechanics methods have been used very successfully in conformational analysis of acyclic and cyclic hydrocarbons. However, when the parametrization (technically described as a forcefield) appropriate for these hydrocarbons was applied to carbohydrates, it became apparent that additional factors were at work. The most important of these is termed the exo-anomeric effect, which has been reviewed in detail by Tvaroska and Bleha (11). The consequence of this effect is that the aglyconic carbon prefers an orientation in which ϕ_H is ~ +60° in β-D-glycosides and ~ −60° in α-D-glycosides, respectively.

The earliest conformational energy calculations on oligosaccharides were pioneered by Lemieux *et al.* (5–7). These workers utilized a simple forcefield which included a Kitaygorodski term (12) to describe the nonbonded interaction between atoms, a torsional potential to describe the exo-anomeric effect, rigid monosaccharide ring geometry, and neglect of partial atomic charges, giving rise to so-called hard sphere exo-anomeric effect (HSEA) calculations. Despite their simplicity, convincing results were obtained concerning the preferred conformation of oligosaccharides related to blood groups (5, 6), and conformational analysis of simpler oligosaccharides such as sucrose (13) has yielded results which are in very good accord with other experimental and theoretical considerations.

The HSEA approach was used by Bock and co-workers to determine the preferred conformation of oligosaccharides related to complex-type N-linked glycans (14), and this represented one of the first studies of the conformation of these compounds. The values of ϕ and ψ which these authors obtained for each glycosidic linkage in the minimum energy configuration of a biantennary oligosaccharide are shown in Figure 3. Also shown in Figure 3 is a stereo diagram illustrating the resulting overall conformation of the glycan. This is shown with the torsion angle ω_H about the C-5–C-6 bond of the Manβ residue set to 180°. Note that the glycan exists in an extended configuration, with no tendency for different regions of the glycan to interact. This appears to be a property of most N-linked glycans studied to date, and in view of their highly branched nature, the surface area which they can span is very large.

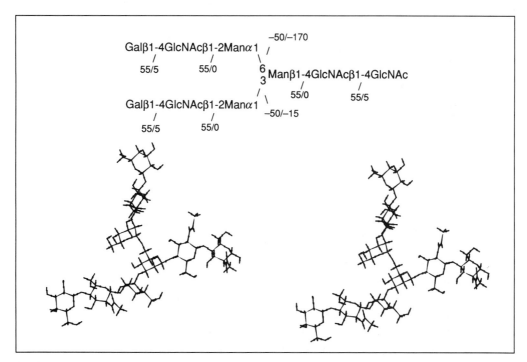

Fig. 3 (a) Torsion angles about the glycosidic linkages of a biantennary complex-type glycan derived from theoretical HSEA calculations (14). These angles are denoted in the figure as ϕ/ψ pairs for each linkage. In addition the value of ω about the 1-6 linkage was found to adopt two preferred conformations, 180° and −60°. (b) Stereo plot of the structure shown in (a). The GlcNAcβ1-4GlcNAc core is shown to the right of the figure, and the 1-6 arm is to the top. The torsion angle ω about the 1-6 arm is shown in the 180° configuration

3.2 Nuclear magnetic resonance

NMR is currently the only technique by which high resolution structural data can be obtained in solution. Two NMR parameters exist which give quantitative information on three-dimensional structure. The first, known as spin–spin coupling, is a through-bond effect which is manifest in the NMR spectrum as a splitting of the resonance line(s) of one nucleus as a consequence of the difference in spin states of another (15). The effect can be transmitted through several bonds, but in proton NMR is very weak if the number of consecutive saturated bonds exceeds three. Spin–spin coupling is quantified in terms of a spin–spin coupling constant (J), and importantly J is dependent upon the dihedral angle between vicinal protons, as shown in Figure 4.

A second NMR parameter, known as the nuclear Overhauser effect (NOE), is not manifest in the NMR spectrum directly. The NOE is a through-space effect which is apparent in the change in intensity of a given proton resonance when the resonance corresponding to a nearby proton (< 5 Å distant) is perturbed (15). Under appropriate conditions, the NOE can be considered to be proportional to the inverse sixth power of the distance between the protons in question, and hence is a very sensitive 'molecular ruler'. The nature of the NMR technique is such that NOEs and Js give imprecise distance and angular information. In general therefore NMR data are supported by molecular mechanical energy calculations in order to refine the data (16–18). Loosely stated, NMR measurements are useful in determining which conformations are possible, whereas energy calculations are useful in determining which conformations are not possible, and hence the two approaches are complementary.

Shortly after the work of Bock et al., high-resolution NMR investigations of solution conformations began on disaccharides and trisaccharides as model fragments of N-linked glycans, together with preliminary studies on the complete glycans. The first detailed analyses incorporating NOE and J measurements together with molecular mechanical energy calculations of both model fragments (19) and complete glycans (20, 21) were reported by Brisson and Carver. In the

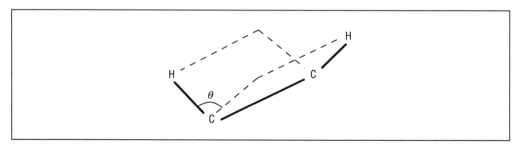

Fig. 4 Stereochemical dependence of the spin coupling constant $^3J_{H,H}$ between two vicinal protons. The magnitude of this coupling can be determined empirically from the Karplus equation which takes the form $J = A\cos^2\theta + B\cos\theta + C$, where A, B, and C are constants which must be determined for the system of interest, and θ is the dihedral angle formed between the two C—H bonds, as shown

model compound Manα1–6(Manα1–3)Manα1–OMe, these authors found ϕ, ψ values for the Manα1–3Man linkage of −45°, −15°, and ϕ, ψ = −60°, 180° for the Manα1–6Man linkage (19), in excellent agreement with the above values reported by Bock et al. In addition, by measurement of $J_{5,6}$ and $J_{5,6'}$ for the Manα1-OMe unit, two rotamers of similar probability were found about the C-5–C-6 bond of this unit, at ω = 180° and ω = −60°. Since the NOE data for the Manα1–3Man linkage could be interpreted in terms of essentially fixed geometry about the glycosidic linkages, these data predicted that oligosaccharides may exist in solution with joint rigidity and flexibility.

In an extension of these studies to a series of N-linked glycans, Brisson and Carver reported ϕ, ψ values (20, 21) for the majority of glycosidic linkages found in glycoproteins (Figure 5), and the joint rigidity/flexibility found in model compounds was also suggested to exist in the intact oligosaccharides. Importantly, the rotamer distributions about the C-5–C-6 bond of Manβ were found to depend upon the primary sequence of the oligosaccharide (21). Using a similar approach, but with semiempirical quantum mechanical energy calculations rather than HSEA calculations, Homans et al. characterized the solution conformations about the Manα1–3Man and Manα1–6Man linkage in a series of biosynthetically related oligosaccharides (22, 23). The reported values of ϕ, ψ about certain linkages were found to differ from those found by Brisson and Carver, and in particular the conformation of the Manα1–3Man linkage was reported to be dependent upon the primary sequence of the oligosaccharide (23). While the reasons for this discrepancy were for some time unclear, it is now known that this derives from motional averaging about the Manα1–3Man linkage, and both conformations are in fact present in solution (see section 4).

A question which often arises in discussions of the solution conformations of oligosaccharides is that of hydrogen bonding. Naively, one would expect that hydrogen bonding predicted from *in vacuo* energy calculations would not exist in solution, due to competition by the vast excess of solvent water. However, recent advances in NMR technology have made possible the direct demonstration of the existence of inter- and intraresidue hydrogen bonds in solution (24), and hydrogen bonds are often observed in crystal structures of model disaccharides and trisaccharides

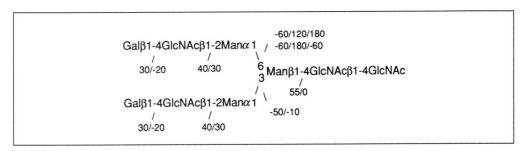

Fig. 5 Values of the torsion angles obtained in a biantennary complex-type glycan according to the NMR studies of Brisson and Carver (20, 21). The torsion angles are denoted as ϕ/ψ pairs for each linkage, except in the case of the 1-6 linkage, where the presence of two stable rotamers about the C-5–C-6 bond of the Manβ residue gives rise to a pair of ϕ/ψ/ω values

(section 3.3), and have been suggested from molecular dynamics simulations (section 4.1). The presence of an inter-residue hydrogen bond would be expected partially to stabilize the torsional flexibility of the glycosidic linkage which it spans.

3.3 X-ray crystallography

X-ray crystallographic data on free oligosaccharides (in contrast to oligosaccharide–protein complexes, section 5) is limited in the main to disaccharides and trisaccharides, since larger oligosaccharides appear not to crystallize readily. Indeed, the largest structure to be investigated of relevance to N-linked glycans is the trisaccharide Manα1-3Manβ1-4GlcNAc (25). This study was of great importance, since it allowed the values of ϕ and ψ for two glycosidic linkages at least, to be compared with those determined theoretically (section 3.1) or by NMR (section 3.2). The observed values for Manα1-3Man (ϕ_H, ψ_H = $-57.6°$, $-19.4°$) and Manβ1-4GlcNAc (ϕ_H, ψ_H = $+48.1°$, $-0.6°$) are in good agreement with those predicted by these two methods. However, such agreement may lead to a false sense of security, since, as mentioned above, we now know that there is significant conformational flexibility about the Manα1-3Man linkage. Each of the monosaccharide residue rings in the crystal structure was found to be in the $^4C_1$ configuration, and the ring geometry of Manβ was distorted slightly due to intermolecular hydrogen bonding. An inter-residue hydrogen bond was observed between the ring oxygen of Manβ and the C-3 hydroxyl proton of the GlcNAc at the reducing terminus.

3.4 Conclusions

Early studies regarded oligosaccharides as having essentially fixed conformations about glycosidic linkages (with the exception of 1-6 linkages), and good agreement was found between conformations predicted from experimental data (NMR or X-ray crystallography) and the 'global minimum' structures derived from energy calculations. However, attempts to extend these studies to more complex structures (26–28) resulted in poor agreement between these approaches, and it became clear that most glycosidic linkages exist with a degree of conformational variability. Strictly, it is therefore meaningless to describe an oligosaccharide in terms of its rigid, global minimum energy conformation. However, this picture is a useful starting point with which to visualize the dynamic behaviour of an oligosaccharide, and becomes very important when we consider the thermodynamics of an oligosaccharide binding to a protein, where again it may be appropriate to think in terms of a highly restrained structure.

4. Dynamics of oligosaccharides

4.1 Theoretical predictions

The interpretation of early NMR data in terms of essentially rigid conformations was supported in part by the depth of the potential wells in HSEA calculations.

However, as molecular mechanical calculations became more refined in carbohydrates by allowing all bonds, angles, and torsion angles to vary during minimization (i.e. 'flexible geometry') (18, 29–34), it became apparent that the degree of motional freedom, or, in technical terms, the area of conformation space available to the oligosaccharide, was counterintuitive to the concept of rigidity about φ and ψ. For a simple disaccharide, the conformational space mapped about φ and ψ can be estimated by calculation of a so-called gridsearch, i.e. φ and ψ are varied independently in small steps (for example 10°), and the geometry is optimized at each step to give a value for the minimum energy at that point. An example is shown in Figure 6a for the disaccharide Manα1-3Man. The area of this gridsearch which the molecule can occupy is temperature dependent, and is governed by the Boltzmann law.

A second method to determine accessible conformation space relies on molecular dynamics simulations (29, 33, 35–37). Since the molecule is essentially a classical object in terms of molecular mechanics, it is possible to derive the motional properties of the system giving suitable starting conditions from Newton's laws of motion. The advantage of the MD approach is that it is useful for much larger structures than can be handled with gridsearch calculations—the number of such calculations required in the latter for a system with a large number of degrees of freedom rapidly becomes prohibitive. The MD data can, however, be plotted in exactly the same manner as a gridsearch, and values of φ and ψ plotted over the

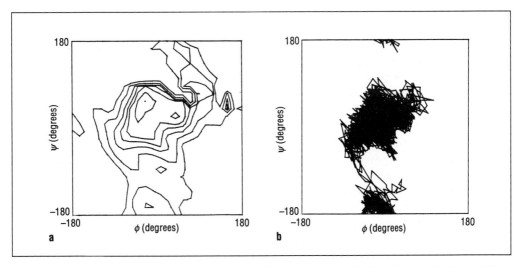

Fig. 6 (a) *In vacuo* gridsearch calculation about the torsion angles φ and ψ for the disaccharide Manα1-3Man. This was generated by adjusting each torsion angle in 10° steps, following by minimization at each point while holding the values of the torsion angles constant. The resulting potential surface is contoured at 0.01, 0.3, 0.8, 1.3, 2.0, 3.5, and 5.5 kcal/mol above the global minimum, which in this example is found at φ, ψ = −30°, +45°. (b) Trajectory of a 500 ps molecular dynamics simulation of the same disaccharide. However, in contrast to the gridsearch, this simulation was generated with explicit inclusion of solvent water molecules. Note how the conformations visited by the disaccharide (black areas) correspond with the low energy regions of the gridsearch in (a)

timecourse of a 500 pS MD simulation at 300 K of Manα1-3Man are shown in Figure 6b. Note how the conformational space mapped in the MD simulation closely follows the low energy contours in the gridsearch.

Initial studies made use of both *in vacuo* gridsearch calculations and *in vacuo* molecular dynamics simulations on simple disaccharide models, and these indicated that more than one minimum may exist about the glycosidic linkage, suggesting that motional averaging was taking place at physiological temperatures. This idea was further strengthened by Cumming and Carver, who found that it was impossible to interpret NOE data for certain glycosidic linkages in terms of a single rigid conformer (26, 27). NMR parameters such as the NOE are time- and ensemble-averaged properties, and since the timescale of build up of the NOE may extend to hundreds of milliseconds, then the measured NOE is an average over the different conformational states which may be sampled in this time. As is seen from Figure 6b, a very large area of conformational space is sampled on timescale many orders of magnitude lower than this.

The realization that a glycosidic linkage is not in fact rigid, has led to intensive investigations by a number of groups in order to characterize the nature of this motional flexibility. Cumming and Carver investigated the nature of dynamics about the glycosidic linkage by generating a series of gridsearch plots for various glycosidic linkages using HSEA calculations *in vacuo* (27). An interesting aspect of this study was the generation of several gridsearch plots for a given glycosidic linkage using various types of HSEA forcefield, for example by inclusion of electrostatic or hydrogen bond terms. The resulting potential surface, not surprisingly, was found to depend upon the precise forcefield employed. In order to gain some measure of the correctness of the various gridsearch calculations, theoretical NOEs were computed from each gridsearch, and compared with the experimental values. The NOEs were computed on the assumption that the rate of internal motion responsible for the motional averaging is slow with respect to the overall tumbling of the molecule, as a result of which the theoretical NOE is a straightforward average overall accessible conformations at a given temperature, weighted by their Boltzmann distribution. By use of this approach, good agreement was found between experimental and theoretical NOEs for certain forcefield parametrizations. However, it is clear that such agreement is not always found, and in a recent study by Imberty *et al.* for the disaccharide Manα1–3Manα using the MM2CARB forcefield, it was not possible to obtain agreement of all experimental NOEs with those found theoretically (34). An inherent inadequacy in the use of HSEA calculations for the simulation of carbohydrates is that not all terms are included in the forcefield. As demonstrated by Cumming and Carver, the nature of the result clearly depends upon precisely which terms are included or excluded. While the HSEA approach is excellent for the generation of approximate static 'minimum energy' carbohydrate conformations, most workers now appreciate that a full molecular mechanical forcefield is more appropriate in the study of dynamics, where the potential surface must be accurate over a wide range of ϕ and ψ values. In this regard, Ha *et al.* produced a full forcefield for carbohydrates, based upon the original CHARMm parametrization (38), which has subsequently been extended by Homans (39) with inclusion of an exo-anomeric torsional potential for simulation of oligosaccharides. An important aspect of these more complete forcefield parametrizations is that it is possible to include solvent water molecules explicitly, which is clearly more comparable with experimental measurements in solution than simulations *in vacuo*.

When applied to model systems, these more detailed simulations have shed further light on the behaviour of oligosaccharides in solution (39, 40). In a recent study of the structure and dynamics of the trisaccharide Manα1-3Manβ1-4GlcNAc both *in vacuo* and with explicit inclusion of solvent water (39), it was discovered that the large torsional fluctuations observed *in vacuo*, which has also been seen in previous studies, were highly damped in the presence of solvent water. This damping results partially from the viscosity of the solvent, but also results from a network of solvent–solute and solute–solvent–solute hydrogen bonds which stabilize the structure in the region of the global minimum. In particular, the inter-residue hydrogen bond observed in the crystal structure was found to persist transiently in solution. However, the structure is by no means rigid, as was implied in the earliest studies (above), and two important conclusions can be drawn from this study. First, the rate of internal motion is on a timescale measurable in tens of picoseconds, and hence is much faster than overall tumbling for an oligosaccharide of moderate size. Second, in the case of the Manα1-3Man linkage, *each* of the conformations observed independently by Homans *et al.* and Brisson and Carver are in fact predicted in solution, thus resolving the discrepancy between the results of the two laboratories.

An inherent problem with dynamics simulations is that they are just that—theoretical simulations with no direct experimental input. It is therefore tempting to attempt to justify their correctness by computation of theoretical parameters from them for comparison with experiment. The computation of theoretical NOEs from dynamics simulations can be achieved in principle by an analogous approach to that described above for gridsearch calculations. However, since the study described above indicates that the rate of internal motion is fast with respect to overall tumbling, the computation of theoretical NOEs is in fact considerably more complex. The precise nature of these calculations is outside the scope of this article, but essentially it is necessary to include terms called spherical harmonics which account for the orientational dependence with respect to the molecular axes of the vector joining the two atoms whose NOE we wish to simulate (41). In a study of the Le^x determinant (42), Homans and Forster demonstrated that reasonable agreement between experimental and theoretical NOEs could be obtained if these factors were taken into account. However, the calculation of theoretical NOEs from MD simulations is itself subject to an important caveat, in that whereas the timescale of buildup of the NOE is on the order of milliseconds to seconds, we cannot hope to simulate over time periods extending to greater than tens of nanoseconds for solvated systems with present computational technology. Hence any agreement between experimental NOEs and theoretical NOEs derived from MD simulations implies that the averaging process is complete in the shorter timescale, which is certainly possible in view of very rapid internal motions. However, since agreement is not good for all linkages, then clearly the averaging process is not always complete on the shorter timescale, and hence computation of theoretical NOEs from MD simulations is currently not entirely satisfactory. An ideal protocol would be to compute theoretical NOEs from a gridsearch calculation on a model disaccharide with explicit inclusion of solvent water, using the formalism appropriate for fast internal motions. However, such a gridsearch is extremely difficult to compute, since the energy of a solvated system is dominated by the solvent, and the global minimum energy at each point in the gridsearch is correspondingly difficult to estimate.

4.2 Nuclear magnetic resonance

From the discussion in the previous section, it is clear that the NOE can be used indirectly as a probe of dynamics by comparison with theoretical values computed either from MD simulations or from gridsearch calculations, but the results are not entirely convincing. The formal possibility remains that we may *never* obtain excellent agreement between theoretical and experimental NOE data, in view of the nature of the molecular mechanical formalism which we employ. It must be remembered that we are attempting to simulate a quantum system via classical mechanics, and whereas this may in itself not present too many problems, the use, for example, of point charges is an approximation, since the partial charge on an atom is dependent on the conformation, as anyone who has performed quantum-mechanical calculations on a molecular system will attest. When approximations such as these are coupled to the fact that the NOE depends upon the inverse sixth power of distance, very small inadequacies in the forcefield can give rise to enormous errors in the computed NOE. However, if we accept the results of molecular dynamics simulations at face value, and the fact that reasonable agreement between experimental and theoretical parameters is often found, we find a picture of oligosaccharide dynamics which is rather different from that predicted from the earliest observations, namely, a population of discrete conformers about the glycosidic linkages, rather than a single, rigid geometry. Why, then, were the early NOE measurements interpretable in terms of rigid geometry? The answer to this question lies in the fact that the NOE is a rather imprecise measure of local geometry, unless a large number of constraints exists. Often only one or two NOEs are measurable across glycosidic linkages, and it is usually possible to compute a conformation which simultaneously satisfies theoretical predictions together with the NOE restraints, when the error of measurement is taken into account. For this reason it has been proposed that NOE restraints applied to oligosaccharides are best employed in the restriction of conformation space during conformational searching, but should not be utilized in the final stages of a conformational analysis strategy (42).

It is therefore perhaps not surprising that more attention is being focused on the generation of new experimental data to confirm theoretical predictions of oligosaccharide dynamics, and NMR is the primary tool for this purpose. Fortunately, the NOE is not the only parameter in NMR which provides information on conformation and dynamics, and these alternative parameters will be discussed below. However, one further study based upon the NOE should first be considered. If the measured NOE is an average over a variety of different conformational states, then its value should change with temperature since the populations of the various conformational states visited is temperature dependent, as discussed above. Alternatively, if the NOE derives from an essentially rigid, single conformer, then the value of the NOE should remain roughly constant with temperature. This question was addressed for the case of sucrose by Poppe *et al.* (43), and temperature dependent NOEs were observed. Taking into account various other factors which might influence the NOE with respect to temperature, these authors concluded that significant motional flexibility exists about the glycosidic linkage in sucrose.

The spin–spin coupling constant, mentioned in section 3.2, can also be utilized as a conformational probe of the glycosidic linkage (15). In particular, the long-range three bond couplings $^3J_{CH}$ between the anomeric proton and the aglyconic carbon, and the anomeric carbon and the aglyconic proton, provide information on ϕ and ψ respectively via the appropriate Karplus relationship (44, 45). However, in the case of internal motion, $^3J_{CH}$ again becomes an average quantity, but in this case the average is over an angular (cosinusoidal) function rather than some function of distance. If long-range couplings can be measured, they therefore provide important additional restraints. That they cannot be always measured in oligosaccharides is a reflection of the fact that ^{13}C is abundant naturally at about 1%, and several milligrams of material are usually required in view of the low sensitivity. This often excludes 'interesting' oligosaccharides which are not available from the biological source in such quantities. However, measurement of long-range couplings in model disaccharides is usually straightforward, and one of the first such studies to the author's knowledge involved a conformational analysis of methyl-β-cellobioside (46). Subsequently, several similar studies have been undertaken (44, 45, 47, 48) and values of $^3J_{CH}$ corresponding to ϕ to have uniformly been measured at around 4 Hz. These data immediately strengthen the argument that torsional oscillation exists about ϕ, since if this angle were restricted to, for example ±60° as required by the exo-anomeric effect, then the predicted value would be around 1.5 Hz. An alternative explanation would be that ϕ is restricted to around ±30°, and further work is required to distinguish between these possibilities. It should also be noted that the parametrization of the Karplus equation itself is unlikely to be erroneous, since one can readily find examples of intraresidue torsional angles which are certainly restricted to around ±60°, and each of these corresponds to a long-range coupling in the 1 Hz range.

The NMR relaxation parameters T_1 and T_2 can in principle provide distance information, but more commonly they are utilized in probing dynamics (15, 49, 50). An excited NMR nucleus depends upon local magnetic fields generated by neighbouring nuclei for its relaxation, and for this relaxation to be efficient, these magnetic fields must fluctuate at the correct frequency. These fluctuations in the magnetic field are generated by motional fluctuations of the nearby nuclei, and hence T_1 and T_2 are a direct measure of mobility. Unfortunately, the characterization of the nature of such motion in quantitative terms requires a model, and the complexity of the torsional fluctuations along an oligosaccharide chain are such that a tractably reasonable model is impossible. However, important qualitative information on oligosaccharide dynamics can in principle be obtained from relaxation measurements, and with the advent of new, high sensitivity techniques for the measurement of T_1 and T_2 for ^{13}C nuclei (51), we can expect that such measurements will be a valuable aid in the study of oligosaccharide dynamics in the near future.

A final example of the use of NMR for probing oligosaccharide dynamics derives not from high resolution solution work, but from a recent solid-state NMR study on frozen solutions of (1-4)-α-D-glucans by Gidley and Bociek (52). ^{13}C-cross polarization-magic angle spinning (c.p.-m.a.s.) NMR chemical shift data for these glucans in the solid state are sensitive to ϕ and ψ, and comparison of the single-pulse excitation ^{13}C NMR spectrum (i.e. the conventional carbon spectrum) of an aqueous 20% solution of α-cyclodextrin at 30 °C with the ^{13}C-c.p.-m.a.s. spectrum of the same solution at $-40°$ shows broad spectral features due to variations in chemical shift. Average chemical shifts of the resolved signals were found to be similar for frozen solutions and at 30°, suggesting that the broadening of the spectral features is due to trapping of a range of conformations. In this regard the broadest signal was found for C-6, which would be expected conformationally to be the most mobile site. In contrast, the widths of signals for the crystalline hexahydrate of α-cyclodextrin were found to be similar at 30° and $-40°$. These data thus strengthen the idea of conformational flexibility about glycosidic linkages, but unfortunately no quantitative information is available since thus far it has been possible to determine a firm relationship between these chemical shifts and ϕ and ψ.

4.3 Conclusions

While the concept of conformational flexibility of oligosaccharides in solution is now well accepted, we are just beginning to understand the details of such dynamics. One should not think of an oligosaccharide in solution as being flexible in the sense of a random coil protein, since the conformational space available about glycosidic linkages at physiological temperatures appears to be restricted to several low energy minima on the conformational surface. The precise populations, or indeed locations of these minima are just beginning to be delineated, as computational power and forcefield parametrizations improve. However, one firm conclusion that can be made at present is that it is dangerous to extrapolate data on the 'global minimum' conformation of an oligosaccharide in predicting how it will bind to a specific protein receptor. The presence of several low energy minima on the potential surface suggests that any one of these may be the bound conformation. In section 5 we shall examine examples where indeed an oligosaccharide appears to be bound to a protein in a conformation which is not the global minimum.

5. Sterical structure of oligosaccharide-protein complexes

5.1 Glycoproteins

While study of the conformational properties of oligosaccharides in solution is interesting and indeed necessary, it is essential at some point to ask whether the conformation of an oligosaccharide when associated with a protein differs from that in the 'free' state. One of the first high-resolution studies on a glycoprotein was the crystal structure of the F_C region of human immunoglobulin G (53), which has a single complex-type biantennary N-linked oligosaccharide at Asn297 on each symmetry related $C_{\gamma 2}$ domain. The observed values of ϕ and ψ (and ω in the case of 1-6 linkages) for each monosaccharide residue for which electron density was available are given in Figure 7.

For the glycosidic linkages in the pentasaccharide core region [Manα1-6(Manα1-3)Manβ1-4GlcNAcβ1-4GlcNAc], the torsion angles are different from those predicted by NMR and theoretical calculations (Figure 3), but not dramatically so. In contrast, the values of ϕ and ψ for the GlcNAcβ1-2Man moiety on the 1-6 arm and for the Galβ1-4GlcNAc moiety on the 1-3 arm are significantly different from the predicted 'free solution' values. In particular, the value of ϕ in each case is not consistent with the exo-anomeric effect. These discrepancies probably reside in the fact that the Gal residue on the 1-6 arm undergoes extensive interactions with the protein, whereas the GlcNAc on the 1-3 arm is proposed to interact with the 1-3 arm of the oligosaccharide unit on the symmetry related domain (Figure 8). To our knowledge this is the only characterized example of a glycoprotein where the oligosaccharide is involved in extensive interactions with the protein. The functional significance of these protein–oligosaccharide and oligosaccharide–oligosaccharide

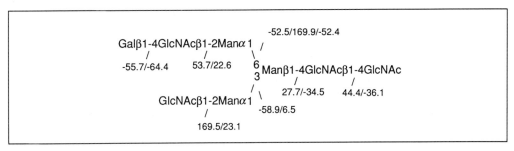

Fig. 7 Values of the torsion angles for the complex-type biantennary glycan attached at Asn297 in the F_C region of human IgG, derived from crystallographic studies (53). These are denoted as ϕ/ψ pairs, or as $\phi/\psi/\omega$ in the case of the 1-6 linkage. The galactose residue on the 1-3 arm is either lacking in the sample used for the crystallographic study, or has sufficient mobility that the electron density is not observed

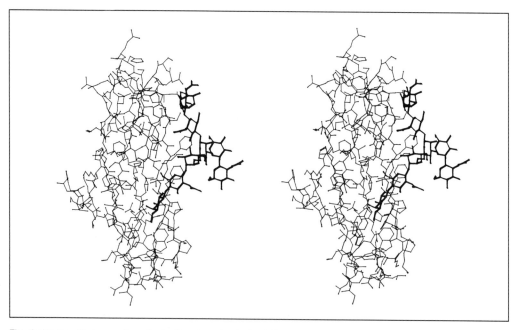

Fig. 8 Stereo diagram of a single C$_{\gamma 2}$ domain derived from the crystal structure of human F$_c$ (53). The oligosaccharide is shown in bold, with the core region attached at Asn297 at the top of the figure. The 1-6 arm is at the left and interacts extensively with the protein, whereas the 1-3 arm is to the right, and interacts with the 1-3 arm on the symmetry-related domain in the intact F$_c$

interactions are not known with certainty. However, several effector functions of IgG are apparently affected if the oligosaccharide is removed (54, 55), and abnormal glycosylation of IgG appears to correlate with various autoimmune diseases (56).

5.2 Lectin–oligosaccharide interactions

Several crystallographic studies have now been reported on the interactions of monosaccharides and their derivatives with lectins. Perhaps the best known of these is the interaction of Concanavalin A with methyl-α-mannoside (57, 58). The mannoside is bound in the chair conformation, with C-4 of the saccharide ring deepest in the binding site cavity and C-1 closest to the surface (Figure 9). A network of hydrogen bonds links oxygen atoms of the sugar to hydrogen bond donors and acceptors of the protein. OH-6 is bonded to OD1 of an asparagine residue (Asp208), and the lone pair of O-6 is bonded to the peptide NH's of a tyrosine (Tyr100) and a leucine (Leu99) residue. The ring oxygen is also bonded through its lone pair to the peptide NH of Leu99. OH-4 is bonded to OD2 of Asp208, and the lone pair of O-4 is bonded to the amide NH of an asparagine (Asn14). The lone pair of O-3 is bonded to the peptide NH of an arginine residue (Arg228), but there is no hydrogen bond acceptor within bonding distance of OH-3.

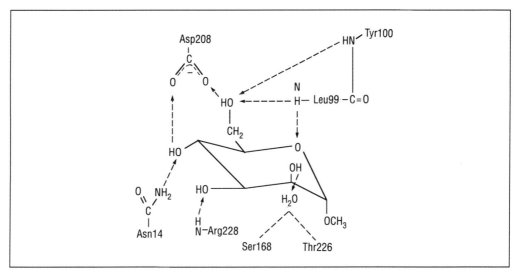

Fig. 9 Diagrammatic illustration of the interactions of methyl-α-mannoside with binding-site residues of Concanavalin A, from the crystal structure of the complex (57, 58). The dotted lines represent hydrogen bonds, with arrows from donor to acceptor. Note that the C-2 OH is hydrogen bonded indirectly via a bound water molecule. See text for further details. Adapted from reference 58

The hydroxyl group at C-2 is not hydrogen bonded to the binding site but linked through a water bridge to an adjacent subunit. Thus, Con A binds saccharide by forming hydrogen bonds with nearly all the donors and acceptors of the saccharide hydroxyls.

A recent study on the specificity of the L-arabinose binding protein (ABP) for several monosaccharide residues has been reported by Quiocho et al. (59, 60). ABP is an initial receptor for the high-affinity active transport or permease system in gram-negative bacteria. L-arabinose and D-galactose are both excellent substrates of ABP ($K_d = 0.98 \times 10^{-7}$ M and 2.3×10^{-7} M respectively), whereas D-fucose is bound less tightly ($K_d = 3.8 \times 10^{-6}$ M). Thus, substitution of the equatorial proton at C-5 of L-arabinose ($^4C_1$ conformation) by a bulky CH_2OH group (D-galactose) does not change the tightness of ligand binding substantially, but substitution by a smaller, apolar methyl group (D-fucose) reduces the binding affinity by ~40-fold. The crystal structures of ABP-Fuc, ABP-Gal, and ABP-Ara complexes were found to share several prominent features. Each monopyranoside fills and is completely enclosed by a cleft between the two globular domains of ABP. The pyranose rings and atoms common to all three sugars are positioned identically within the sugar-binding site. The sugars are held in place by strong hydrogen bonds with mainly planar, polar side chains and by van der Waal's interactions (Figure 10). Identical hydrogen bonds are associated with OH1, OH2, OH3, and OH4 of the monopyranosides in all three complexes, and a hydrogen bond is found between a water molecule and the sugar ring oxygen in both ABP-Ara and ABP-Fuc complexes but not ABP-Gal. A hydrophobic patch created by the C3-H,

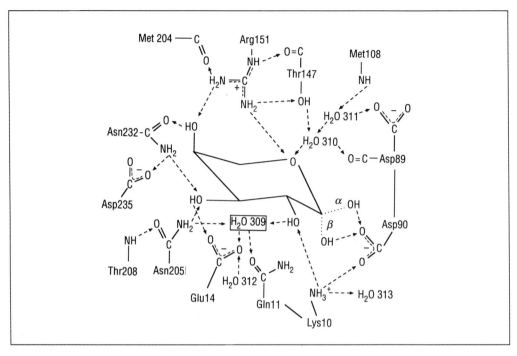

Fig. 10 Diagrammatic illustration of the interactions of arabinose with binding-site residues of the arabinose binding protein (ABP), from the crystal structure of the complex (60). The dotted lines represent hydrogen bonds, with arrows from donor to acceptor. See text for further details. Adapted from reference 60

C4-H, and C5-H on the β face of the pyranose ring, is partially stacked with the β face of the indole ring of a tryptophan residue (Trp16), and the exposed C2-H on the α face is close to the ε-methyl group of a methionine (Met204). The methyl group of fucose causes a rotation of a second methionine residue (Met108) relative to arabinose binding, which puts the methionine methyl group within 3.7 Å of the methyl group of fucose, creating a favourable nonpolar interaction similar to that involving Met204. The hydroxymethyl group of galactose induces many more local changes, which are particularly large for two bonded water molecules. These observations demonstrate several roles for bound water molecules in substrate binding and specificity. These stabilize a particular complex by mediating hydrogen bonds between the functional groups of the protein and the sugar and by filling in 'voids' inside the binding site. The mobility, polarity, and hydrogen bonding potential of the water molecules are major factors in enabling the binding site of ABP also to accommodate galactose. It is of interest that the D-galactose-binding protein does not bind L-arabinose of D-fucose, because the protein lacks the equivalent of the hydrogen-bonded water molecules found in ABP.

Further insight into the mechanism of specificity of oligosaccharide protein interactions is provided by a recent crystallographic study of lactose bound to the heat-labile enterotoxin from *Escherichia coli* (61). Recognition of the oligosaccharide portion of ganglioside G_{M1} in membranes of target cells by this toxin is the crucial first step in pathogenesis, and this study reveals the location of the binding site of the terminal galactose of G_{M1}. The specificity of the galactose binding stems from the fact that every hydrogen-bond donor and acceptor of galactose, except the ring oxygen and anomeric oxygen, is engaged in hydrogen bonding. These hydrogen bonds involve polar residues such as glutamic acid (Glu51), Glutamine (Gln61), Asparagine (Asn90) and lysine (Lys91), together with three well-defined water molecules. There are extensive van der Waal's interactions between hydrophobic groups of galactose and a tryptophan residue (Trp88), and thus overall the types of interaction bear a distinct similarity with the ABP (above).

Another interesting example of a lectin-oligosaccharide interaction is the crystallographic study of Bourne *et al.* on the complex between Isolectin I of *L. ochrus* and the trisaccharide Manα1-3Manβ1-4GlcNAc (62). Isolectin I consists of two subunits composed of two heavy β and two light α chains of 181 and 52 amino acid residues, respectively. The molecular entity is a dimer formed by the association of two αβ monomers. Each monomer requires Ca^{2+} and Mn^{2+} ions for carbohydrate binding and contains one carbohydrate binding site. The crystal structure of Isolectin I in the absence of carbohydrate is very similar to other lectins specific to mannose and glucose such as concanavalin A (57, 58), pea lectin (63), and favin (64). The interest in the structure of the trisaccharide/isolectin I complex derives from the fact that the conformation of the trisaccharide in free solution has been well-studied as a fragment within N-linked glycans, and also a good crystal structure is available (25). The trisaccharide adopts an extended conformation when complexed to the lectin, as found in the study of the isolated trisaccharide. However, a detailed comparison of the two structures shows that the Manβ residue and in particular the GlcNAcβ residue both adopt different positions in the complex than in the isolated oligosaccharide. The observed values of the torsion angles were $\phi, \psi = 69.7°, 123.7°$ for Manα1-3Man and $\phi, \psi = -69°, -116°$ for Manβ1-4GlcNAc, as compared with those found in the free trisaccharide; $\phi, \psi = 60.5°, 97°$ and $\phi, \psi = -75.7°, -130.6°$ respectively (25). This torsional variance is generated primarily by displacement the Manβ and GlcNAc residues toward the lectin, not via direct binding but rather via lectin-water-sugar hydrogen bond bridges. However, the complexed trisaccharide is still found in a low energy region of the potential surface according to the relevant adiabatic ϕ, ψ plots for each glycosidic linkage (65). The only *direct* lectin-sugar interaction involves the binding of the Manα residue in a pocket near the Ca^{2+}, in a site known to bind monosaccharides. Several residues in this site hydrogen bond to Manα as shown in Table 1, and it is of interest that all potential hydrogen-bonding sites on the Manα residue are exploited, with the exception of O2 and HO2 which face outwards from the protein. The only protein movement upon trisaccharide binding involves residues in the monosaccharide-binding site.

From these studies it is again clear that the conformation of an oligosaccharide is modified when complexed with a protein. However, in contrast to the example of IgG described in section 5.1, torsional strain in the oligosaccharide is generated not

Table 1 Hydrogen bonding network between the Manα residue of Manα1-3Manβ1-4GlcNAc and residues within the binding site of isolectin I

Protein atoms	Trisaccharide atoms	Distance (Å)
Gly99 NH	O3 (Manα)	2.6
Gly99 NH	O4 (Manα)	2.9
Asn125 NHD2	O4 (Manα)	2.8
Asp81 OD2	HO4 (Manα)	2.8
Ala210 NH	O5 (Manα)	3.0
Asp81 OD1	HO6 (Manα)	2.9
Ala210 N	O6 (Manα)	3.0
Glu211 N	O6 (Manα)	3.3
Gly209 N	O6 (Manα)	3.4

from direct oligosaccharide–protein interactions, but from sugar–water–protein hydrogen bonds acting upon regions of the oligosaccharide *distal* to the primary binding site. In this context it is interesting to note that related lectins such as Con A, although having a rather broad specificity for mannose and glucose, bind larger oligosaccharides with varying affinity, suggesting that binding of these distal regions of the glycan may be an important element governing fine specificity.

5.3 Antibody–oligosaccharide interactions

In a recent NMR study, Glaudemans *et al.* have described the conformation of an oligosaccharide derivative both in free solution and when bound to an intact antibody molecule, IgA X24 (66). The detailed determination of the conformation of an intact antibody with bound antigen is not feasible by NMR, in view of the large size of the complex. However, if the ligand binds to the antibody only weakly, a technique termed transferred nuclear Overhauser effect (TRNOE) spectroscopy (15, 67–69) can be used to investigate the conformation of ligand in its bound form. As with conventional nuclear Overhauser effect spectroscopy, TRNOE measurements exploit the dependence of magnetization transfer between two protons close in space ($< \sim 5$ Å) upon the inverse sixth power of the distance between these protons. For such protons on a small, unbound ligand which tumbles rapidly in solution, the rate of magnetization transfer is very slow, and is opposite in sign to the very fast transfer which occurs with a ligand bound to a slowly tumbling protein. If the ligand is bound to the protein for a sufficiently short time, then this rapid exchange can be exploited to observe the effect of NOEs in the bound state by observing the much stronger and sharper resonances of excess free ligand present in solution. Effectively, when the bound ligand leaves the protein to become free, it carries with it the 'memory' of the NOEs which were present in the bound state. Using this approach, Glaudemans *et al.* were able to demonstrate that the conformation of a derivative of Galβ1-6Galβ-OMe was different in the bound and free states. Values of ϕ, ψ, and ω for the disaccharide in free solution were reported as −120°, 180°, and 75° respectively, whereas the TRNOE data for the bound disaccharide were consistent with two possible conformers at −53°, 154°, −173° or −152°, −128°, −158° respectively, which could not be distinguished in view of the limited accuracy of the TRNOE data. Nevertheless, each of these conformers is again quite different from the solution conformation, indicating that the antibody combining site constrains the disaccharide into a higher energy conformation. This increase in energy, in common with the other examples considered above, is most likely more than compensated by favourable binding interactions between the sugar and the antibody.

A second example of an antibody–oligosaccharide interaction is given by the crystallographic study of Cygler et al. (70) on the recognition of a cell–surface oligosaccharide of pathogenic Salmonella by an antibody Fab fragment. The antigen is comprised of repeating units containing D-mannose, D-galactose, L-rhamnose (Rha), and 3,6-dideoxy-D-galactose (abequose, Abe); {3Galα1-2(Abeα1-3)Manα1-4Rhaα1-}$_n$. The complex of Fab with three repeating units of the saccharide showed that the complementarity determining regions (CDRs) in the Fab create a pocket approximately 8 Å deep and 7 Å wide, which is occupied by the abequose residue. This residue becomes totally buried, whereas mannose and galactose lie on the protein surface, and are partially exposed to solvent. The shapes of the oligosaccharide and antibody surfaces that are in contact are highly complementary. Both hydrogen bonds and hydrophobic interactions are important for binding. All sugar hydroxyl groups, with the exception of those at galactose C-6 and mannose C-6 participate in protein-carbohydrate hydrogen bonds. A buried water molecule, which accepts hydrogen bonds from O-4 of abequose and donates to the ring oxygen atom, enhances Fab-antigen surface complementarity. In contrast to the arabinose binding protein (above), aromatic amino acid side chains dominate the Fab hydrogen bonding scheme. The conformation observed for the bound trisaccharide epitope lies within 14 kJ/mol of the calculated global minimum energy conformer for the free oligosaccharide (71). The glycosidic torsion angles of the Abeα1-3Man linkage in the bound sugar agree reasonably well with predicted values (ϕ, ψ = 80°, 91° in the crystal versus 66°, 106° for the free glycan), but the torsion angle of the Galα1-2Man linkage adopts a ϕ value that is shifted about 40° (05, ψ = 105°, 38° in the crystal versus 69°, 101° for the free glycan) from the value of 60° predicted from consideration of the exo-anomeric effect. Hence in this second example of an antibody–oligosaccharide interaction the conformation of the oligosaccharide when bound again differs appreciably from that observed in the uncomplexed oligosaccharide in solution.

6. Functional implications

6.1 Oligosaccharide biosynthesis

The study of the solution conformations and dynamics of oligosaccharides in free solution is of considerable practical importance concerning oligosaccharide biosynthesis, since any oligosaccharide is potentially an acceptor substrate for a given glycosyltransferase. Oligosaccharide biosynthesis appears not to be template directed, and yet certain controls clearly operate during biosynthesis (72, 73). For example, different oligosaccharide chains are often found at different glycosylation sites on the same glycoprotein. Since different chains on the same glycoprotein rules out the possibility of indirect genetic control by modulation of glycosyltransferase activity (although this mechanism may well explain tissue-specific glycosylation), then such differences must arise by differences in processing *in situ*. An obvious explanation would be that certain glycosylation sites are inaccessible to the processing enzymes by virtue of the steric constraints imposed by the three-dimensional structure of the protein. This is consistent with observations that the extent of processing is correlated with the extent of sensitivity of an oligosaccharide to digestion by endoglycosidase H, at least in certain systems (74, 75). However, since glycoproteins exist where different sites on a glycoprotein have structures representing products on different branches of the processing pathway, then this explanation is incomplete. A novel alternative explanation has been suggested by Carver and co-workers, termed 'site-directed processing' (76). This model arose from work on a human myeloma IgG(Hom) which was found to contain oligosaccharides at Asn107 of the light chain which were all of the Neu5Acα-6 terminated, bisected-biantennary complex type (Figure 11), whereas those at Asn297 of the heavy chain were found to comprise primarily unbisected complex types terminating in Gal or Neu5Ac. Since GlcNAc transferase III must act before galactosylation occurs (77), then absence of bisected structures at Asn 297 due to inaccessibility should also result in lack of galactosylation and sialylation since the relevant transferases act later in the pathway. This apparent paradox can be resolved by noting that the protein constrains the oligosaccharide with the 1-6 arm fixed at ω =

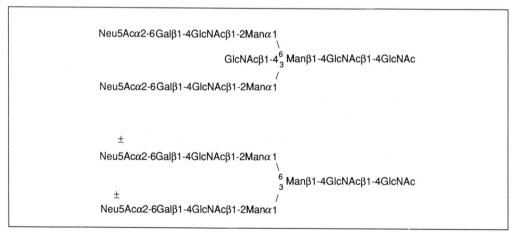

Fig. 11 Diagrammatic illustration of (a) the Neu5ACα2-6 terminated, bisected-biantennary oligosaccharides found at Asn107 of IgG(Hom), and (b) the biantennary complex type oligosaccharides found at Asn297 of the same protein. See text for further details

$-60°$, whereas in free solution $\omega = 180°$ is also a highly populated rotamer. If the latter is the conformation recognized by GlcNAc transferase III, then the oligosaccharide in association with protein may not be a suitable substrate for the transferase. This concept requires further study of the precise substrate specificities of glycosyltransferases, but it presents intriguing and far-reaching possibilities regarding the control of biosynthesis of N-linked glycans.

A more recent study suggests that glycan conformation may modulate the activity of the oligosaccharyltransferase which is responsible for the transfer of the protein N-glycosylation precursor $Glc_3Man_9GlcNAc_2$ from dolichol to nascent protein (78). This study indicated that the three-dimensional structure of the Glcα1-2Glcα1-3Glcα1-3 moiety on the precursor was of a flattened trimeric unit in solution, and it was proposed that this unit may modulate the activity of the oligosaccharyltransferase by binding to a hydrophobic cleft in the enzyme. In support of this notion is the fact that yeast strains which are unable to glucosylate the precursor produce proteins which are underglycosylated. If correct, this would represent the first example to the author's knowledge where an N-linked glycan could act as an allosteric effector.

6.2 Oligosaccharide-mediated recognition phenomena

Of the many functions which have been proposed for oligosaccharides, none has been more intriguing than their possible roles as recognition signals. While a great deal of circumstantial evidence has accumulated, only in a few key systems have the precise oligosaccharide ligands been identified. A primary difficulty with the concept of 'oligosaccharide-mediated recognition' is the phenomenon of microheterogeneity, i.e. the presence of a manifold of biosynthetically related structures at a given glycosylation site. This presents a paradox, since any recognition phenomenon by definition must be specific, and yet microheterogeneity in primary sequence would appear to defeat this specificity. In an attempt to rationalize this problem, it has been proposed that microheterogeneity is of great importance in diversifying the function of a protein, such that a given protein molecule to which a given oligosaccharide primary sequence is appended may be functionally distinct from another to which a different oligosaccharide is appended (79). This hypothesis neatly avoids the problem of maintenance in specificity, since it implies that the same protein molecule may be recognized specifically by a different receptor depending upon the precise oligosaccharide sequence. In the view of the author, such functional significance of microheterogeneity is unlikely, and instances where the nature of the oligosaccharide primary sequence appears to modulate protein activity might easily be accounted for by small coincidental changes in the physicochemical properties of the molecule. Pragmatically, if one considers the biosynthetic pathway of N-linked glycans, one finds a nontemplate directed synthesis which must potentially be influenced by various factors in addition to those already mentioned, and it is difficult to envisage a scenario where one would *not* expect to find heterogeneity in sequence.

If one accepts this alternative view, then one still must tackle the heterogeneity/specificity paradox. Since recognition is a 'shape' problem, then it follows that the important determinant of specificity is the three-dimensional structure of the oligosaccharide, and not its primary sequence. In this regard, a study of several biosynthetically related oligosaccharides of differing primary sequence has demonstrated that these can be categorized into a much smaller subset of related three-dimensional structures (22). The majority of monosaccharide substitutions or deletions were found not to affect the overall conformation of the oligosaccharide. These were designated 'type I' structural changes. Other substitutions or deletions resulted in conformational changes in linkages proximal to the site of the primary sequence change, and were designated as 'type 2p' changes, whereas others, surprisingly, gave rise to conformational changes distal to the site of the primary sequence change, and were designated as 'type 2d' changes. Thus, on the assumption that primary sequence heterogeneity is an unavoidable penalty of a biosynthetic pathway which is not template directed, this could be accommodated by common three-dimensional motifs (members of the 'type I' structural class) which can be recognized by the same receptor, whereas specificity can be modulated by 'type 2' structural changes. This hypothesis explains why certain lectins appear to recognize several monosaccharide residues within a given oligosaccharide and are specific for a rather wide range of oligosaccharides with different primary sequences (80), and yet are intolerant to apparently trivial differences in primary sequence. These sequence differences would be predicted to be those which would give rise to 'type 2' changes. With techniques which we now have at our disposal for the conformational analysis of oligosaccharides, this hypothesis is readily testable.

The possibility that the overall conformation of an oligosaccharide can be modulated by the addition or removal of an oligosaccharide residue has implications which may be important in the regulation of recognition. For example, carbohydrates have long been implicated as 'differentiation antigens' during development, on the basis of observed changes in oligosaccharide primary sequence during embryogenesis (81, reviewed in references 82, 83). A 'key player' in these phenomena is a structure known as Lewis X (Le^X). During embryogenesis, the sequence Galβ1-4GlcNAc found in the I, i differentiation antigens, when fucosylated to generate the Le^X structure Galβ1-4(Fucα1-3)GlcNAc, generates a new antigenic determinant known as the SSEA-1 antigen. We have recently undertaken a detailed investigation of the structure and dynamics of Le^X (42), and it is of interest that the presence of the fucose appears to modify the conformation of the Galβ1-4GlcNAc unit from $\phi, \psi = 55°, 0°$ in its absence, to $\phi, \psi = 55°, 30°$ in its presence, which is a conformational change of type 2p in the above nomenclature. It is at present too early to tell whether conformational changes such as these are an inherent requirement for modulation of recognition. They may simply represent extreme examples of a general principle whereby the addition of a monosaccharide residue to an existing oligosaccharide chain creates a new determinant or hides one that previously existed, or both. A possible example of this latter principle may be the structure known as sialylated Lewis X (SLe^X, Neu5Acα2-6 Galβ1-4(Fucα1-3)GlcNAc), the proposed ligand for the selectin ELAM-1 (84, 85), where the presence of the sialic acid does not alter the conformation of the underlying Le^X moiety (Homans, unpublished observations).

References

1. Burnet, F. M. (1951). Mucoproteins in relation to virus action. *Physiol. Rev.*, **31**, 131.
2. Springer, T. A. (1990). Adhesion receptors of the immune system. *Nature*, **346**, 425.
3. Springer, T. A. and Lasky, L. A. (1991). Sticky sugars for selectins. *Nature*, **349**, 196.
4. Homans, S. W. (1990). Oligosaccharide conformations: Application of NMR and energy calculations. *Prog. NMR Spectrosc.*, **22**, 55.
5. Lemieux, R. U., Delbaere, L. T. J., Koto, S., and Rao, V. S. (1980). The conformations of oligosaccharides related to the ABH and Lewis human blood group determinants. *Can. J. Chem.*, **58**, 631.
6. Thogersen, H., Lemieux, R. U., Bock, K., and Meyer, B. (1982). Further justification for the exo-anomeric effect. Conformational analysis based on nuclear magnetic resonance spectroscopy of oligosaccharides. *Can. J. Chem.*, **60**, 44.
7. Sabesan, S., Bock, K., and Lemieux, R. U. (1984). The conformational properties of the gangliosides $G_{M2}G_{M1}$ based on ^{1}H and ^{13}C nuclear magnetic resonance studies. *Can. J. Chem.*, **62**, 1034.
8. Paulsen, H., Peters, T., Sinnwell, V., Heume, M., and Meyer, B. (1986). Konformations analyse der verzweigten pentasaccharid-sequenz der "bisected" struktur von N-glycoproteinen. *Carbohydr. Res.*, **156**, 87.
9. Biswas, M., Chandra Sekharudu, Y., and Rao, V. S. R. (1987). The Conformation of glycans of the oligo-D-mannosidic type, and their interaction with concanavalin A: A computer-modelling study. *Carbohydr. Res.*, **160**, 151.
10. Bock, K. (1983). The preferred conformation of oligosaccharides in solution inferred from high resolution NMR data and hard sphere exo-anomeric calculations. *Pure and Appl. Chem.*, **55**, 605.
11. Tvaroska, I. and Bleha, T. (1989). Anomeric and exo-anomeric effects in carbohydrate chemistry. *Adv. Carbohydr. Chem. Biochem.*, **47**, 45.
12. Kitaygorodsky, A. I. (1961). The interaction curve of non-bonded carbon and hydrogen atoms and its applications. *Tetrahedron*, **14**, 230.
13. Bock, K. and Lemieux, R. U. (1982). The conformational properties of sucrose in aqueous solution–intramolecular hydrogen bonding. *Carbohydr. Res.*, **100**, 63.
14. Bock, K., Arnarp, J., and Lonngren, J. (1982). The preferred conformation of oligosaccharides derived from complex-type carbohydrate portions of glycoproteins. *Eur. J. Biochem.*, **129**, 171.
15. Homans, S. W. (1992). *A dictionary of concepts in NMR* (revised edition). Oxford University Press.
16. Wuethrich, K. (1986). *NMR of proteins and nucleic acids*. Wiley, New York.
17. Clore, G., Brunger, A., Karplus, M., and Gronenborn, A. (1986). Application of molecular dynamics with interproton distance restraints to three-dimensional protein structure determination. A model study of Crambin. *J. Mol. Biol.*, **191**, 523.
18. Scarsdale, J. N., Ram, P., Prestegard, J. H., and Yu, R. K. (1988). A molecular

mechanics-NMR pseudoenergy approach to the solution conformation of glycolipids. *J. Comput. Chem.*, **9**, 133.
19. Brisson, J.-R. and Carver, J. P. (1983). Solution conformation of α-D(1-3)- and αD(1-6)-linked oligomannosides using proton nuclear magnetic resonance. *Biochemistry*, **22**, 1362.
20. Brisson, J.-R. and Carver, J. P. (1983). Solution conformation of asparagine-linked oligosaccharides: α(1-3)-, β(1-2)-, and β(1-4)-linked units. *Biochemistry*, **22**, 3671.
21. Brisson, J.-R. and Carver, J. P. (1983). Solution conformation of asparagine-linked oligosaccharides: α-(1-6)-linked moiety. *Biochemistry*, **22**, 3680.
22. Homans, S. W., Dwek, R. A., Boyd, J., Mahmoudian, M., Richards, W. G., and Rademacher, T. W. (1986). Conformational transitions in N-linked oligosaccharides. *Biochemistry*, **25**, 6342.
23. Homans, S. W., Dwek, R. A., and Rademacher, T. W. (1987). Tertiary structure in N-linked oligosaccharides. *Biochemistry*, **26**, 6553.
24. Poppe, L. and van Halbeek, H. (1991). Nuclear magnetic resonance of hydroxyl and amido protons of oligosaccharides in aqueous solution: Evidence for a strong intramolecular hydrogen bond in sialic acid residues. *J. Am. Chem. Soc.*, **113**, 363.
25. Warin, V., Baert, F., Fouret, R., Strecker, G., Spik, G., Fournet, B., and Montreuil, J. (1979). Crystal and molecular structure of O-alpha-D-mannopyranosyl(1-3)O-β-D-mannopyranosyl(1-4)2-acetamido-2 deoxy-α-D-glycopyranose. *Carbohyd. Res.*, **76**, 11.
26. Cumming, D. A., Shah, R. N., Krepinksy, J. J., Grey, A. A., and Carver, J. P. (1987). Solution conformation of the branch points of N-linked glycans: Synthetic model compounds for Tri'-antennary and teraantennary glycans. *Biochemistry*, **26**, 6655.
27. Cumming, D. A. and Carver, J. P. (1987). Virtual and solution conformations of oligosaccharides. *Biochemistry*, **26**, 6664.
28. Cumming, D. A. and Carver, J. P. (1987). Reevaluation of rotamer populations for 1,6 linkages: Reconciliation with potential energy calculations. *Biochemistry*, **26**, 6676.
29. Homans, S. W., Pastore, A., Dwek, R. A., and Rademacher, T. W. (1987). Structure and dynamics in oligomannose type oligosaccharides. *Biochemistry*, **26**, 6649.
30. French, A. (1989). Comparisons of rigid and relaxed confromational maps for cellobiose and maltose. *Carbohyd. Res.*, **188**, 206.
31. Ha, S. N., Madsen, L. J., and Brady, J. W. (1988). Conformational analysis and molecular dynamics simulations of maltose. *Biopolymers*, **27**, 1927.
32. Homans, S. W., Edge, C. J., Ferguson, M. A. J., Dwek, R. A., and Rademacher, T. W. (1989). Solution structure of the glycosylphosphatidylinositol membrane anchor glycan of Trypanosoma brucei variant surface glycoprotein. *Biochemistry*, **28**, 2881.
33. Yan, Z.-Y. and Bush, C. A. (1990). Molecular dynamics simulations and the conformational mobility of blood group oligosaccharides. *Biopolymers*, **29**, 799.
34. Imberty, A., Tran, V., and Perez, S. (1989). Relaxed potential energy surfaces of N-linked oligosaccharides: The mannose-α(1-3)-mannose case. *J. Comput. Chem.*, **11**, 205.
35. Brady, J. W. (1986). Molecular dynamics simulations of α-D-glucose. *J. Am. Chem. Soc.*, **108**, 8153.
36. Brady, J. W. (1987). Molecular dynamics imulations of β-D-glycopyranose. *Carbohydr. Res.*, **165**, 306.
37. Madsen, L. J., Ha, S. N., Trans, V. H., and Brady, J. W. (1990). Molecular dynamics simulations of carbohydrates and their solvation. In *Computer modeling of carbohydrate molecules* (eds A. D. French and J. W. Brady). ACS Symposium series. 430, pp. 69–90.
38. Ha, S. N., Giammona, A., Field, M., and Brady, J. W. (1988). A revised potential-

energy surface for molecular mechanics studies of carbohydrates. *Carbohydr. Res.*, **180**, 207.
39. Homans, S. W. (1990). A Molecular mechanical force field for the conformational analysis of oligosaccharides: comparison of theorectical and crystal structures of Manα1-3Manβ1-4GlcNAc. *Biochemistry*, **29**, 9110.
40. Edge, C. J., Singh, U. C., Bazzo, R., Taylor, G. L., Dwek, R. A., and Rademacher, T. W. (1990). 500-picosecond molecular dynamics in water of the Manα1-2Manα glycosidic linkage present in Asn-linked oligomannose-type structures on glycoproteins. *Biochemistry*, **29**, 1971.
41. Tropp, J. (1980). Dipolar relaxation and nuclear Overhauser effects in non-rigid molecules: The effect of fluctuating internuclear distances. *J. Chem. Phys.*, **72**, 6035.
42. Homans, S. W. and Forster, M. (1992). Application of restrained minimization, simulated annealing and molecular dynamics simulations for the conformational anlysis of oligosaccharides. *Glycobiology*, **2**, 143.
43. Poppe, L. and van Halbeek, H. (1992). The rigidity of sucrose — just an illusion? *J. Am. Chem. Soc.*, **114**, 1092.
44. Mulloy, B., Frenkiel, T. A., and Davies, D. B. (1988). Long-range carbon–proton coupling constants: Application to conformational studies of oligosaccharides. *Carbohydr. Res.*, **184**, 39.
45. Tvaroska, I., Hricovini, M., and Petrakova, E. (1989). An attempt to derive a new Karplus-type equation of vicinal proton-carbon coupling constants for C–O–C–H segments of bonded atoms. *Carbohydr. Res.*, **189**, 359.
46. Hamer, G. K., Balza, F., Cyr, N., and Perlin, A. S. (1978). A conformational study of methyl β-cellobioside-d_8 by ^{13}C nuclear magnetic resonance spectroscopy: dihedral angle dependence of $^3J_{C-H}$ in ^{13}C–O–C–^{1}H arrays. *Can J. Chem.*, **56**, 3109.
47. Poppe, L. and van Halbeek, H. (1991). ^{1}H-Detected measurements of long-range heteronuclear coupling constants. Application to a trisaccharide. *J. Magn. Reson.*, **92**, 636.
48. Poppe, L. and van Halbeek, H. (1991). Selective, inverse-detected measurements of long-range ^{13}C, ^{1}H coupling constants. Application to a disaccharide. *J. Magn. Reson.*, **93**, 214.
49. Lipari, G. and Szabo, A. (1982). Model-free approach to the interpretation of nuclear magnetic resonance relaxation in macromolecules. 1. Theory and range of validity. *J. Am. Chem. Soc.*, **104**, 4546.
50. Levy, R. M., Karplus, M., and McCammon, J. A. (1981). Increase of ^{13}C NMR relaxation times in proteins due to picosecond motional average. *J. Am. Chem. Soc.*, **103**, 994.
51. Kay, L. E., Torchia, D. A., and Bax, A. (1989). Backbone dynamics of proteins as studied by ^{15}N inverse detected heteronuclear NMR spectroscopy: Application to staphylococcal nuclease. *Biochemistry*, **28**, 8972.
52. Gidley, M. J. and Bociek, S. M. (1988) ^{13}C-C.p.-m.a.s. NMR studies of frozen solutions of (1-4)-α-D-glucans as a probe of the range of conformations of glycosidic linkages: the confromation of cyclomaltohexaose and amylopectin in aqueous solution. *Carbohydr. Res.*, **183**, 126.
53. Deisenhofer, J. (1981). Crystallographic refinement and atomic models of a human Fc fragment and its complex with fragment B from staphylococcus aureus at 2.9 Å and 2.8 Å resolution. *Biochemistry*, **20**, 2361.
54. Leatherbarrow, R. J., Rademacher, T. W., Dwek, R. A., Woof, J., Clark, A., Burton, D., Richardson, N., and Feinstein, A. (1985). *Molec. Immunol.*, **22**, 407.

55. Jefferis, R., Lund, J., Mizutani, H., Nakagawa, H., Kawazoe, Y., Arata, Y., and Takahashi, N. (1990). A comparative study of the N-linked oligosaccharide structures of human IgG subclass proteins. *Biochem. J.*, **268**, 529.
56. Parekh, R. B., Dwek, R. A., Sutton, B. J., Fernandes, D. L., Leung, A., Stanworth, D., Rademacher, T. W., Mizuochi, T., Taniguchi, T., Matsuta, K., Takeuchi, F., Nagano, Y., Miyamoto, T., and Kobata, A. (1985). Association of rheumatoid arthritis and primary osteoarthritis with changes in the glycosylation pattern of total serum IgG. *Nature*, **316**, 452.
57. Yariv, J., Kalb (Gilboa), A. J., Papiz, M. Z., Helliwell, J. R., Andrews, S. J., and Habash, J. (1987). Properties of a new crystal form of the complex of concanavalin A with methyl-alpha-glucopyranoside. *J. Mol. Biol.*, **195**, 759.
58. Derewenda, Z., Yariv, J., Helliwell, J. R., Kalb (Gilboa), A. J., Dodson, E. J., Papiz, M. Z., Wan, T., and Campbell, J. (1989). The structure of the saccharide-binding site of concanavalin A. *EMBO J.*, **8**, 2189.
59. Quiocho, F. A. (1989). Protein–carbohydrate interactions. Basic molecular features. *Pure Appl. Chem.*, **61**, 1293.
60. Quiocho, F. A., Wilson, D. K., and Vyas, N. J. (1989). Substrate specificity and affinity of a protein modulated by water molecules. *Nature*, **340**, 404.
61. Sixma, T. K., Pronk, S. E., Kalk, K. H., van Zanten, B. A. M., Berghuis, A. M., and Hol, W. G. J. (1992). Lactose binding to heat labile enterotoxin revealed by x-ray crystallography. *Nature*, **355**, 561.
62. Bourne, Y., Rouge, P., and Cambillau, C. (1990). X-ray structure of a (α-Man(1-3)β-Man(1-4)GlcNAc)-lectin complex at 2.1-Å resolution. *Biol. Chem.*, **265**, 18161.
63. Einspahr, H., Parks, H. E., Suguna, K., Subramanian, E., and Suddath, F. L. (1986). Crystal structure of pea lectin at 3 Å resolution. *J. Biol. Chem.*, **261**, 16518.
64. Reeke, G. N., Jr. and Becker, J. W. (1986). Three dimensional structure of favin–saccharide binding–cyclic permutation in leguminous lectins. *Science*, **234**, 1108.
65. Imerty, A., Delage, M.-M., Bourne, Y., Cambillau, C., and Perez, S. (1991). Data bank of three-dimensional structures of disaccharides: Part II, N-acetyllactosaminic type N-glycans. Comparison with the crystal structure of a biantennary octasaccharide. *Glycoconjugate J.*, **8**, 456.
66. Glaudemans, C. P. J., Lerner, L., Daves, G. D. Jr., Kovac, P., Venable, R., and Bax, A. (1990). Significant conformational changes in an antigenic carbohydrate epitope upon binding to a monoclonal antibody. *Biochemistry*, **29**, 10906.
67. Clore, G. M. and Gronenborn, A. M. (1982). Theory and applications of the transferred nuclear overhauser effect to the study of the conformations of small ligands bound to proteins. *J. Magn. Reson.*, **48**, 402.
68. Clore, G. M. and Gronenborn, A. M. (1983). Theory of the time dependent transferred nuclear overhauser effect: applications to structural analysis of ligand-protein complexes in solution. *J. Magn. Reson.*, **53**, 423.
69. Feeney, J., Birdsall, B., Roberts, G. C. K., and Burgen, A. S. V. (1983). Use of transferred nuclear overhauser effect measurements to compare binding of coenzyme analogues to dihydrofolate reductase. *Biochemistry*, **22**, 628.
70. Cygler, M., Rose, D. R., and Bundle, D. R. (1991). Recognition of a cell-surface oligosaccharide of pathogenic salmonella by an antibody Fab fragment. *Science*, **253**, 442.
71. Bock, K., Meldal, M., Bundel, D. R., Iversen, T., Pinto, B. M., Garegg, P. J., Kvastrom, I., Norberg, T., Lindberg, A. A., and Svenson, S. B. (1984). The conformation of

salmonella O-antigenic oligosaccharides of serogroup A, serogroup B, and serogroup d1 inferred from ^{1}H and ^{13}C NMR. *Carbohydr. Res.*, **130**, 23.
72. Pollack, L. and Atkinson, P. H. (1983). Correlation of glycosylation forms with position in amino-acid sequence. *J. Cell Biol.*, **97**, 293.
73. Schachter, H., Narasimhan, S., Gleeson, P., and Vella, G. (1983). Control of branching during biosynthesis of Asn-linked oligosaccharides. *Can J. Biochem. Cell Biol.*, **61**, 1049.
74. Hsieh, P., Rosner, M. R., and Robbins, P. W. (1983). Selective cleavage by endo-H at individual glycosylation sites of sindbis virus envelope glycoproteins. *J. Biol. Chem.*, **258**, 2555.
75. Trimble, R. B., Maley, F., and Chu, F. K. (1983). Glycoprotein biosynthesis in yeast-protein conformation affects processing of high mannose oligosaccharides in carboxypeptidase-Y and invertase. *J. Biol. Chem.*, **258**, 2562.
76. Carver, J. P. and Cumming, D. A. (1987). Site-directed processing of N-linked oligosaccharides: the role of three-dimensional structure. *Pure Appl. Chem.*, **59**, 1465.
77. Narasimhan, S. (1982). Control of glycoprotein synthesis 7. GlcNAc transferase III, and enzyme in hen oviduct which adds GlcNAc in beta-1,4 linkage to the beta-1,4 linked mannose of the trimannose core of *N*-glycosyl oligosaccharides. *J. Biol. Chem.*, **257**, 235.
78. Alvarado, E., Nukada, T., Ogawa, T., and Ballou, C. E. (1991). Conformation of the glucotriose unit in the lipid-linked oligosaccharide precursor for protein glycosylation. *Biochemistry*, **30**, 881.
79. Rademacher, T. W., Parekh, R. B., and Dwek, R. A. (1988). Glycobiology *Ann. Rev. Biochem.*, **57**, 785.
80. Yamashita, K., Hitoi, A., and Kobata, A. (1983). Structural determinants of phaseolus vulgaris erthroagglutinating lectin for oligosaccharides. *J. Biol. Chem.*, **258**, 14753.
81. Gooi, H. C., Feizi, T., Kapadia, A., Knowles, B. B., Solter, D., and Evans, M. J. (1981). SSEA-1 involves alpha-1,3 fusocylated type-2 blood group chains. *Nature*, **292**, 156.
82. Feizi, T. (1981). Carbohydrate differentiation antigens. *Trends in Biol. Sci.*, **6**, 333.
83. Feizi, T. (1991). Carbohydrate differentiation antigens: probable ligands for cell adhesion molecules. *Trends in Biol. Sci.*, **16**, 84.
84. Phillips, M. L., Nudelman, E., Gaeta, C. A., Perez, M., Singhal, A. K., Hakomori, S.-I., and Paulson, J. C. (1991). ELAM-1 mediates cell adhesion by recognition of a carbohydrate ligand, sialy-Le^X *Science*, **250**, 1130.
85. Berg, E. L., Robinson, M. K., Mansson, O., Butcher, E. C., and Magnani, J. L. (1991). A carbohydrate domain common to both sialyl Lea and sialyl Lex is recognised by the endothelial cell leukocyte adhesion molecule ELAM-1. *J. Biol. Chem.*, **266**, 14869.

Appendix to Chapter 3

Note added in proof

The following papers relevant to the topic of this chapter appeared after submission of the manuscript to the printers.

Several new glycosyltransferase genes have been cloned. All of these genes encode proteins with a putative Type II integral membrane protein domain structure.

(i) UDP-GlcNAc:GlcNAc:GlcNAcβ1-2Manα1-6Manβ-R (GlcNAc to Manα1-6) β1,6-N-acetylglucosaminyltransferase V (GlcNAc-T V; EC 2.4.1.155) was purified from rat kidney (1) and a human lung cancer cell line (2). Oligonucleotides based on partial amino acid sequence data were used as primers in polymerase chain reaction (PCR) amplification of cDNA to yield partial CDNA clones for mouse and rat GlcNAc-T V (3, 4). A 4.8 kb rat cDNA clone was isolated containing a 2220 bp open reading frame encoding a 740 amino acid protein; this is the largest lumenal Golgi glycosyltransferase reported to date. GlcNAc-T V activity was obtained on transient expression of this clone in COS-7 cells. There was no sequence homology to any previously cloned glycosyltransferases.

(ii) Human UDP-GlcNAc:R_1-Manα1-6[GlcNAcβ1-2Manα1-3]Manβ-4R_2 (GlcNAc to Manβ1-4) β4-GlcNAc-transferase III (E.C. 2.4.1.144) has been cloned and expressed and the gene has been localized to human chromosome 22q.13.1 (5).

(iii) UDP-GalNAc:polypeptide N-acetylgalactosaminyltransferase (polypeptide GalNAc-T, EC 2.4.1.41) was purified from bovine colostrum and PCR was used as above to obtain the bovine cDNA clone (6, 7). The cDNA contains an open reading frame encoding a 559 amino acid protein. GalNAc-T activity was obtained by transient expression in COS-7 cells and with the baculovirus/Sf9 insect cell expression system. The recombinant GalNAc-T transferred GalNAc much more effectively to Thr than to Ser residues. There was no sequence homology to any previously cloned glycosyltransferases.

(iv) Sialyltransferase genes. It has been suggested that there are at least two different CMP-sialic acid:Galβ1-3/4GlcNAc-R α2,3-sialyltransferases (ST3N), differing in their preference for Galβ1-3GlcNAc (type 1) and Galβ1-4GlcNAc (type 2) acceptors (8). Although the type-2-specific sialyltransferase was reported to occur in human placenta, a sialyltransferase gene cloned from this tissue was a type-1-specific ST3N (9). Expression cloning was used to clone the gene for a human CMP-sialic acid:Galβ1-3/4GlcNAc-R α2,3-sialyltransferase that the authors believe may be different from the previously cloned type 1-specific rat liver ST3N and may be a type 2-specific ST3N involved in the formation of sialyl-Le^X *in vivo* (10). Mouse brain CMP-sialic acid:Galβ1-3GalNAc-R α2,3-sialyltransferase (ST30) has been cloned (11). Homology screening based on similarities in sequences between sialyltransferase genes (sialylmotif) have resulted in the isolation of a putative sialyltransferase cDNA which is expressed only in newborn rat brain (12); all acceptors tested to date with the recombinant protein were ineffective as substrates.

References

1. Shoreibah, M. G., Hindsgaul, O., and Pierce, M. (1992). Purification and characterization of rat kidney UDP-N-acetylglucosamine:α-6-D-Mannoside β-1,6-N-acetylglucosaminyltransferase. *J. Biol. Chem.*, **267**, 2920.
2. Gu, J., Nishikawa, A., Tsuruoka, N., Ohno, M., Yamaguchi, N., Kanagawa, K., and Taniguchi, N. (1993). Purification and characterization of UDP-N-acetylglucosamine:α-6-D-mannoside betal-6-Nacetylglycosaminyltransferase (N-acetylglucosaminyltransferase V) from a human lung cancer cell line. *J. Biochem.*, **113**, 614.
3. Shoreibah, M., Perng, G. S., Adler, B., Weinstein, J., Basu, R., Cupples, R., Wen, D., Browne, J. K., Buckhaults, P., Fregien, N., and Pierce, M. (1993). Isolation, characterization, and expression of a cDNA encoding N-acetylglucosaminyltransferase V. *J. Biol. Chem.*, **268**, 15381.
4. Miyoshi, E., Nishikawa, A., Ihara, Y., Gu, J., Sugiyama, T., Hayashi, N., Fusamoto, H., Kamada, T., and Taniguchi, N. (1993). N-acetylglucosaminyltransferase III and V messenger RNA levels in LEC rats during hepatocarcinogenesis. *Cancer Res.*, **53**, 3899.
5. Ihara, Y., Nishikawa, A., Tohma, T., Soejima, H., Hiikawa, N., and Taniguchi, N. (1993). cDNA cloning, expression, and chromosomal localization of human N-acetylglycosaminyltransferase III (Gnt-III). *J. Biochem.*, **113**, 692.
6. Homa, F. L., Hollander, T., Lehman, D. J., Thomsen, D. R., and Elhammer, Å. P. (1993). Isolation and expression of a cDNA clone encoding a bovine UDP-GalNAc:polypeptide N-acetylgalactosaminyltransferase. *J. Biol. Chem.*, **268**, 12609.
7. Hagen, F. K., Vanwuyckhuyse, B., and Tabak, L. A. (1993). Purification, cloning, and expression of a bovine UDP-GalNAc–polypeptide N-acetyl-galactosaminyltransferase. *J. Biol. Chem.*, **268**, 18960.
8. Nemansky, M. and Van den Eijnden, D. H. (1993). Enzymatic characterization of CMP-NeuAc:Galβ1-4GlcNAc-R α(2-3)-sialyltransferase frome human placenta. *Glycoconjugate J.*, **10**, 99.
9. Kitagawa, H., and Paulson, J. C. (1993). Cloning and expression of human Galbeta1,3(4)GlcNAc α2,3-sialyltransferase. *Biochem. Biophys. Res. Commun.*, **194**, 375.
10. Sasaki, K., Watanabe, E., Kawashima, K., Sekine, S., Dohi, T., Oshima, M., Hanai, N., Nishi, T., and Hasegawa, M. (1993). Expression cloning of a novel gal β(1-3/1-4)GlcNAc α2,3-sialyltransferase using lectin resistance selection. *J. Biol. Chem.*, **268**, 22782.
11. Lee, Y. C., Kurosawa, N., Hamamoto, T., Nakaoka, T., and Tsuji, S. (1993). Molecular cloning and expression of gal-β-1,3GalNAc-α-2,3-sialyltransferase from mouse brain. *Eur. J. Biochem.*, **216**, 377.
12. Livingston, B. D. and Paulson, J. C. (1993). Polymerase chain reaction cloning of a developmentally regulated member of the sialyltransferase gene family. *J. Biol. Chem.*, **268**, 11504.

Index

N-acetylgalactosaminyltransferase 258
ABO blood group antigens 17, 19–21, 92, 97–8
N-acetylglucosaminylphosphotransferase 10, 56
N-acetylglucosaminyltransferase 216
   N-acetylglucosaminyltransferase I 8, 90, 138–43
   N-acetylglucosaminyltransferase II 8, 90, 143–4
   N-acetylglucosaminyltransferase III 90, 144, 250, 251, 258
   N-acetylglucosaminyltransferase IV 8, 90
β-1,6-N-acetylglucosaminyltransferases 38
   core 2 β-1,6-N-acetylglucosaminyltransferase 38, 144–5
   core 4 β-1,6-N-acetylglucosaminyltransferase 32–3, 91, 96
   I-branching enzyme 39, 145–6
   N-acetylglucosaminyltransferase V 9, 37, 40, 258
N-acetyllactosamine 7
N-acetyllactosamine repeats 7, 11
activation-dependent shedding 172
adression 25
   GlyCAM-1 25, 170–2
animal lectins 53–87
antennary 5, 90
   biantennary 5–6
   triantennary 5–6
   tetrantennary 5–6
   pentantennary 5–6
antibody–oligosaccharide complex 248, 249
arabinose binding protein 245
asialoglycoprotein receptor 22, 58–60
azido sugars 212

bisecting N-acetylglucosamine 10
blood group A α-1,3-N-acetylgalactosaminyltransferase 19, 137–8
blood group B α-1,3-galactosyltransferase 19, 116–17
blood group H α-1,2-fucosyltransferase 19, 129
branch specificity 13
   of β-1,3-N-acetylglucosaminyltransferase 13
   of α-2,6-sialyltransferase 14
branching enzymes 215
   core 2 branching enzyme 32, 34–8, 144–5
   I-branching enzyme 12, 17–19, 145–6

carbohydrate ligands 185–6, 206
carbohydrate–protein interaction 52, 73
carbohydrate–lectin interaction 53, 73
carbohydrate–protein recognition 53, 243
carbohydrate recognition 54, 70–4, 230
carbohydrate recognition domain (CRD) 22, 54, 70–4
catalytic domain 99–100
cell-type specific expression of carbohydrates 17–38
cell-type specific transcription of glycosyltransferase genes 119, 122–4, 148
changes in O-glycans, increase of core 2 β-1,6-N-acetylglucosaminyltransferase in
   choriocarcinoma 37
   immunodeficiency 36–8
   leukaemia 34, 36, 38
   metastatic tumour cells 37
   T-cell activation 34, 35
chemical–enzymatic synthesis 222
chemoattractants or activating substances 24, 193
chimeric protein with human IgG1 heavy chain Fc domain 170
cleavage by endo-β-galactosidase 13
collectins 65
complement receptor repeat, CR repeat 23, 168, 175–6, 187
complex-type oligosaccharides 4
concanavalin A 244
conformation of oligosaccharides 230
conformational energy calculations 231
conformational flexibility 231

core structure of O-glycans 32–3, 91
   core 1 33
   core 2 33
   core 3 33
   core 4 33
crystallography 231, 236

diversity of N-glycans 4–8
dolichol-linked intermediate 8

E-selectin, ELAM-1 23, 63, 174–86
endocytic receptors 59–63
enzyme substrates 218
epidermal growth factor-like domain 167, 175, 187
Escherichia coli enterotoxin 246

fucose-specific receptor 61
α-1,2-fucosyltransferase H-enzyme 1 92, 129
α-1,3/4-fucosyltransferase III, Lewis enzyme 130–1, 135–7
α-1,3-fucosyltransferase IV, myeloid type 131–3, 135–7
α-1,3-fucosyltransferase V 133–4, 135
α-1,3-fucosyltransferase VI, plasma type 134–7

Galβ1→3(4)GlcNAc→R α-2,3-sialyltransferase 127–8, 258–9
Galβ1→3GalNAc→R α-2,3-sialyltransferase 126–7
Galβ1→3GalNAc→R α-2,6-sialyltransferase 91
Galβ1→4GlcNAc→R α-2,6-sialyltransferase 118–26
galactose-binding proteins 59, 74–7
galactosyltransferase-associated p58 protein kinase 110
α-1,3-galactosyltransferase 19, 111–17
β-1,4-galactosyltransferase 98–111
GlyCAM-1, Sgp^{50} 170–1

INDEX

N-glycan conformation 233
N-glycans 3–29
O-glycans
  mucin-type glycoproteins 29–38
glycosidases 206
glycosyl bromides 215
glycosyl donor 209
glycosylation 209
glycosyltransferase acceptors 215, 217
glycosyltransferase assays 215
glycosyltransferase inhibitors 220, 221
glycosyltransferase gene family 147
glycosyltransferases 88–162, 206
GM3/GD3 β-1,4-N-acetygalactosaminyl-
  transferase 138
Golgi retention signal 107–10, 124–6

HEV, high endothelial venules 164
high mannose-type oligosaccharide 3–4
HSEA calculation 233
hybrid-type oligosaccharide 3–4
hydrogen bonding 235, 245

I/i antigens 12, 17–19, 145–6
IgG oligosaccharide 243

L-14 lectins
  binding to polylactosamine 76
L-30 lectins
  binding to IgE 77
  as ribonuclear protein components 77
L. ochrus lectin 247
L-selectin 63, 164–74
  LEC-CAM1 63, 167–74
  lymphocyte homing receptor 164–74
  MEL-14 63, 164–74
lectins, animal 54
  C-type lectin 58–74
  P-type lectin 55–8
  S-type lectin 74–9
leukocyte rolling 24, 192–4
leukosialin (CD43) 30–2
Lewis^x, Le^x, X-antigen 17, 23, 177
ligands—potential as anti-inflammatory
  pharmaceuticals 185–6
lysosomal enzymes 10, 56

mammalian glycosidic linkages 208
mannose-6-phosphate 56–9
mannose receptor 62
metastatic tumor cells 26
molecular dynamics 236–40
molecular mechanics 232
monogamous bivalency 19
mucin-type glycoproteins 30–8

neighbouring group participation 211
nonreducing terminus (end) 6
nuclear magnetic resonance (NMR) 234,
  240

oligosaccharide analogues 206
oligosaccharide biosynthesis 249
oligosaccharide conformation 230
oligosaccharide primers 222
oligosaccharide synthesis 206

P-selectin 22, 63–4, 186–92
  PADGEM 22, 63–4
  GMP-140 22, 63–4
  CD62 22, 63–4
P-type lectin
  mannose 6-phosphate receptors 55
  cation-dependent and cation-
    independent 55
peripheral lymph nodes 164–6
Peyer's patches 164
PNHEV, the post-capillary high
  endothelial venules within
  peripheral lymph nodes 165–6
poly-N-acetyllactosamines 11–29
  preferential site for sialyl Le^x
    formation 7
poly-N-acetyllactosamine formation
  carrier protein such as band 3, 14
  in membrane protein 15
  time necessary for passages to Golgi
    complex 16
PPME, a yeast polysaccharide of
  mannose and mannose-6-
  phosphate 165
processing of N-glycans 8

protecting groups 207, 211
pseudogene 114

rolling leukocytes 24, 192–4

Se α-1,2-fucosyltransferase 129
selectins 230, 253
  E-selectin 22–4, 63–4, 174–86
  L-selectin 22–4, 63–4, 164–74
  P-selectin 22–4, 63–4, 186–92
sialic acid binding lectins 75
sialyl Le^a
  ligand for E-selectin 27, 183–5
sialyl Le^x 206, 253
  ligand for E-and P-selectin 179–83
side chains 4–8
signal/anchor domain 39, 99, 108
single-exonic versus multi-exonic genes
  146–7
solid state NMR 242
spin–spin coupling 234
Stamper–Woodruff assay, L-selectin
  binding 164
stem region 99–100
S-type lectins 74–7
sulphate-binding protein, common
  motif 78
sulphatide 191
synthetic inhibitors 206

thio-glycosides 215
three bond C–H coupling 241
three-mannosyl core 3–4
transmembrane domain 99–100
trichloroacetimidates 215

uromodulin 78

variation in terminal structures
  the degree of branches (antennary) 5–6
  variation in the nonreducing terminus
    6–8
VIM-2 antigen 132–3

Weibel–Palade bodies 186